Innovation Engineering:
The Power of Intangible Networks

Innovation Engineering: The Power of Intangible Networks

Edited by
Patrick Corsi
Simon Richir
Hervé Christofol
Henri Samier

British Library Cataloguing-in-Publication Data
A CIP record for this book is available from the British Library
ISBN 10: 1-905209-55-X
ISBN 13: 978-1-905209-55-2

Printed and bound in Great Britain by Antony Rowe Ltd, Chippenham, Wiltshire.

Table of Contents

Chapter 3. From Knowledge to Business: Virtual Encounters Propagate Innovation
Patrick CORSI and Barnabas TAKÁCS

Chapter 4. Value Management's Creative-Destruction via Digitalized Innovation: The Winning Plan
Jean MICHEL and Roy WOODHEAD

Chapter 5. Research, Innovation and Technological Development   85
Mélissa SAADOUN and Lin YANNING

Chapter 6. Sustainable Innovation through Community Based Collaborative Environments
Marc PALLOT and Kul PAWAR

Chapter 7. New Spaces for Innovation, New Challenges 123
Hiroshi MIZUTA, Victor SANDOVAL and Henri SAMIER

Chapter 12. C4 Innovation Method: A Method for Designing Innovations   223
Olaf MAXANT, Gérald PIAT and Benoît ROUSSEL

Chapter 13. Creativity World   239
Michel SINTES

PART 3. Innovation Management: Which Factors Underpin Success? . . .   251

Chapter 14. Psychology of Innovation and Change Factors   253
Laurent DUKAN

Chapter 15. Intellectual Property for Networks and Software   267
Sylvain ALLANO

PART 1

The Global Innovation World:
Which Visions Ahead?

Introduction

This first part introduces the historical basis of innovation as well as the relationships with foresight with a view to understand what levers to act upon in order to create a new wealth. Such wealth lies in human resources, changes in individual and collective behaviors, and management styles that are associated to networked organizations and finally new creation and collaboration spaces.

Each chapter stresses some theoretical foundations that are required for a deeper understanding of innovation and is illustrated with practical cases and applications. We state that diversity in innovation always rests upon a duality between "theory" (the concepts) and "practice" (applications). The variety of the seeds to innovation, be they human, affective, technological or organizational, means it is necessary to create a method on how to put into use the proposed steps within enterprises and organizations.

We introduce foresight and innovation in order to analyze how these two disciplines cross-fertilized themselves throughout their history. Then we explain that innovation results from the interaction of societal, human, managerial, organizational, scientific and technological components.

We develop the notion of collaborative networks made of individuals, projects and enterprises in a way similar to communities of practices based on the evidence that an optimal functioning of a technological network is founded on individuals and their competencies first. On a side account, the systemic propagation of innovation will lead us towards new concepts through an analysis of enterprise cases.

We then discover new realms of innovation based on information technologies that own their own laws and therefore are characterized differently from classical innovation areas. We develop networks of innovation through their modeling, organizational and information technologies aspects while taking care of analyzing the existing and future impact on employment and remote working relationships.

Finally we shed light upon value management and the enabling the notion of "valorization" that bridges working methods and enterprise goals.

In so doing, this first part delivers a number of realistic views about innovation while decoding the intrinsic complexity of a discipline that is resolutely multi-dimensional, pluridisciplinary and, above all, intensely compelling.

Chapter 1

Inventing the Future

"Tomorrow will not be like yesterday. It will be new and will depend on us. It is less to discover than to invent. The future of the ancient man had to be revealed. The future of the 19th century scholar could be forecast. Our future is to be built by invention and work. We have been progressively freed from material job by our machines, only to be asked to provide more and more intellectual work, really human work, that is, invention" [BER 64].

When reading this quotation from Gaston Berger, father of the French "prospective", one immediately understands the very close link between futures thinking and innovation, thus breaking with a future-oriented thinking, which is traditionally more retrospective (projecting the past onto the future) than "prospective" (imagining new futures).

What are we talking about? Fashionable notions today, innovation and future thinking are in fact very complex objects that are not easy to categorize; the effort to explain them before describing them is seldom taken. That is why we will first undertake to define some concepts and then explain some of the basics of futures thinking.

An innovative look through futures thinking on innovation and a future-oriented contribution of innovation to futures thinking: the cross-fertilization of these two attitudes towards the future – indissolubly linked – can restore meaning and purpose to the shaping of our future.

Chapter written by Fabienne GOUX-BAUDIMENT and Christopher B. JONES.

So, first of all, we will precisely define the notion of innovation and show the profile of the innovator; then we will introduce the field of futures thinking and the notion of change. Finally, we will show what futures thinking can bring to innovation and how the former contributes to the latter in order to invent the future.

1.1. Innovation

"The problem of the future transforms itself and, to some extent, simplifies itself when, rather than over-emphasizing the prospective discoveries, one thinks on the basis of manifested needs or satisfaction of deep expectations" [BER 60].

What are we talking about when we speak of innovation today? Let's define the nature of innovation itself before we turn to the more human-oriented profile of the innovator.

1.1.1. *How should innovation be designed?*

Three distinctive approaches help to encompass the topic and reveal its main points.

1.1.1.1. *A change*

First of all, an innovation is a change. As such, it directly engages futures thinking, which is a field of studying, creating and leading change.

The word "innovation" comes from the verb "to innovate" which means to "introduce something new" or to introduce "a new idea, method, or device".

The introduction of this novelty goes through various different processes according to its domain. In the economy, this is the introduction within the process of production or sale of a new product, equipment or process, which presupposes a phenomenon of integration of the novelty into the existing process. In sociology, innovation is defined as a process of influence that leads to a social change and whose effect is the rejection of the existing social norms and the adoption of new ones. Within this framework, the problem is less about integrating innovation with what already exists than substituting a new system for the previous one.

Alongside these definitions are two fundamental approaches to innovation. The first one helps to distinguish between innovation and invention; the second one between two different natures of innovation: incremental innovation and radical innovation.

1.1.1.2. *A contextualized process*

Innovation is different from invention, although it also manifests itself in change. Yet a change occurring at the level of the object itself creates only a change "in itself", independently of specific contexts, while the change induced by innovation modifies a set of strongly differentiated processes (e.g., from the assembly line to the final use of the product). For if invention is defined as "the action to imagining, inventing, creating something new" or "the faculty to find something, to create by imagination", then innovation, especially in the economy, defines itself as "the whole process proceeding from the beginning of an idea until its materialization (the launching of a new product), through market research, the development of the prototype and the first steps of the production".

Moreover, innovation can change the modes of distribution, of consumption, even the recycling of the innovative object. In doing so, innovation can extend its ramifications, induced impacts, even to its modes of payment, transportation or interpersonal communication. This is how it constitutes a process, at the opposite end of invention which is only a specific moment whose effects are limited to the object of invention.

Indeed, this makes innovation a lot more complex, much more so than invention. Because innovation is not only the expression of the emergence of change (as invention is), but is also the expression of adequacy to this change in the world, it can only exist in conjunction with the social and economic acceptability of change. Thus, if invention can be considered as disconnected from time and space, innovation is, on the contrary, the reflection of its time and a specific space through the culture of this location.[1]

1.1.1.3. *From incrementation to rupture*

The generic word "innovation" encompasses two distinct phenomena: an incremental change and a radical change. One often forgets to remember this fundamental distinction, thus erasing a cleavage intrinsic to the very notion of innovation.

Incremental innovation concerns a change brought to an already existing product (in the broad sense of the word). It improves the product, according to a specific use, or attaches complementary functions to it, transforming it into a slightly different object.

Radical innovation creates a product that is rarer and very different from those which existed before. This is not only because it must be the fruit of an invention in

---

1 As demonstrated by Thierry Gaudin in [GAU 78].

rupture with what has been already existing before – which is the most difficult because it comes from scarce effort of imagination – but above all because the environment will accept less easily a whole novelty as opposed to a simple improvement, as novelty often induces a chain reaction of change. So the advent of a real novelty and its economic and social acceptability is an infrequent phenomenon.

Considering the current pressure coming from the need to reduce the "time to market" and from the shortening of return on investment, incremental innovation is most favored by companies. It usually provides fewer benefits, but does so more quickly, and it is generally less risky than radical innovation whose parameters, in addition, are less well understood and less easily controlled.

Indeed, incremental innovation can be guided thanks to methods such as functional analysis or morphological analysis [REY 93] or more specific methods like TRIZ, for example. Radical innovation is less amenable to such an analytical and systematic approach (see below).

1.1.2. *Profile of the innovator*

Whether an independent innovator (innovating almost by chance) or a researcher within an industrial research center (innovating by professional duty), cognitive phenomenon related to innovation is not well known. It is often said that innovation is the fruit of the marriage between invention and its market. However, the skills of the innovator are generally due to some features of their personality profile.

1.1.2.1. *The liberating role of ignorance*

Most innovators share unique, perhaps strange, similarities which suggests that some qualities are correlated to the faculty of innovating.

Among them, ignorance plays a special role. In fact, too much knowledge would reduce imagination, learning substituting itself for invention, the mind closing itself over what it has already gained, refusing to imagine solutions which, filtered by the current theories, would not appear to conform to the body of knowledge. Moreover, one observes some intellectual laziness over building novelty from a certain level of learned knowledge.

It is easy to test this on students for example: to ask them to work on a topic they do not know anything about. At the end, you will always get some nuggets from smart brains that have entirely rethought the problem according to new criteria. Doing so, they have gone beyond the usual analysis of most of the well known experts, simply because they have considered the problem from a new and more innovative approach. However, if you ask them to work on a topic they know something about or about which they can access information, the best result will be a good compilation with the least personal contribution.

Researchers, writers and other intellectuals know well the phenomenon of the "white paper" whereby, after a very intensive period of documentation, everything seems have been said on the topic and nothing new can be added. Only when enough time has passed for this information to have been forgotten can the brain work again by itself.

This "distancing" from knowledge or information is often seen as a capacity for critical judgment, an aptitude for discernment. By taking a critical look backwards at acquired formal knowledge, the innovator opens the door to other kinds of knowledge which is more intuitive and more subconscious.

1.1.2.2. *The quality of the listening for signals*

So, although he should be ignorant – at least partially – the innovator must be attuned to societal needs and expectations in order to differentiate himself from the inventor. That is why he usually possesses an ability to "listen for signals". This intuition allows him to read the weak signals hidden within the informational noise of our societies, to distinguish between what is the real and structural, and what are only mass media constructions or "lifestyle" fashion effects.

This listening ability expresses itself through a capacity of problematization, a means of transforming scattered, often ill-assorted data into a coherent whole carrying meaning or significance. Innovation then comes from the research of an answer to a problem, such as the Tetrabrik® system replacing the traditional glass bottle.

The innovator's ability to listen for signals does not limit itself to intuition of the societal expectations. It is also tuned, even unconsciously, on his environment: colleagues, hierarchy, personal relations, etc. So the innovator can mobilize his network for the benefit of his idea – to test it, or for its diffusion – to achieve it.

Thus, while the inventor is rather solitary, enclosed in his garage, the innovator is an integral part of the thickness of the world: he thrusts his offshoots, his tendrils, his extensions deep into it. It is as if the quality of his listening for signals would give him access to a new dimension within which his mind can easily build new solutions.

1.2. Futures thinking

Moore's Law extends computer memory capabilities; "nomadic objects" (things are built to be easily moved everywhere); electronic objects perform ever more functions without an end in sight; the Internet every day spins the McLuhan global village web; the effects of an acceleration of the pace of change are felt everywhere, even in our everyday life, jamming our bearings and perceptions of time.

Time, change, novelty, future: the scene is set. As Janus, futures thinking presents many facets: "interdisciplinary discipline" to study the future, "science for action", "science of change", "philosophical attitude" toward the future; futures thinking is all this and much more, hence the urgent need for some definition.

1.2.1. *Futures thinking: a tool to build the future*

As is the case with every complex object, futures thinking is very often sliced into various sections in order to be better understood. Industrial futures thinking (the French *prospective industrielle*) is different from State futures thinking. Strategic futures thinking is different from organizational or managerial futures thinking. Exploratory futures thinking is dedicated to the exploration of the future, while the normative futures thinking is dedicated to the building of the future. Global futures thinking (whether industrial or strategic) contrasts with territorial futures thinking (used to build or plan a territory or community project), regional futures thinking (also called "regional foresight"), urban futures thinking (also called "urban planning"), technology futures thinking ("technology foresight"), thematic futures studies (according to economic sectors or resources, such as food sector or energy), etc. Futures thinking is a simple food that can be eaten with various spices. However, it has a history and a corpus, which are not well known, that make it a rightful discipline.

1.2.1.1. *A French orientation*

Both a philosopher and head of a company, then head of the Higher Education at the French Ministry of Education, Gaston Berger (1896-1960) formulated the notion of "futures-oriented anthropology" as early as 1955, followed by the concept of "*prospective*" in 1957, which we translate today as futures thinking [BER 57].

He defined futures thinking as field of study; it is different from forecasting as it only concerns the very short term, it must be very precise to be useful, and it is built on quantitative data. In contrast, futures thinking is oriented toward the mid- and long-term (10 to 20 years ahead); it must scan the comprehensive environment very broadly, be "free and bold" in order to help the decision-maker to understand the transformations happening in front of them, and give more importance to qualitative information and analysis.

Since the 1960s, futures thinking has deeply influenced the captains of industry and most of the senior civil servants and government officials in France (Louis Armand, Pierre Masse, Jerome Monod, etc.), organizations that institutionalized futures thinking, as well the public sector (Commissariat Général au Plan (1946), Délégation à l'Aménagement du Territoire-DATAR (1963), Ministère des Armées (1964), etc.) and in large corporate companies (CDC, Ciments LAFARGE, KODAK, SAINT GOBAIN, SNCF, SNECMA, etc.).

1.2.1.2. *A discipline in expansion*

Since this golden age, several generations of futurists (namely the prospectivists) have followed one another, each one bringing its own contribution to the corpus of futures thinking. The first generation of these pioneers grew up within the spirit of the 19th century scholars (G. Berger, P. Masse, J. Fourastie, B. de Jouvenel, etc.). The second generation (1970s) was that of the engineers, providing a large toolbox for futures thinking (from American methodologies, such as DELPHI, to made-in-France methods, such as MICMAC, MACTOR, etc.). The third generation (1990s) has reconnected itself with the values of the first generation: multidisciplinary, global thinking, and humanistic (sustainable development, democracy, etc.).

While the American orientation of futures thinking (forecasting) looks for the "colonization of the future" [BAR 93], based on a very deterministic vision of the future, the French orientation has shown the way of the "*futuribles*" – the possible futures that one can create if one is willing to do so. A large number of developing countries, especially in Africa and South America, have adopted this "French prospective" as a tool to invent their own, desired futures.

1.2.1.3. *Operational thinking about change*

Futures thinking can be defined by several characteristics: it is global, systemic,[2] taking into account both the object of the study and its environment (context); it puts the person at the core of its work, taking an interest in the relationship between the

---

2 A system is a complex of interacting elements. The elements are open to, and interact with, their environments. In addition, they can acquire qualitatively new properties through emergence and thus are in a continual evolution. System thinking is both part-to-whole and whole-to-part thinking about making connections between the various elements so that they fit together as a whole.

person and the object studied; it looks "far ahead and far away",[3] adopting a critical distance thanks to the practice of macro-history[4] and far futures scenarios.

Futures thinking goes through a logical and rigorous three-step process. The first step makes it possible to acquire the information needed to produce a dynamic diagnostic (diachronic) of the studied system and also an anticipation of the possible (trends, breakthroughs). The second step helps to formulate the problem that justifies the study: what is the problem (subjective approach), what are its components (objective approach), why is it a problem (values- and outcome-based approaches)? The third step aims to elaborate the most desirable solutions and to discuss them from a strategic point of view (return on investment, mid- and long-term impacts) and a operational point of view (about implementation: who, what, when, where, how). Then the decision-maker has all the cards in hand to make the correct decision.

1.2.2. *Profile of the futurist*

When a discipline is not frozen, its "orthodoxy" is not clearly defined or recognized, and the role of those practicing it is crucial. Although most of the concepts and methods of futures thinking can be learned,[5] the real value of a futurist usually dwells in what cannot be learned: cognitive behaviors and approaches that education does not usually teach.

1.2.2.1. *A behavior "in and outside the world"*

Like the innovator, the futurist needs a critical distance from knowledge, especially because it is too often built on a snapshot, a state of the art at a very precise moment in time. Indeed, the specific contribution of a futurist is both his fresh look (an outside look) and his dynamic (non-static) approach that, whilst deeply anchored in time, is also well beyond the apparent source of the studied facts. That is why the futurist is often an efficient macro-historian, able to identify the pattern of evolution over millennia.

Like the innovator again, the futurist is continuously listening to the world, less to perceive the immediate expectations than to grasp the "big picture", to see the structure of the final image of a puzzle, the pieces of which could never fit together. He spends a significant part of his life listening to, searching, scanning, rummaging, etc., in his quest for evolving social and cultural mutations and their understanding,

---

3 As it was prescribed by his "inventor" Gaston Berger [BER 57].
4 Johan Galtung and Sohail Inayatullah (eds), *Macrohistory and Macrohistorians*, Greenwood Press, 1997.
5 It is taught in several universities around the world.

for developing, declining or stagnating trends, for hidden weak signals, for probable ruptures and breakthroughs, and for all the consequences of these elements on the future of humanity or of a very specific population, a city or a firm, for example.

1.2.2.2. *A "post-industrial" way of thinking*

If one admits that modern thinking is characterized by processes analogous to industrial processes (products/tasks assembly line, products/ideas mass production, reduction of complex processes/tasks into their simplest versions, products/graduates, standardization), then one can call it "post-industrial" thinking, thinking that uses complex approaches [MOR 99], systems approaches, methodologies such as spiral dynamics, multi-layered analysis, futures wheels and various other methods usually very different from methods taught in traditional training, education or learning.

Futurists and innovators are a product of this very post-industrial way of thinking. This way of thinking presents, amongst other characteristics, the following four characteristics.

A distancing approach to knowledge: in a world where the most important thing is the accumulation of information, even if the information is already obsolete, creativity is very often curbed by this intellectual formatting. To escape this, the futurist looks for knowledge that is synthetic rather than analytic, comparative or applied rather than *in abstracto*, within which the critical analysis can find its best place.

An unfailing curiosity is essential because it makes it easier to absorb multiple sources of information, including those that have nothing to do with the studied topic. The futurist's work is based on intentionally broad general knowledge. This curiosity also allows the futurist to progress because his universe undergoing a rapid evolution: evolution of the discipline itself that must adapt itself to the various problems encountered; evolution of change itself which is in ongoing transformation. That is why the futurist must constantly evolve and adapt himself as quickly as possible, in order to keep pace with the evolution of change.

The alternative thinking relates back to the assessment that there is no longer a unique truth, but a large number of roads by which one can reach the same point. Alternative thinking often encounters a form of totalitarianism of thought that forbids the alternatives and unique ways of thinking, a type of "intellectually correct" thinking.[6] For example, some very innovative systems of thinking, smarter

---

6 As we sometimes see at school when children are forbidden to recite a lesson in their own words rather than in the words of the book; or when it is decided, in higher education, that

than most of the intellectually-correct ones, such as the general system theory [BER 68] and the theory of general semantics [KOR 33], have been prevented from spreading due to such resistance. However, the alternative thinking approach, which enables pluralistic thinking, imagining possible alternatives, is the closest to the specific human capability to grasp facts, to interpret them, and then to create the most adaptable solution, according to the context and all the differential elements (culture, values, etc.). This is the same alternative thinking that the innovator displays.

Last but not least, inductive thinking has been proven to be as rich, if not more so, than deductive thinking: in the evolution of humanity, practice has usually come before theory, and not the other way round. Technical change, for example, comes more often from innovators' initiative than from the implementation of scientific theories [GAU 98]. This is likely the reason why there is a correlation between the Anglo-Saxon pragmatism and the American leadership in the field of patents, etc. [PER 81].

Critical and alternative thinking, curiosity and induction are many of the complex tools that the futurist uses to understand the world and its various evolutions, hence his proximity to the innovator who shapes the world by transforming it.

1.3. Change and network

What kind of relationship do innovation and futures thinking have? As seen in the definitions and profiles above, each deals with change. Innovation creates change and futures thinking studies it and advises on it. Innovation is a breeding ground that feeds futures thinking, while futures thinking helps to facilitate innovation, both as an individual act and as a social practice.

1.3.1. *When innovation feeds futures thinking: the study of change*

Innovation literature has flourished in France, English-speaking countries and the rest of the world, and has been covered widely by the mass media. A large number of clubs, meetings, workshops and conferences are organized on the subject. Specific training and degrees are established and attract many students.

This general interest in innovation raises questions for those who observe the evolution of society: why has innovation come so much into demand [GIG 98]?

---

deductive thinking is the only rigorous thinking and that inductive thinking should not be used.

The futurist's first answer – used to scrutinize the signals of any kind of change, especially the weal signals anticipating the future – will not be "because of current global competition", but "because change itself is needed". But why is change needed? And what kind of change are we looking for?

1.3.1.1. *Why should we change?*

As early as the 1950s, Gaston Berger, when inventing the French futures thinking, explained to the French public and private-sector leaders that the reconstruction of the country after World War II was a great opportunity to fundamentally change France in order to modernize both minds and organizations; this conviction was the source of French futures thinking [BER 59, 61].

This vision of French futures thinking was rooted in the observation, on the one hand, of the acceleration of the pace of history – already pointed out at the end of 19[7] century by the evolutionists[7] – and, on the other hand, of a growing gap between this accelerating pace of change and the slowness in adaptation of French institutions, especially educational institutions. Consequently, Gaston Berger suggested a change of organizations and ways of thinking, towards greater openness, improved action and efficiency, so that they could more quickly and efficiently adapt to the new needs of humanity as a whole.

This approach is still valid 50 years later, because the gap is still there. In France, the gap is all the more obvious due to the resistance to change in institutional structures traditionally oriented to centralized state planning, and also, probably, to the weigh of a strongly rooted, 1,000-year-old culture, which leans towards conservatism rather than towards novelty[8].

Yet this is the same French society, so easily denounced as fossilized, that has welcomed radical innovations such as the first department store in 1869, with the first hypermarket, almost one century later in 1963, the smart card (1974) which is used today for money and health care, and the participatory democratic processes used to build territorial strategic projects, at the end of the 1980s, before the fall of the Berlin Wall that paved the way for participatory democracy and citizenship. It is also an important source of futures thinking itself, introduced in France at the same time as in the USA (1950s), and which other European countries only began to discover during the 1990s. Thus, France enjoys a high potential for invention and innovation, but too many restraints prevent it from benefiting from this potential. That is why a change in organizations and ways of thinking is still needed.

---

7 See [MEY 48, 54, 74].

8 That is why we needed a devastating revolution to end up with a monarchy, rather than to evolve quietly towards more freedom.

The case of France is not unique; many countries are in the same situation, which creates a global state of emergency. Not only does the pace of history still continue its acceleration, increasing the stress on society within which the gaps grow in number and depth and making it more and more difficult for each individual to keep the pace with it, but the paradigm shift that has occurred during the last half of the 20th century is hardening the conditions of globalized competition between firms and states. States feel a growing need to intervene in the market in order both to smooth the consequences of this evolution and to protect their own economy and firms, which, in turn, creates tensions at the geopolitical level.

"In the last resort, however, it is always a system of values, of ideas, of ideologies – choose whatever word you like – that is decisive."[9] If a country, an organization or an individual wants to change, the reason is that they implicitly believe in progress, whether it is human, technological, social, political or economic; they hope that tomorrow will be better than today, thanks to the introduction and diffusion of new inventions[10]. This thinking is not so obvious as it might seem to be. In fact, it is the fruit of a very difficult balance of the recognition of the state of the current situation (diagnosis), the assessment that this situation is not good and needs to be improved, the belief that this improvement will not be the result of a mechanical adjustment (such as the market laws, for example) but of an adjustment that demands human will and action for implementation, and this deep inner hope or confidence in the fact that "progress" exists (versus a technological doom, for example). If you are too optimistic, you will not be aware of the disaster early enough; if you are too pessimistic, you will not mobilize your energy to act against the trend. Moreover, you must also cultivate a special relationship with openness in order to accept assessment as well as uncertainty; assessment of the current situation and uncertainty of the future (from the futurist's point of view) and of the consequences of the novelty's introduction (from the innovator's point of view).

Inventing the future means includes self-confidence, openness, ability to integrate novelty and to adapt oneself to it, and, above all, ambition for humanity and a willingness to secure a better future.

1.3.1.2. *What change to look for?*

With this perspective, change has value only if it is a catalyst leading to improvements, such as modernity, progress, and improvement of the general welfare of humanity. Change is a means, not an end.

---

9 Ludwig Von Bertalanffy, "The world of science and the world of value" in *Teachers College Record* 65: 244–55 at p 245 (1964).
10 Which is the characteristic of the thermodynamic societies according to Claude Levi-Strauss.

Innovation is nested in the core of this vision of change, which, unfortunately, is not unique, thus explaining the different attitudes towards change. If innovation is indeed the process by which invention can meet the economic and social acceptability that enables it to spread itself, it can emerge only at the moment when it is ready to be heard, or is expected, often in a subconscious way [GAU 98].

This societal dimension of innovation, the fact that "the seed will germinate where the soil is favorable" is often missing in innovation theory. Heads of companies, inventors, discoverers, creators, marketers, etc., are used to thinking that a good idea will impose itself on the market whatever the needs or the expectations of the market. From a short-term point of view this is not wrong as supply has pulled demand for at least half a century. Yet such a demand is neither sustainable nor truly "progressive".

Concerning product innovation, this is exactly where the shoe pinches in the innovation race between firms. A firm will not look specifically for an innovation that brings an improvement to the social welfare or that corresponds to an implicit societal expectation. Its aim will rather be to promote an innovation that will give an acceptable return on investment and that has a competitive benefit that will lead to a larger market-share or even that will just keep its market position for longer. The short life of the product is not a problem, considering the quick turn-over of the market. This explains why the quick launching of innovative products (the time to market) is more important than the degree of innovation within the product itself, the marketing budget often being privileged over the R&D budget.

Process innovation, being organizational or technical, comes from a different change objective. Its main aim is to improve the efficiency of a given system (whether technical, economic, social or political), usually in relation to improved productivity. As it answers expectations in terms of improvement, this kind of innovation is often more durable than the product innovation. Process innovation aims to deliver a product with a higher quality than the previous one on the market, at the same price (than the previous one) and even, sometimes, a higher quality with a lower cost of production than the previous product, for the greater satisfaction of the organization and its clients, all of which is significant in such a competitive world. This is perhaps the reason why one can observe a lack of innovation in the public services when they are in a monopolistic situation.

Beyond these two basic needs for change (making money and keeping pace with the global competition), what other motives can justify innovation?

In most cases, innovation is dedicated to greater efficiency and is a means of reaching the aim with the maximum of gain and the minimum of loss, not only for the firm, but also for the clients (hence, the notion of "service"). This is why the postal service must not only quickly and safely take your letter from one point to another, but also provide the shortest possible waiting time at the post office. Further, innovation means that the need to improve energy efficiency cannot be limited to acquiring a high-performance generator, but also extends to all the systems that protect against energy loss, such as pipe insulation, the use of specific devices, building materials and "smart" architecture.

Finally, technological innovation can be motivated by the will to access the same – even better – products (or services) than other countries or clients. See, for example, the difference between France and the USA, in the 1980s, the advantage of France in the field of telephony and at the advantage of the USA in the field of the Internet, opening the road to innovations such as the Minitel and the Bi-bop (the first cell phone).

Futures thinking learns from these lessons so as to inform the various futures scenarios it must create or assess: to identify what is sustainable or not; to advise specific forms of change; to assess the impact of specific forms of change.

Yet, it is not enough to have a new idea, even if it meets an expectation: it must arrive when the right person is able to implement and/or sell it. Futurist and innovator share the problem that the resistance to change is deep-rooted and powerful, as well as the question of how to diminish the resistance.

<table>
<tr><td>

Sidebar: on innovation

Much work has been devoted to explain why things are as they are, but very little work has been undertaken in order to understand how things change, and even less work to learning how to accompany their transformation.

How can we explain that in such a well-educated community of business and civil service officials, who are able to adapt to the vicissitudes of power, instinctive resistances appear in contradiction with the discourse on principle in favor of novelties.

I think it comes from deep reflexes, acquired at school, which give priority to conformist mental attitudes and mistrust creative processes. In a way, these reflexes assume that knowledge emanates only from the institution. The institution is already there, overhanging human beings in a transcendent way.

Experience shows that creation proceeds from the opposite movement. Creative flux moves up, like the sap of the tree, toward light. Its chemistry is not one of transcendence, but the one of immanence.

From Thierry Gaudin (with the assistance of Jean-Eric Aubert), *De l'innovation*, 1998, La Tour-d'Aigues: Editions de l'Aube.

</td></tr>
</table>

1.3.2. *When futures thinking helps innovation: opening the road to change*

It is easy to believe that the use of futures-oriented methodologies is enough to open the road to change, especially scenario-building. Yet, much more than exploratory futures thinking, the real driver to innovation is the normative approach of the future. Indeed, the leading of change goes through a preliminary awareness of the need to change. From the simple refusal to change one's habits to the aggressive anxiety provoked by the fear of possible dangers, including the traditional inertia regarding every fashion, resistance to change is significant.

The role of futures thinking, in its daily use, is to put in evidence the present and future difficulties that could influence the future in a significant way. It shows the possible ways of change that can be taken. Usually problems are well known, generating diagnostic after diagnostic, but old, obsolete models that are no longer adequate are applied to solve them.

Futures thinking is the right tool to promote novelty, to encourage the development of new ways of thinking, to observe and act over the world, whatever the geographical scale. For the conditions of implementation, the context is much more important than the scale.

Therefore, futures thinking and innovation are on the same side. If innovation appears today as a key factor in business and society, this is because we are aware more than ever of our need to change. This one comes from a dissatisfaction of the societal expectations – in advance of its institutions and organizations – as well as from the imperative of ongoing adaptation imposed by technological and economic competition. [WEL 98]. Another emerging factor that also motivates this need for innovation is the creation of organizations and products able to satisfy the trends peculiar to the current society, such as mobility, hedonism, and the increasing speed of our activities. Therefore, the close relationship between futures thinking and innovation opens the way to change that is most likely to improve human life (chosen change) and to help novelty insert itself into common practices without causing too much damage.

To do so, futures thinking acts on innovation at several levels. Indirectly, it favors the development of a favorable context, for example by raising awareness of the need to create and maintain "agents" of innovation, such as venture capital, or the conditions to create a climate of trust. Directly, it can intervene via various creativity methods or radical ways of thinking. Although innovation cannot be conjured up on demand (you cannot say to someone "innovate now" and them innovate immediately; it is not an order that can be given and obeyed instantly!), it is possible to assist its appearance by altering the context within which it will occur.

1.3.2.1. *Promoting a favorable context*

Consider, for example, two elements of the context essential to the transformation of invention into innovation: venture capital and trust.

Money is the sinew of war. Indeed, product innovation is a long process, which begins with a pilot product to test the feasibility of the product, followed by the adjustment of the industrial chain that will produce, then pack, distribute and sell the final product. Organizational innovation is slightly less expensive: the difference is in hidden costs, direct or indirect (workers' training, time lost in transition from the old to the new system, initial errors, and the struggle against resistance or inertia). The classic firm, in a time of very high competition, hesitates to launch risky initiatives. When the innovator is also the entrepreneur,[11] the situation is even more critical: most importantly, he must be able to protect his discovery by a patent (which can be expensive), then develop a persuasive argument to convince either a firm to buy his patent or an investor to help him put it on the market.

Taking into account this difficulty and the decreasingly favorable context coming from the growth in such demands (hyper-requests) and the consequences of a speculative financial failure (e.g., when Silicon Valley start-ups ran short of capital because of the speculation on the stock exchange), alternative solutions have been developed.

In the USA, venture capital has become an institution with its "venture-capitalists", ready to bet on the future of specific innovations, its own financial stock exchange (NASDAQ), its criteria and assessments, its business angels and all the interested professionals those who look for investment opportunities to finance their projects, as well as those who look for projects to maximize their investments.

In France, venture capitalism is less developed. Admittedly, it has existed since 1996 as a "new market" (NM) for innovative enterprises or for those enterprises that have a high potential for growth, as well as for networks of business angels. However, there seems to be less openness to innovation than in the USA, mainly for cultural reasons, such as the weakness of entrepreneurship, bureaucratic constraints, and a very reticent attitude toward risk-taking. This has prompted the development of mechanisms of substitution such as the National Agency for the Promotion of Research (ANVAR) or the local government agencies that help to fundraise for innovation or to facilitate spin-off development.

---

11 The innovator is either an independent innovator, seeking to produce and market his own invention, or an entrepreneur having bought a patent and who is trying to develop a new product based on the patent.

The European Union supports innovation within the framework of mega-projects, including, in the main, large corporations, government agencies and universities, or within the framework of actions dedicated to SMEs.

Slowly, venture capital is strengthening itself, perhaps too slowly considering the fast pace of evolution.

The second very important element of the context is the climate of trust.

"We believe in all our stakes (joint venture), otherwise we could not invest!"[12] Whatever the operations to support the fundraising needed to implement an innovation, trust is always crucial.

The notion of trust is defined as "a positive anticipation attached to an assumed risk: a trusting actor assesses favorably the intentions and the capability of another person to achieve a given action and estimates that the non-achievement of this given action would be prejudicial to him" [ALI 98].

This trust manifests itself in various forms that can be brought together in two categories: rationalized trust and intuitive trust.

Rationalized trust is based on a series of indicators that help the "business angel" to form his opinion of a given project [MAI 99]. Those indicators (business plan, investment, size of the market ratio, etc.) give a more objective view of the innovator and his capability to run a business, thus avoiding the possibility that his business community (investors, bankers, etc.) will be seduced by an idea or a person (the project-holder). Consequently, the choice and evaluation of the indicators are major parameters in the building of this trust, a rationalized, reasonable and justifiable trust.

It would be a negation of the reality – as sometimes management theory or economic theory does – to limit the phenomenon of trust to this rational aspect because, whatever its justifications, trust is fundamentally subjective and, as such, includes a part of the irrational (emotion, belief, etc.). Even a business plan, read by two different people, will have two interpretations, a rather optimistic one and a rather pessimistic one. Each of reader of the plan will consolidate his own initial feeling – trust or distrust – spontaneously felt towards the man or his project. We call this feeling the intuitive trust because it comes before any rationalization and escapes to justification.

---

12 Bernard MAITRE, President of Galileo Partners, in an interview available at http://www.neteconomie.com/perl/navig.pl/neteconomie/infos/article/20000317010555.

If trust plays such a role within the process of innovation [ALI 98], how can it or the factors that give rise to it be detected? What are its drivers?

Three structuring factors of this trust mechanism are linked all along the process of innovation: the innovator's self-confidence; the support of "trust intermediaries"; and the care and maintenance of "trust-capital" (this is, capital that is in trust rather than in monetary form).

The first element, preliminary to the process itself, is the self-confidence the investor must feel and communicate to others, not only because it is very likely to be one of the sources of his innovation [EME 41], but also because it forces him to undertake the long road from the idea to the market. Self-confidence is the strength of his conviction, which will ultimately feed his energy, despite all the obstacles he will meet, and will mobilize his partners around him (self-confidence being part of the innovator's charisma).

The second element lies in the transitivity that characterizes the phenomenon of trust: it is easier to trust someone recommended by an acquaintance than another person not known by any means, even by the intermediary of an acquaintance. Hence the importance of the personal networks within which the innovator must be immersed; these networks are part of his credibility. This is the reason why the innovator looks for trusted, well known persons, and includes them in the project as president, associates, sponsors, members of the scientific committee, etc. Well known organizations are also "seals of trust", for example government agencies, university laboratories, business angels, and financial organizations. Trust is transitive, spreading from one player in the innovator's environment to another.

Once acquired, this trust-capital must be maintained and nurtured. This is most important as trust is likely to weaken during the evolution of the project. Causes of this weakening are numerous: interpersonal tensions arise as the number of actors involved in the process increases; conflicts of interest occur when the product is about to be realized; conflicts arise when associates feel envy, jealousy, pride, greed for gain, distrust; and some partners, tired of difficulties, quit the project. Also, as the implementation moves ahead, the entrepreneur will take the lead over the innovator and his attention will switch from his partners to his team.

Therefore, rules for a healthy trust are required in order to avoid most of these difficulties. Among the most important rules, two are provided by technical principles and two by ethics.

The two first rules are technical and can be considered as principles. The first one, "formalization", can be summed up as follows: whatever the degree of trust between the innovator and members of his business community (associate, investor, member of the team), the expression of this trust must be formalized, usually as a contract. This formalization should be considered as a symbolic step to strengthen the agreement, if it can be made easily. However, above all, the formalization is a guarantee against the violation of trust on the part of one or other of the parties.

The second rule is more difficult to define: we will call it "donation/counter-donation", following a very well known anthropological principle. It concerns information that the innovator must communicate to promote his "trust-intermediaries" when he introduces his project. Indeed, if an innovator has such people around him, he should promote the fact in order to increase the trust-capital he needs, this maximizing the benefit of having trust-intermediaries. Therefore, communication about trust-intermediaries will be crucial. However, this process must not be one-sided: the trust-intermediaries must also reciprocate. Each partner of this exchange must find an interest, a motivation, for mutual satisfaction. The challenge is to maintain this equilibrium so that no party feels injured; that's the logic of donation/counter-donation.

The two other rules are of an ethical nature. The first one is transparency because trust is by nature deeply rooted in it. Suspicion about things that are hidden, even by simple omission, is the most powerful encourager of systematic distrust, which can never be wholly erased. This is the breakdown of a moral "trust agreement". The danger in the process of innovation is that the innovator will naturally tend to maximize the positive aspects and to minimize the negative ones, in order not to scare his partners; but nobody will forgive him if the slightest accident or setback happens during the course of the process.

Finally, the last rule is ethics itself. Innovators' speech and actions must be guided by strong ethics. What can appear obvious in a routine situation is far less obvious when the stakes are very high; not only financial stakes, but also – and sometimes more importantly – the stakes linked to credibility. The emotive factor then tends to prevail over any other considerations: "the end justifies the means". However, when the process of innovation is made possible only by the interconnections of various partners for whom the network amongst them is based on trust, the ethics concerning the means is as important as the end itself.

Therefore, futures thinking tends to identify the innovator as the central point of the network that spins around him in order to successfully handle the process of innovation. As such, the innovator is a captor of the societal expectations, an entrepreneur, and a promoter of change. As a keystone to this network, trust reveals both the tremendous fragility of innovation – this subjectivity – and the amazing power of the human will as soon as it manifests itself in a synergy of participants who are together ready to take the same gamble with the future.

1.3.2.2. *Thinking the novelty*

Beyond this context, can futures thinking facilitate innovation itself? There is no certainty; however, futures thinking can at least contribute through two approaches: specific methodologies and specific ways of thinking.

Futures thinking draws from the creativity toolbox methods, such as TRIZ[13] (Theory of Inventive Problem Solving), morphological analysis, multi-criteria analysis, and other instruments.[14] These methods are part of a common corpus, in a more or less derivate form, adapted to each aim. In technology firms where these tools are well known, futures thinking brings no specific added value. On the other hand, in service-based firms or local government (including agencies and all the public sector organizations), futures thinking can bring innovative thinking, thanks to these situation-adapted tools.

Beyond these tools, the main asset of futures thinking in this field is probably its capability to build a collective intelligence. The systems approach, whatever the method used, sparks off a synergy in which the intelligence of several people is greater than the intelligence of only one, which in turn favors introduction and appropriation of novelty. The elaboration of dystopian scenarios, the designing of trees of competences, and the utilization of "future histories" are some ways to explore paths likely to produce novelty. Futures thinking does not interest itself in the present: it extends the research of innovation by moving the problem or the given data in the flow of the time in order to broaden the "possible" and the "visible".[15]

By the same logic, working with the "desirable" (meaning desirable futures from the point of view of a firm's social body or a territory's stakeholders, for example), futures thinking tries to orientate innovation toward improvement or novelty as expressed by visions of desirable futures.

At the same time, as an observatory of emerging trends, futures thinking performs the function of an early detection system because it listens for weak signals. As such, innovations can occur by associating existing problems with means in embryonic form, or by hybridizing new developments in radically different fields.

---

13 See a description at http://www.mazur.net/triz.

14 See an overview of these methods at http://erwan.neau.free.fr/outils_innovation_3.htm.

15 The "visibles" are problems hidden in the present, but that can be seen, or are made visible, when you look at them from the future or from a broader perspective.

Thus, futures thinking offers a way of thinking more adapted to this new paradigm we have entered during the last half century: a paradigm where complexity, global thinking, and the role of complex interactions are becoming dominant. Unfortunately our traditional ways of thinking do not evolve as fast. It always aims to reduce a complex phenomenon to its simplest elements, as if the sum of the solutions brought to each of these elements would be the solution for the whole [GAU 03]. This is to reckon without the synergies produced by the interactions of the different components; in fact, the whole is greater than the sum of the parts. Henceforth novelty can be drawn from this new source.

Finally, the most original introduction of futures thinking to the process of innovation, as a creation of novelty, is the "time shift". What does that mean? Futures thinking works on change as a material to build the future, whether desirable or not. Consequently, the future interests the futurist if it is different from the past. In order to recognize this difference, one does need to know the past well.

The futurist is not interested only in the factual past, but also in the scope of social evolution, the changes it reveals or contains, the causal chains it gives us to understand. To study this human evolution, the historical scale is nothing less than millennia, as macrohistory shows.

Thus, the macro-historical synthesis teaches us that our current state of evolution is one of integration: integration of men and machines, integration of different cultures, economies, nations and disciplines. Interdisciplinarity and multidisciplinarity are at the core of the developments to come and, consequently, of the underlying innovations.

In the same way, with a mirror effect, futures thinking works well in macro-futures; these futures are so far away that they free the creative imagination. This could be an essential skill especially for fields and industries in crisis and chaos, for example the technology research and video game industry,[16] where innovation is mainly incremental. Exploration of far futures is a better way to test various configurations of possible scenarios that the current constraints prevent us perceiving. By opening the mind to various novelties, futures thinking promotes a state of creativity which facilitates innovation.

Mobilizing all the disciplines to integrate the long term into our thinking, anticipating the evolutions to come, identifying the current ones, and grasping the possible breakthroughs are the objectives of futures thinking.

---

16 For the evolution of the video game industry, see http://www.gdconf.com/archives; for the European program NEST, see http://www.futura-sciences.com/sinformer/n/news3204.php.

Therefore, futures thinking aims to give everyone – be they head of a company, civil servant or citizen – the means to be responsible for their own future, both individual and collective, a responsibility that is shared by the innovators, as inventors of the future.

Chapter 2

Innovation Management:
How to Change the Future

2.1. The innovation, beyond technique

2.1.1. *The fiction of the linear model*

In an era which stands out compared to all others for its continuous technological progress, the adjective "technical" goes almost automatically with the word "innovation". And even if they deny it, most people have the old linear model in their minds, according to which the fundamental research gives rise to knowledge which nourishes more practical research: these make it possible to carry out industrial research leading to applications and innovations. The scientist, the engineer, and then the industrialist and the distributor – each has their place. It is therefore sufficient to invest in research upstream in order to reap the benefits of inventions and innovations downstream.

In fact, things were never like this before. If we had waited to understand the laws of nature before creating and using the bow, wheel, bronze, iron, and power of steam, we would all still be only at the prehominid stage. Innovation originates from a new idea which is further worked upon to obtain a result and to get it adopted by the society and by the market in case it is a product – in a broader sense, commercialization. Invention is a new idea to obtain a result, but not necessarily an idea validated and adopted by the users. It can be said that innovation is an invention

---

Chapter written by André-Yves PORTNOFF.

ratified by success. In reality, the innovative idea can be basically old. It is possible that the idea is old and that it has not drawn to the attention of the public or that its promoter has not been able to take it to the application stage. This chapter will examine the conditions of the routing in the society until success is achieved.

2.1.2. *Technically and societally viable*

It is obvious that at least two basic conditions have to be fulfilled: the idea should be technically viable and applicable in the context of the society. "Technically viable" has two meanings which are often misunderstood: one is the difficulty of putting into practice an idea which involves going against the laws of nature, and the other is the availability of a certain number of scientific and technical tools. These tools are: knowledge, know-how and equipments. This does not boil down merely to the necessity of progress in research. It will be observed below that erroneous knowledge can impede the exploration of a path but does not deter innovations based on wrong concepts or on purely empirical knowledge, as is the case with the Montgolfier brothers. Know-how and equipments have an essential role to play. Pascal and Leibnitz had designed calculating machines which practically could not be used because they did not use the know-how in the field of delicate mechanics used by good clockmakers of that period and the machines were not user-friendly. Almost a century later, Thomas gave a marketable form to his arithmometer. This instrument, being a sustainable innovation, was sold for 30 years till the second half of the 19th century.

The other field of possibility corresponds to the conditions of the society: will such a function be accepted, desired and adopted? It will be noticed that it involves anticipating the reactions of a series of intermediate actors and not just the potential customers. All this depends on the economic and socio-cultural conditions, the rules and regulations and the forced relationship between competitors. Other non-material factors like cultural elements, mental framework and values determine the behavior and the decisions have a major role to play in whatever is technically and societally possible. These two fields of possibilities are separated to facilitate the presentation: apart from the respect to the principles of physics, all other technical conditions connected to knowledge, beliefs, know-how and equipments are indicative of a certain state of a society.

Under these conditions, it is obvious how absurd and restricting it is to talk only about the technological innovations and how simplistic and false it is to combine innovation with industrial research. I[1] have written that a good innovation is possible

---

1 Portnoff, André-Yves, "Sentiers d'innovation" (Pathways to innovation), *Bilingual Perspectives Futuribles*, 2004.

without being a researcher, without having done any research or having missed a research; just as it is possible to miss opportunities to innovate with thousands of researchers around. Xerox, IBM or the Thomson group have proved it in the past. The concept of innovation spread in France in the early 1970s for faulty reasons. It was used to clear the conscience of the advisors to the authorities who sacrificed easily the research budget...

2.1.3. *Technical and societal futuribles*

The relation between innovation and "prospective" (future studies) is obvious. Future studies explore the field of possible futures, *futuribles* (contraction of *futurs possibles* in French). It differs from forecasts, extrapolation of the past and the present, and more so from prophecies, belief in a future which is already sealed but just needs revealing. The persistent success of the prophetic sales talks amongst the educated senior executives with a so called superior education and who believe in these talks more than the workers and the peasant farmers prove that there is a lot more to do in the field of promoting scientific spirit. Prospective is useful for the person who wants to be a role player in his future, says Hugues de Jouvenel[2]. The complaisant attitude towards prophecy is also contradictory to innovation, an act which basically makes possible that which is impossible to others at a given time and context.

The pioneer needs a vision regarding that which is possible technically and societally. The prospective of techniques that could develop and become available and the prospective of the futuribles of the society are two tools essential for enforcing the innovative ideas and helping the innovator to build his plan of action. The evolution of values is a very useful guide in anticipating the reactions of a society towards a proposition. The European studies substantiated by Futuribles (a French center for prospective studies and foresight) constitute a mine of data to be explored[3] in this regard. Surprisingly prospective is practiced rarely by the research organizations who seem to be scared of being asked to carry out programs based on the results of prospective studies which while providing them a larger field of vision could help them justify the choice of new and original subjects.

Prospective is all the more useful to the innovator as the factors which prevent the perception of weak signals are generally also those that oppose innovations and those that get them rejected. It is therefore a tool which not only highlights the tough path followed by the innovator but also helps reduce incredulities, the errors of vision of other actors who need to be convinced to arrive at the innovation.

---

2 Director of the group for prospective Futuribles.
3 Values of the Europeans, long-term tendencies, *Futuribles* special issue, July-August 2002, 216 pages.

2.2. Innovations in an era of digital networks

2.2.1. *More and more power*

The originality of this book is its consideration for innovation in the context of a society marked by the explosion of networks and other applications of digital technology. Let us see in what way these transform the context of the "technically possible" and even of the "societally possible"[4]. I leave aside the more general framework of the intellectual revolution to the post-World War II work that is accelerated by the digital mutation[5].

The spread of digital technology and the resultant convergence of domains separated till then from mainstream informatics, telecommunication and electronics give a common characteristic to the evolution of these sectors while still offering creative opportunities for innovation combining their properties: microcomputers, telephones, printers or audio-visual equipments with Internet terminals, communicating cameras, etc.

The main factors for evolution of technical origin can be summarized into two general trends: a continuous rise in the available power, and a reduction in the communication and transaction costs that develop the network at all levels and force the specific properties of these to be taken into account[6].

The rise in the available power is symbolized by the so-called Moore's Law, after the name of one of the founders of Intel. The prospectivist observes that it is a good example of self-fulfilling prophecy. Over the last 30 years, the wealth of Intel has enabled it to go ahead with investments which also follow the famous law. By not innovating the basics and by continuing with the same architecture, it was possible for Intel to dissuade less opulent competitors and double its declared performance every 18 months. But a limit was being reached, the electrical consumption and the heat generated, two main disadvantages for the portable and the domestic applications, also follow this law, whereas the performances useful for the users progress rather slowly as explained by Jean-Paul Colin[7]. Faced with aggressive competitors, Intel is forced to innovate.

---

4 Dalloz, Xavier and Portnoff, André-Yves, "The digital proliferation", *Futuribles* no. 266, July-August 2001, pp. 23-40. and Portnoff, André-Yves and Dalloz, Xavier, "The e-novation of companies", *Futuribles* no. 266, July-August 2001, pp. 41-66.
5 Portnoff, André-Yves and Arlette (ed.), *Bureaucratic Societies against Revolution of the intelligence*, L'Harmattan, 1994.
6 Portnoff, André-Yves, "The Challenge of the intelligence, the chips, the mice and men", *Col. Perspectives, Futuribles*, 2004.
7 Colin, Jean-Paul, "After Moore?" *Futuribles* no. 294, February 2004, pp. 5-16. and "The Law of Moore, what limits?" *Futuribles* no. 278, September 2002, pp. 49-56.

Moore's Law or not, a combined progress in hardware and software is going to offer increasing possibilities of computing power at affordable prices to small enterprises and to individuals. Bigger organizations can always process their customers easily and individually and also trace those individuals eventually violating their privacy if proper measures are not taken.

2.2.2. *Cost of organizational transaction and innovation*

The fall in transaction and communication cost has led to two interconnected logic, of ubiquity and networks. The cost and difficulty in transacting, communicating, working, inquiring and collaborating at a distance have reduced. Wireless communications enable the user to stay connected to his resources and to his correspondents wherever he is. It is obviously not the physical ubiquity but that of action: in order to act, the physical presence is not necessary. Networks develop between persons, organizations and objects. Now networks have properties of complex systems: by multiplying they have caused an explosion of synergies[8] that help all those desirous of collaborating between themselves by mutualizing the resources. The competitive spirit is disturbed; the networks of small groups can stand out against those of huge centralized pyramidal groups. That is why a new field, that of innovative organization, opens up. Its challenge is to release more and more synergies[9] in-house amongst the personnel and external between the companies and their representatives. Henceforth the major innovations are organizational[10], are difficult to copy because they are based on participative management and on a win-win partnership strategy that is beyond all catchphrases. This strategy of a real extended business venture explains the mind-boggling success of Wal-Mart and Dell, which their competitors have seen but have neither known nor want to imitate. The same synergic effect can influence one actor to take a hegemonic position such as Microsoft and Intel. There has to be a break at a certain point of time so that the innovators can question the established monopoly. That is what is happening to the two associates with the advent of new mainstream markets that demand new characteristics.

---

8 See footnote 6.

9 Portnoff, André-Yves with Lamblin, Véronique, "The real value of organizations", *Futuribles* no. 288, July-August 2003, pp. 43-62.

10 Portnoff, André-Yves, "Organisational innovation", *Futuribles*, pp. 92-94. no. 281, December 2003.

2.3. Shortsightedness against innovation

2.3.1. *Credibility of the message and the messenger?*

No change occurs without forerunners, also known as signs that foretell the future or weak signals. The innovator is one who grasps the new possibilities and opportunities that escape other actors. Prospective provides a checklist to identify signs of change at an early stage. A signal can be difficult to see or to decipher. It could also be disturbed and as a result not noticed. It is therefore necessary to take into account the objective perceptibility of the signal that physicists refer to as its height compared to the background noise, and its psychological receptivity, mainly subjective and related to every actor in this context at a given point of time.

The following equation which is explained later is proposed to the amateurs of formula:

$$\text{Level of perception} = \frac{\text{intensity of the signal}}{\text{background noise}} \times \frac{\text{messenger's credibility}}{\text{questioned by the message}}$$

Certain signals are themselves important whether they are noticed or not. We have difficulty in foreseeing the evolution of the climate but this evolution will nevertheless influence us whether we want it to or not. On the other hand, the emergence of a possibility of technical or commercial innovation will have no practical consequences if no actor notices it. This is the reason why many potential innovations are either not achieved or are concretized only with a delay of so many years or centuries.

According to Pascal, judging the credibility of the carrier of a signal and not its content constitutes one of the main misleading forces. IBM has treated with disregard the works of the University of Hawaii that were to be the creator of Ethernet, but for IBM, Hawaii had more authority in surfing competitions than in informatics... In the same way, the importance of the AIDS epidemic was ignored for a long time because the warnings issued by doctors were considered marginal because they were interested in risk groups who were also considered marginal like homosexuals and prostitutes. The responsible people of the Academy of Medicine were not vigilant.

2.3.2. *Outdated evidence*

Misleading evidence is often due to the power of a concept overshot by the changes in the context. It is always a known fact that a well guarded manufacturing secret is a guarantee of force. IBM and DEC, like Apple, have existed for long time with this logic that led them to favor the so called proprietary systems for a variety of computers which were non-compatible with those of others. Customers equipped with IBM had to be loyal to their supplier so as not to be stuck with piecemeal machines and programs. Thus the manufacturer thought that he was protecting his commercial interests by rejecting the Unix operation system developed in the 1970s by Bell Labs. In the eyes of the company, Unix had the disadvantage of being an open system. It was precisely this aspect of Unix that attracted university researchers and who in turn influenced NASA to adopt it for the Apollo program. From that time onwards the success of Unix was assured for years and IBM had to follow suit.

In the same way, Apple was blinded by the evidence of the profits that it maintained by banning the cloning of Macintosh. As a result, it neglected one of the consequences of the law of networks according to which the success is only at the cost of a certain sharing even if there is an opportunity to impose a standard and to create a self-maintained success phenomenon. What it earned in excess margins on its sale prices was lost dramatically in volume sales and the isolation to which it was condemned almost killed it. With its system of on-line sale of music, Apple did not commit the same mistake again as it had accepted to produce the drives for Hewlett-Packard.

2.3.3. *A too narrow vision*

If IBM considered the computer network proposed by the University of Hawaii in the 1970s as marginal, it was also because the new solution consumed a lot of bandwidth. Now the fiber optics and compression techniques have almost eliminated this disadvantage and the Hawaiian works have created the Ethernet network which have won over the "token sharing" networks of IBM.

This is an example of a cultural compartmentalization of specialists who do not see that their sector could be disturbed by the arrival of an outside technology. Many other errors have been due to the blinkers which make one neglect a factor that is becoming more and more influential. Thus, the supporters of centralized computing or large photocopiers have ignored the increasing desire for autonomy of the actors. That is why until the beginning of the 1990s, IBM refused to take seriously the micro-informatic phenomenon and Xerox presented Canon with the market for medium-sized photocopiers.

2.3.4. *False proofs*

The concepts that are simply false are the cause for a number of forecasting errors. It is because they had a false vision of science and technology that the director of the American Bureau of Patents proclaimed in 1875 that "there is nothing more to invent" and that Marcelin Berthelot claimed seven years later "the universe without mysteries". Scientific errors have led brilliant minds to prove the theoretical impossibility of the electric motor a few years before Gramme from Liège, a self-taught man, made it. Simon Newcomb became popular by mathematically proving that objects heavier than air cannot fly. When the Wright brothers ridiculed him, it made him say that the airplane was not beneficial.

2.3.5. *Significances ignored*

Lack of imagination goes with a lot of forecasting errors caused by non-perception of the significance of a new possibility. An interesting case is that of optical telegraph analyzed minutely by Patrice Flichy[11]. In 1684, the British astronomer Robert Hooke described "the way to make his thoughts known at long distances" through light signals and in 1690 the physicist Guillaume Amontons made a concluding demonstration of information transformation through optical semaphore at the Luxembourg garden in Paris. The potential utility of rapid communication was blanked out for almost a century because of the fact that the proposed media was new, even though it was clearly known that communication was important and strategic. Ultimately the Jacobins had to come to power in order for the telegraph reinvented by the Chappe brothers to be used. The use of mobile telephones and the Internet by the public was also largely underestimated, even contested, for a long time which resulted in a serious handicap to Alcatel and France Télécom.

Non-perception of potential needs is very frequent in public areas because of professionals with atypical consumer attitudes. This is an error committed frequently in the field of electronic commerce. The problem is to have enough empathy to imagine what people different from us can accept or reject. The reason for this is that people do not act as rational automats but make decisions based on their personality and subjectivity. The binary spirit also leads to asking questions in a wrong way: it is not a question of knowing whether "the French will buy clothes on-line" but whether "enough French people will buy so that the investments towards building up an on-line offer becomes cost-effective in a reasonable space of time", which is totally different.

---

11 Flichy Patrice, *A History of the Modern Communication*, La Découverte, 1991.

2.3.6. *Under-estimation of evolution potential*

In technology assessment the essential and the most difficult involve appreciating the development potential of emerging solutions, almost always less impressive than those already in place due to the proven veracity of these and the experience of the people using them. In France, the majority of commentators, mainly among intellectuals with a non-scientific background, prefer to go wrong due to pessimism for fear of "allowing oneself getting carried away by technical illusions", a tradition established by Rousseau who recommended permanent ban on printing, probably for authors other than himself. This good mannered skepticism explains to some extent the dramatic refusal to face facts related to digital phenomenon from a part of the French elite. Even today, the Internet is a target for innumerable criticisms, and the experts have dared to write until the last few years "that nobody has ever earned money in e-commerce". As a result there are imperfections, Intranet networks are often badly designed and cause loss of time. The arrival of every new communication tool gives rise to such failures and justifies the criticisms, albeit temporarily. The novelty of a vector is that the counter-measures to its perverse usage are still not in place and that there is a tendency to pin it to an outmoded organization, initially creating costly complications.

2.3.7. *Dare to imagine breaks*

It is difficult to foresee breaks in evolution with respect to the past. In this, prospective becomes precious in relation to the extrapolations of the forecasters. The imposing nature of the present having an undisputed power of that which exists prevents us from considering a future disconnected from what is known. In 1956, when Ampex marketed the first professional video cassette players at $50,000 a unit, it made little effort to derive from there a version suitable for the public because it sounded so incredible to be able to divide the price by 100. At JVC, the chief of projects Masaru Ikuba had the audacity to stick with the product and succeeded in 20 years. In the beginning his company was so difficult to convince that he decided to act without notifying its hierarchy.

In prospective, one of the difficulties is to make the specialists consider the hypothesis which appears impossible at the outset but which has a strong impact, be it positive or negative, when accomplished. What would be the consequences of a multiplication by 100 of the performances of electrical batteries? Which technical locks have to be broken for this? Who would be well placed to collect the necessary skills? These are the questions which can open the innovative perspectives hidden until then by the experiences of the past evolutions.

2.3.8. *Blinding arrogance*

Arrogance, the feeling of superiority which often existed but is disappearing, has prevented many leaders from evaluating the real extent of the changes which were being carried out under their eyes. From 1980 the publication of Japanese articles describing a new type of powerful electronic component phasing out all thyristors used up to then did not cause any reaction in France. The railway industry has been the undisputed champion of electric traction in the world for the previous 20 years and it had nothing to learn from anybody... This superiority was questioned barely five years later when the Japanese took over from the French the market for renewal of the French metro from Barcelona. Their major asset was the GTO, a component more compact than thyristors and better adapted to the input control of the electric trains.

IBM against the works from Hawaii and France Télécom denying the Internet phenomenon for years are examples of arrogance. The reasoning given by the excellent professionals to the national French operator was simple: in the field of telematics, France had the best experience in the world owing to Minitel which even though free had penetrated only a third of French households; conclusion, "the French are not Americans, they are not interested in interactivity and had nothing to learn from others". What is remarkable is that often the obvious is not visible due to the place where one lives, due to pressure of the present context and surroundings. The same colleagues placed elsewhere would have noticed the dangers and the opportunities of the Internet. Their attachment to past big achievements of their company and their loyalty blinded them. As a result France Télécom has missed the possibility to rapidly upgrade the knowledge obtained through its research team which is amongst the best in the world – a powerful illustration of the possible gap between research and innovation. Know-how is good but in practice the only thing that counts is the *knowledge of what to do*...

It is advisable to disregard an excuse given by those who hesitate to innovate. "It is better to let the others live in a brand new house". Regarding the Internet, many say "the digital delay is immaterial, it can ultimately be reached". Apart from the fact that the catching up is not always obvious, one knows in prospective that the most important thing is not *where* we shall be in 10 years but *by which path* we are going to arrive. In 10 years a company can achieve the same turnover in a new market according to a graph which takes off very steeply or otherwise. The cumulative turnover in the same period shall be higher in the first case. Worse, the pioneering company is going to attain the point of equilibrium and accumulate experience faster than its followers; it will also observe that its cost is reducing and hence could force others to sell at a loss if they really want to enter its market.

Let us return to the attitude that makes it such that by loyalty or attachment one refuses to contemplate on an event which is harmful to the organization that one serves, and to the persons whom one likes. A good question to ask at the beginning of all exercises of prospective is the following: make a composite drawing of your possible assassin, of the competitor who could destroy your company, the technology which can make yours obsolete! If a futures group has the courage to enter into such an exploration, it will look at the situation differently, discover the faults, the measures to be taken, the opportunities to be exploited and it would find itself stronger probably on the road to innovation because to innovate often means considering differently a situation which can appear to be dedicated to a future already programmed. Hugues de Jouvenel notes that prospective put to use in time gives maneuvering margins to the actors, who very often proclaim that they *do not* have it, whereas in reality due to lack of foresight they *do not have it anymore.*

2.3.9. *"The situation is under control"*

The fear of expecting the worse, even of admitting that one has made a mistake is a cause of blinding that kills many innovations and also people. If importance of AIDS was not seen by responsible people, it was because they failed to recognize with their own eyes that the medical body was wrong to have considered itself fit to master any illness. This attitude made it relax efforts in preventive medicine and public health over the last few decades[12]. At the beginning of the 1990s, Europeans and the Japanese insisted on financing their HD-TV programs based on the analogical techniques, at least two years beyond the moment when it became obvious that digital technology was going to become important. The big programs are subjected to such issues, because the inertia is considerable, the responsibilities are dispersed and the effects of publicity are dear to politicians. But how can this be stopped without reversing one's decision? Few directors have the courage to be self-critical like Bill Gates who in 1996 operated a complete reversal vis-à-vis the Internet which he had previously ignored and in which he has invested ever since. He even analyzed how he could have missed the right track.

Confronted with crisis situations, the same reflex plays a role. The first worry of the authorities is to reassure the population, that is, to protect their image of efficiency and mastery in the face of public opinion. This explains the Chinese censorship against Sras, the soothing reactions of the French political and administrative heads during the dog-days of the summer of 2003, who for several days decided to spare the population "unwanted panic". This step prevented them from detecting a situation that would have been obvious for them through their good

---

12 Lacronique Jean-François, "Prevention, care, two integrated functions?" in *Bureaucratic Societies against Revolution of the Intelligence*, pp. 101-110.

sense in any other position of responsibility even with much less information. This has led Bertrand de Jouvenel to conclude that "for equal intelligence, foresight is minimal for the person who is in power"[13].

2.4. Innovation as a process of creation of values

2.4.1. *Sell the training with the product*

We have developed an evaluation method[14] for an organization which can be summarized as follows: an organization is viable only when it produces for its main representatives more value than it consumes. In a broader perspective the value is that which represents a value for an actor at a given time in a given context. It has a strong psychological dimension (see Chapter 12) and not just financial. The representatives are mainly the shareholders, financiers, customers, staff, suppliers and other actors of the society like key influencer, opinion makers, administrations, associations, etc. At every moment the organization produces through the dynamic network of interactions between the talent and the aspirations of its members, its main collective intelligence. Dynamism also has the other essential element, namely relational capital, which is the capacity to establish and maintain flow of interactions with the external actors, partners who complete the internal collective intelligence with skills and complementary and competitive methods that have to be respected, and finally the customers.

In order that an innovation is accepted, it is necessary to establish a relation with the potential customer, a relation that will allow communication, confidence building and creation of emotion. Communication is the basic condition. It is rich if it allows mutual understanding to use the customized and personalized production possibilities introduced through digital techniques. At the communication level these digital techniques bring interactivity that informs the offer about the expectations of the prospective customer and this may be used by the marketing department upstream and the design office. At this stage the two parties, the offer and the demand, should have enough skills to communicate. If the prospective customer does not understand the product or the proposed service, if he does not know how to use it, the offer should give him the necessary skills, or else there will be refusal. It is particularly the case when the offers are innovative and imply a new behavior and learning. It is a brake which is essential to the penetration of the professional or private digital applications, especially since the usage is new, its advantages and its challenges appear uncertain or even unclear. All studies show that the most delicate

---

13 Bertrand de Jouvenel, "The art of the conjecture". *Sedeis*, 1964.
14 Portnoff, André-Yves with Lamblin, Véronique.

groups – people with fewer qualifications, the aged and isolated – are the most reticent to face innovative products, and more so the Internet.

In these conditions, the pedagogy of the offer is as determining a factor as its content. It can also be useful to build a relation of confidence without which there will be no transaction. The confidence is even more necessary in a new and expanding context like that of on-line commerce. The actors of the classic economy, through their known brand names and their physical layout, have the necessary asset to reassure. It is that which contributes to the efficiency of the "click and mortar" couplings. Finally the emotion has a decisive role to play because it conditions all decisions. The multimedia can be used to show, demonstrate and bring through images and movements an excess of pedagogy and emotion. Virtual reality is going to be used more and more to explain and convince, be it trying a dress on-line, using a tool for tinkering, assembling furniture, choosing a hotel room facing the sea or setting up a building on a site.

Emotion should be sufficient to help take decisions to adopt the innovative offer despite the inconveniences and the uncertainties of this choice even though it makes the investments obsolete. These investments are financial and material and they correspond to the equipments which will become useless and to invested capitals, but the deciding psychological aspect is often obscured: if new skills are necessary to use the innovative offer, certain actors may fear some loss of time in learning, in finalization but also some prestige, authority because they will no longer be the ones who know and because their powers will be reduced. Many small managers delayed the introduction of the Internet and email to retain the illusory power of checking all the mail, which was for a long time their privilege... the rule therefore is that an innovative offer should not only offer a benefit visibly higher than those offered by established solutions but should be perceived as they are by the prospective customers. The expected advantage should seem sufficiently high to compensate for the loss of investment and the uncertainty over the expected results. Thus, one comes under an economy of hope.

2.4.2. *Network, creator of value*

The process which leads from idea to innovation does not boil down to a confrontation between the offer and the potential user. The first path to take is the one that will go from the first formulation of the idea to the construction and then the distribution of the offer which then reaches the prospective customer. When somebody has an idea which seems innovative, it has to be enforced, protected, completed, improved and given a form. All operation processes described above apply: if isolated, the innovator cannot achieve. It is necessary to constitute a small formal or informal organization regrouping a collective intelligence and a relational capital around his idea and his desire to achieve. It could be just a start-up or, if one is already in an established set-up, a project team.

The building up of collective intelligence is indispensable to complete the creative skills as well as its character. One can be a good technician or a brainstormer without having all the necessary technical knowledge, without being a mass leader, an organizer, a financier or a negotiator. The relational capital will allow association of indispensable services to move to the realization stage, legal officers to financiers and to internal or external partners who ensure production and marketing. The team that carries out the project also needs to be associated with people, services and organisms that for various reasons will support, help to succeed, find other supports to and release the resources for the team.

Thus, innovation will transform into action only if a group of representatives are interested in it more than just financially. This approach helps estimate whether an organization is favorable or resistant and hostile to innovations. The appearance of an innovative idea can be subjected to self-censorship if everyone knows that it will be looked upon as disruptive in the environment. The first condition is the willingness to experiment which involves a real right to err including when one makes a mistake. It is possible only if the relationship between people, teams and departments is not conflictual but made so by a management having a divide and rule policy in the name of an unhealthy competition that inhibit the participants. The territories and responsibilities should also not be defined as a feudal private preserve, otherwise any new innovative idea will be considered as an effort to encroach upon the responsibilities of others, more so if the person who introduces that idea participates in a domain which do not correspond to his official functions, whereas the innovative ideas come into existence just at the frontiers... including the borders of the company or beyond its limits. The first question is to know whether the company is sufficiently porous, open to notice the signs of change and the famous weak signals mentioned earlier or whether it is bogged down by its own convictions. Will it ignore, as was the case with Alcatel, the rise in the use of mobiles and the Internet to the major advantage of Cisco?

2.5. Conclusion

The culture of the company or the territory has a deciding role in the possibility of the appearance and evolution of a novel idea. The way to manage personnel, to recognize, compensate and reward its work is a key element which is often ignored. The individual contributions to a collective performance are not compensated sufficiently. The promoter of an offer for telematic services explained to me that the marketing of its baby was handicapped by the fact that the sellers preferred to offer the public the well known accessories rather than spending more time to explain the benefits of an innovative appliance. The solution could probably be through a recalculation of commissions taking into account the difficulty and the time spent by the seller in introducing the new product, thereby making it financially viable for him.

Strengthening the innovation potential of a company – but this can also apply to a territory – can only be achieved through consolidation or adoption of a management method which itself is innovative. Those who are successful in managerial innovation acquire a comparative decisive advantage because their competitors will have lot of trouble copying them: a culture cannot be copied! In our method we examine several indicators which try to identify important cultural characteristics. The level of technological independence, the permeability and the existence of a risk management are generally the revealing criteria. The organization which is least dependant should find out whether this advantage is due to navel-gazing which reduces the risks in the short-term but puts it in danger of missing the innovations vital for its development.

Chapter 3

From Knowledge to Business: Virtual Encounters Propagate Innovation

Introduction

This visionary chapter introduces the notion of a unique, digital self that mimics our behaviors and represents our identity through collaborative networks via virtual humans, also known as avatars. We argue herein that such artifacts, in fact, enable the construction of new trustable business models and represent a major force in the mechanism of innovation itself.

The issue of systemic synergies is raised through the playing of networks and their laws. The discussion starts with a knowledge-based view of the issues at hand and asserts that the new medium that sustains innovation must accommodate huge bandwidths, hence the request for user's immersion, virtual self-representation, and 3D remote realities, i.e. presence. Humans have long learned to consider time and space as webbed in a now somewhat obsolete compilation. However, thanks to the virtual environment's emerging technologies, it is possible to interpret situations with continuous attitudinal patterns through time and space – those no longer resort to old computing schemes, i.e., compiling procedures with documents.

Our key thesis stipulates that current network-based interfaces may be greatly improved by recognizing the long-term need for a novel communication layer that can serve as an intermediary between the user and today's systems of ever-increasing complexity. To fulfill the demand for such a dedicated layer is a task for interdisciplinary research involving many branches of science reaching far beyond

---

Chapter written by Patrick CORSI and Barnabas TAKÁCS.

the boundaries of computer architecture and interface design. Yet it may be considered as the key element of a human-centered protocol that determines how users and their computers interact with one another and use this medium to reach other people for decades to come. Failing to implement such a layer and the corresponding system architecture necessary to support it will eventually hinder our ability to keep up with the amount of information our brains have to cope with on a daily basis; and a true knowledge society may not come within reach.

There is thus a need for a novel and advanced human-computer interface that combines real-time, reactive animated virtual humans (avatars) with perceptive capabilities and communicative intelligence. These self-representing avatars form the foundation needed to implement a closed-loop interaction model between users and complex computer systems which in turn are connected to other users with avatars in the network. To implement such a device, we need a virtually transparent intermediate layer, a tool that eventually will be, and must be, seamlessly integrated into our computer culture. Virtual presence anytime, anywhere and on any device is not just a vision, it is the way to make networks vibrant, operational and useful. This chapter explains why and how the vision can become a reality within the next few years.

3.1. Where information society mixes up our linear and local schemes

It is customary to say that we live in an era of networks. What are the links, if any, between networks, innovation and knowledge? Our dated view of experiencing only local, linear and causal changes has undergone a dramatic shift. It has converted to a strange universe that is non-linear, global and based on accelerated, echoing feedback. It is networks – the behavioral escalation provoked by people connecting to people – that make the non-simple, non-linear alive for us [MAS 06]. Problems become complex, urgent and interruptive; perhaps these are three manifestations of a same process? Also, why has it become so urgent to solve complex problems, and so complex to solve urgent problems?

Within this competitive run up of companies and organizations of all venues and sizes, since the Internet bubble the deepening of business models has becomes a key requirement. Only a knowledge-based description level can explain the structural and factual relationships from which meaning can emerge from apparent chaos. A knowledge-based description level brings added value to the economic models; and in the quest toward building added-value in a more integrated society, virtual assets and artifacts stand as a primordial ingredient. Why is this?:

– For one thing, the Internet itself does not bear the market value anymore [COR 05]. Facelessness is one of the many sources of this problem. On the Internet, there is no easy identification of players. You may know an IP address, a mother card number, or even a cartoon avatar, etc., but who is really behind those things? No one can tell – unless you enter into the "banking trilogy" of ensuring identification, authentication and security of a transaction.

– Then, you have to look for content providers that lend meaningful experiences. Surfing on the Web is a form of entertainment, but hardly an easy way to find *ad hoc* and meaningful information. Since there is no direct way of verifying the source or the information provider, trust becomes a key element in the interaction process. eBay is the modern example of a trust-based society, whereby sellers are assessed by buyers. It represents, in a sense, a Virtual Web Society in which its content draws the users closer together.

– The third problem is that the description of events (we call these "data") is often confused with information. Unfortunately, these two are not the same and replacing them causes many misunderstandings. Data only becomes information after it passes through verification and an editorial process that arranges, catalogs and structures it according to some guiding principle. It is this editorial step that inserts trust (and accountability) into the process and allows information to propagate.

– Lastly, the primary goal of networks is to extend the individual beyond his or her physical limits and increase their efficiency in turning information into knowledge and experiential situations. This may be best achieved by building an affective bond that employs the method of emotional modulation in order to increase the efficiency of information and knowledge transfer. Such a method helps the network user to better understand and later remember the content presented. In other words, emotions are the key catalyst that helps to turn information into knowledge and knowledge into business practice. Virtual encounters offer that unique ability to create personalized relationships which bridge the ever-increasing gap between people and their innovations. In marketing terms, virtual encounters offer a unique and sustainable differentiation power for networks-based business models.

Above are four reasons that compel us to narrow the gap between virtual reality benefits and everyday business practices. It is acknowledged that the products that companies design and produce express the way they embody solutions to the problems tackled by their people acting within such systems. On one hand, we strive to deconstruct (dematerialize) the solidification of past business schemes and practices and to free the knowledge from yesterday's peculiar constructs. On the other hand, it is widely accepted that virtual effects can be used further, practically becoming part of our daily routine through better immersive technologies. They can help systematic thinking, which in turns helps to make people creative, as they perceive the world as a unit and as a system.

This process creates novel opportunities by extending business models, while the latter enables a new ability to communicate with more intensity, speed and impact. We all know how much the Quality of Presence (QoP) contributes to task effectiveness, even if we have not much further explored the conditions to enhance it. A given QoP uniquely combines global (remotely balanced) and local (localized person) aspects: I have a "local" eMe who represents myself, and which encounters another "local" eYou representing other people, that then interrelates with global communities of eUs which bear a collective meaning. Theodore Levitt's [LEV 96] notion of a global village bears new characterization, as embodying behaviors, intelligence and emotion where individuals become really "glocal".

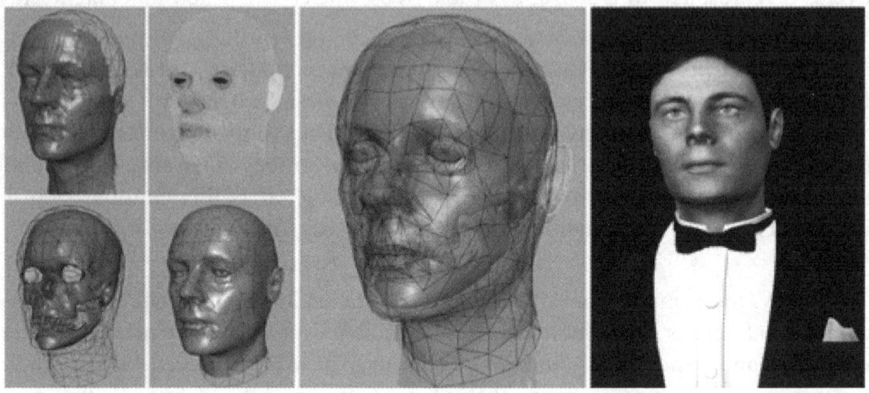

Figure 3.1. *An immersive personalized situation based on photo-realistic avatars can bring new intensity of experience, which in turn changes the underlying model that underpins the rapport*

3.2. Knowledge on the move through networks: examples of innovation processes

We believe that our societies will witness a dramatic innovation as we learn how to narrow the gap between virtual environments (VE), networks and users. Experts on innovation consider certain characteristics to be essential for the successful fielding of systems based on virtual reality. These are as follows:

– sharing the same immersive contextual environment;

– taking the corpus of human emotional states into consideration;

– providing a communication layer to express non-verbal signs of interaction (emotions, body language, hand gestures, etc.);

– adhering to the rules of social dialogs and turn-taking.

Whenever we can associate humans to evaluate situations by bringing in their own emotional state, we add a virtuous benefit. "*Key systems [that] are critical to successful innovation are formal idea evaluation systems that separate creation from evaluation...*" (James Higgins [HIG 95]). We consider a kind of innovation to structure the complex interplay between users and decision-makers; and we would postulate that it is the way forward towards sustainable innovation, whereby innovation is fueled by *"intelligence [that] populates the periphery of networks i.e. the people who inhabit and use the network"* (George Gilder).

We believe that the combination of networks and immersive emotion-based communication gives rise to a new perception of simplicity, usage, practice, and behavior levels. We are still used to considering networks as logistical chains, but they represent far more. When experts don't know the answer, networks do.

Networks enable the making mutual of the expertise and gather it around zones of uncertainty or ignorance. Whenever complete models of a problem are lacking, networks can be used to map out the intrinsic complexity, hence the usefulness of enhanced communication channels with virtual humans and their emotionally-charged rendering. In any case, the tangible outcome should be a solution that is easy to use and to interpret, i.e., to eventually assess. At user level, evaluation is inseparable from usage.

Hence, innovation can be dubbed *"knowledge on the move"*, know-how, practices, heuristics, and codes that flow through the enterprise networks, a new efficiency that energizes staff through intense sharing. At a time when knowledge management has become a challenge and a bottleneck in the securing and evolution of companies' assets, we must strive to free knowledge through new interfaces.

The process of innovation itself may be viewed as the creative resolution of that peculiar tension between the status quo and the escape route, between nominal and lateral thinking. Portioning it into small incremental steps, however, is often vital for success. Specifically, when innovation is too weak it is not keen on building new and competitive values or status. On the other hand, innovation that is too strong may disconnect and rupture companies from their established markets. There is a very narrow path – the width of a razor's edge – between lack of muscle and burning out too soon. This is a path that requires the constant interplay of humans, as users, prescribers, developers, testers and integrators. Therefore, valid modern business models are not simply business-to-business (B2B) or business-to-consumer (B2C), but rather should express the link between a supplier and the clients of its clients, as an integrated form that may be expressed as a combination of the two models, such as B2B2B or B2B2C.

Often, innovation is also nourished by emerging collective usages, as individuals become less adept at eliciting "needs and requirements". This is why networks can far better express the latter. Communities behave, through behavioral patterns, codes of conduct and experience, as one integrated whole that has a meaning. It is the network that knows, rather than its individual members who just cannot keep up with all evolving data and information and *evaluate* the global meaning at the same time. Teenagers using free-of-charge SMS in the mid-1990s quickly assessed the value that SMS had for their own needs, but did not reach the systemic meaning it had for telecommunication operators everywhere; this took longer to establish and the operators did not "ask the network".

3.3. Three laws underpinning technological evolution

Let us recapitulate the three basic "laws" governing the evolution of modern computing:

– *Moore's law* is the engine that substantially makes computers and digital devices more powerful every newly-released version (the density of transistors on a single chip shall double every year and a half). Network-based problems, be they problems in games or grand challenge applications – NP-complete or not – need just that amount of raw CPU power.

– *Metcalfe's law* states that the usefulness of a network grows with the square number of its nodes (users, in our case). As an example, when the Internet reached its critical mass in terms of the number of users having access to it, it became "the" network.

– *Gilder's law* states that bandwidth doubles every six months; hence the consequential communication capacity growth is three times faster than computer power. Of course, the usage bottleneck in a network is always represented by the last mile closest to each user.

Cumulatively adding to the effects of Gordon Moore's observation and conjecture, Gilder's law becomes appreciable (the total bandwidth of communication systems triples every Moorean generation). It opens the new economy with buzzwords that bear new value: collaboration technologies and cooperative work (of which the weblogs, Wikis, and other social computing systems are just the beginning), groupware, virtual communities, concurrent engineering, wireless, interactive displays, integrated value chains, enterprise networks, etc., with each proving that the real value is in the knowledge that it moves and transmits around. However, it is Robert Metcalfe's value proposition that makes networks endorsed by communities transact with each other through models such as P2P, etc.

Therefore, the underlying drivers of the new economy resemble a trilogy made up of the three rulers, *broadband, connectivity* and *usage*. Broadband is a main trend as bandwidth is the engine powering Gilder's law. People in technology often say bandwidth will soon become unlimited. Bandwidth drives usage. The second term, connectivity, is also very important as mobility depends upon it. The end of this spectrum closest to us is the mobile workforce, while the end that is furthest away is called ubiquitous computing. The term *"mobiquity"*, coined by Xavier Dalloz, captures and best expresses the depth and range of the concept. Finally, the last part of the new equation is the ease of usage. Usage is the implicit feature that is always pre-supposed by users, yet frequently remains misunderstood. Simply put, we look for things that are easy to understand and operate, but development still obeys its own rules and methods.

There is one other aspect of network benefits. Individual skills become obsolete quickly, but thanks to their access (or, more accurately, presence) within expanded (and now virtual) worlds, that risk is lowered. Yesterday, we used to say that innovation was the *"anywhere, anytime on any device"* paradigm, but this description has become insufficient. To the notion of access, which really means "access to content", we need to add the QoP, which means, depending on the case, empathy, listening, etc., i.e., emotion. The lack of such emotion only resorts to a sort of "indifference" – mostly the loss of context – created by abolishing distance and time in virtual systems. The half-a-century-long story of computing systems can be summed up as the story of creating indifferent systems.

In today's usage of computerized networks, it is individuals who carry the memory of contexts with them and make it flow freely. This is called experiential know-how. How could you not want to absorb it and inject it into new domains of interest? Competitive behavior within networks requires not only a procedural mindset and the routine to document those procedures, but a better tribal ability to accelerate the cycles of becoming.

So how is it possible that we seemingly miss the proper channel that can install the technically available bandwidth at the user's fingertips? Call it interface if you wish. What are the conditions that a new medium must offer? Before we provide a solution to this question, we shall revisit the definition of a technology lifecycle.

3.4. How do virtual encounters ride the technology lifecycle curve?

One of the many representations of the cyclic nature of technology is shown in Figure 3.2. As time progresses, the technology phases evolve from a basic technology trigger which enables innovation, and goes through rapid and inflated expectations that peak without fulfilling the potential initially anticipated. At this point, capital often withdraws and the long and dark period of disillusionment follows where only the companies and individuals with a vision continue with the process. Eventually, the landscape of business and technology changes, and a slope of enlightenment provides a gradual but expectable pace whereby the innovation reaches its productive phase and eventually plateaus out at a level (usually lower than was initially hoped for) where it becomes a truly useful contribution to society.

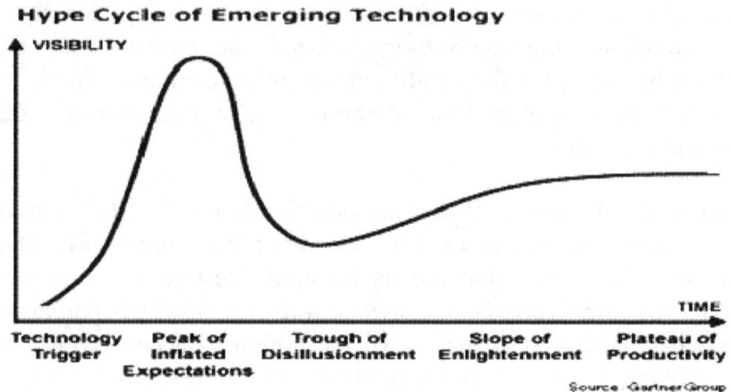

Figure 3.2. *One of the many representations of the cyclic nature of technology (Source: Gartner Group)*

So where are the fields of virtual reality and virtual environments today in this process and what are the events in the business environment that need to happen in order to start pushing an innovation up the slope of enlightenment?

To answer the first question is relatively easy. Virtual reality (VR) and virtual environments (VE) are probably on the slope, halfway towards productivity. The basic tools (head-mounted displays, rendering screens, trackers, etc.) and the associated software methods have been available for decades. Yet they were too expensive to enter the homes of mainstream consumers due to the professional nature of market drivers behind the technology (military, arcade entertainment, scientific visualization and industrial design markets).

It is only recently that a chain of events happened that opened up the possibility of brand new applications of VE. Thus, to answer the second question one must look at the side waves of other revolutions (such as the dotcom era) and build upon those by adopting the values they represent and taking a fresh look at how the existing elements (large bandwidth, computer graphics, "prosumer", i.e. professional consumer level, digital video cameras) may be combined into something unique.

The loss of space and time has been central to the advent of the Internet and has accelerated economic pace. With VR, one salient market value is perhaps in the control of time processes (or timings, a notion that mostly speaks of time structures). The resulting loss of personal engagement (emotions) through the Internet must be now resolved to restore human meaning. Novel innovations should result in re-centering systems on humans. Emotions extend the person to person (P2P) operations model to many to many (M2M) modes whereby each node of a network enjoys the bandwidth of the whole and expresses individuality. They transcend the personal computing era of the last 30 years (1975-2005) into the personalized networking era to come.

When we widen virtual encounters through networks, we build a grid of brains and experiences that is greater than the sum of individual brains and experiences. Today, however, grid architectures stumble on the need to specify interfaces and through such issues as synchronization, problem decomposition and parallel processing. This is where virtual reality and virtual humans come in. Through their immersive paradigm they offer an ultra-wide band of connectivity in a most natural way. People have long experience in natural communication; computers do not. For the first time in the history of modern technology, virtual environments decisively center on people and bring technology to people, rather than the reverse. We can mesh business models based on enterprise memory and network memory in an effective way. From such practice new "killer applications" are expected to emerge that in turn will drive the wheel of innovation further.

Virtual encounters therefore bear that particular ability to augment the innovation space, to weave relationships with meaning, and to extend the P2P model to a fully-implemented M2M model.

3.5. The virtual human interface (VHI) brings a new meaning to communication

To mimic the quality of everyday human communication, future computer interfaces that implement virtual environments must combine the benefits of high visual fidelity animated human agents with conversational intelligence and the ability to modulate the emotions of their users in a personalized manner. High fidelity virtual selves are enlisted to utilize the natural means of interaction, e.g., words, gestures, glances and body language, instead of traditional computer devices, such as the keyboard and mouse.

We created a novel and advanced human-computer interface that combines real-time, reactive animated virtual humans with perceptive capabilities (primarily artificial vision and communicative intelligence) to implement a closed-loop interaction model between users and complex computer systems. The system, called the VHI [TAK 03, TAK 05], was designed to build an affective bond and use the method of *emotional modulation* to increase the efficiency of information transfer, and to help the user to better understand, and later remember, the content being presented. Users entering the VHI may enter a networked VE in which they may meet other people.

To achieve its goals, the VHI builds upon a psychological model of human interaction, attention mechanisms and non-verbal communication that considers the process of human computer interaction as an active dialog. Its goal is to mimic the qualities of everyday human communication and information exchange by uniquely combining advanced computer animation with perception and artificial intelligence. Specifically, a high fidelity (photo-real) virtual agent extends verbal content by conveying information beyond the capabilities of traditional human computer interfaces (HCI) in the form of a complex dictionary of non-verbal signals that include body language, hand gestures and subtle emotional displays that all rendered in a reactive manner driven in response to the real-time actions of the user. These actions may be controlled by intelligent processes (artificial intelligence) or by another user representing him or herself in the virtual environment. The primary cues of social interaction allow the virtual human representation to identify and track faces, maintain eye contact, follow the actions and reactions of another user, etc.

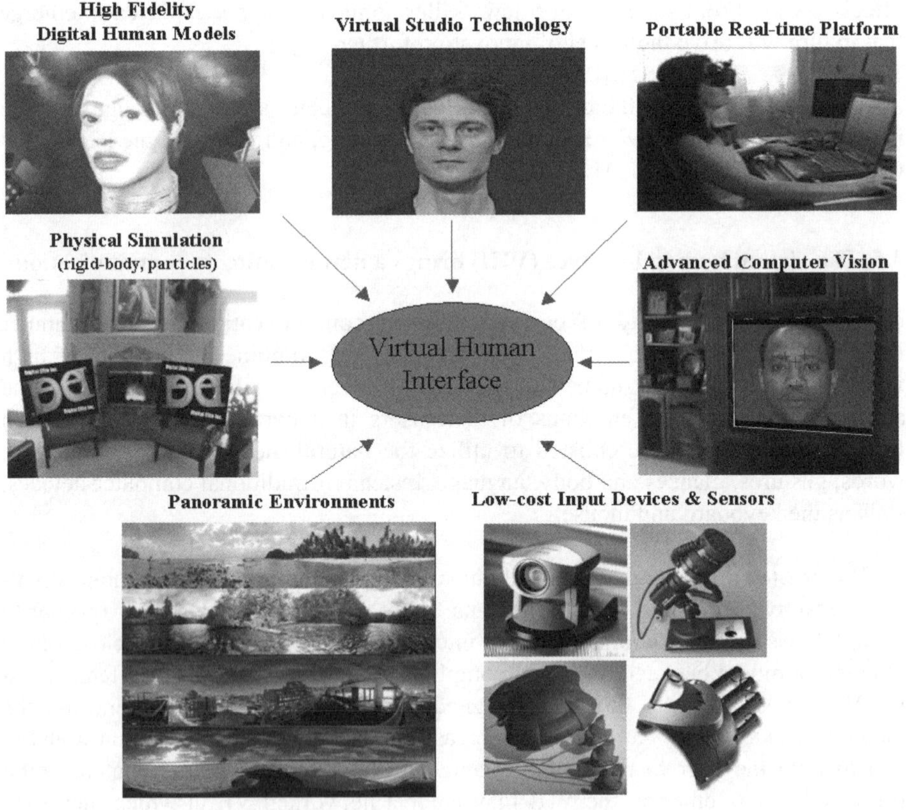

Figure 3.3. *Key modules of the virtual human interface virtual environment*

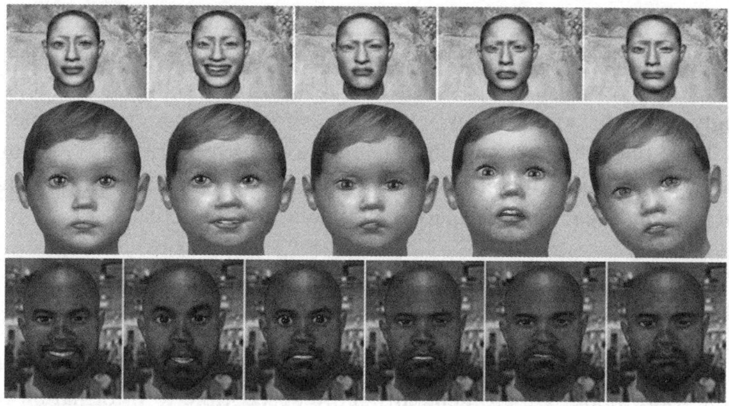

Figure 3.4. *The virtual human interface system uses the innovative method of emotional modulation to increase the efficiency of information and knowledge*

One quickly realizes (without going into the details of how such a system operates) that to fulfill the demand for such a dedicated layer, the tools of animation and perception must be combined into a single closed-loop model that mimics the processes of human interaction and communication. Photo-real virtual humans (digital replicas of living people animated by computers) – who users consider as their equal counterparts – achieve this goal by provoking direct emotional responses via facial gestures and body language. It is precisely those emotions, continuously modulated by the non-verbal cues of the virtual human representation in a reactive manner, which lay the foundation to open an extremely powerful information and knowledge gate and throughput to our brains and thus help users to communicate and learn faster and better.

"Information consumes the attention of its recipients, hence the wealth of information creates a poverty of attention, and the need to allocate that attention efficiently among the sources that might consume it" (Herb Simon, 1995 Nobel Prize winner). Thus, the primary problem with any form of communication is the challenge to direct attention to one or another particular aspect and maintain a level of interest that is high enough and lasts for long enough for the other party to be fully able to absorb it.

Emotional modulation is a technique used by the VHI system that helps to transform *information* to *knowledge* by using our own emotions as a catalyst. Our world is filled with *data*. Data in itself is not a piece of information, but rather it holds the potential of becoming one. For data to become useful information it first must go through a verification or editorial stage that guarantees and qualifies its value in a given context. In essence, when evaluating the usability of information from a business perspective we place our trust in the editors. The purpose of learning is to turn this information into *knowledge* and later move it from the knowledge domain to the business domain. However, as the amount of information grows, the brain finds it increasingly difficult to keep up with this ever-increasing demand and function properly. Thus, one of the greatest challenges of today's modern society is how we can enable people to increase the efficiency of the learning process itself, in general, and their own learning capacity and capability to interact, in particular.

3.6. The emotional modulation opens up new business spaces

The idea of emotional modulation stems from a simple everyday observation. When in a good mood, we are generally more susceptible to information presented in a positive fashion, and when we are sad or feeling down, we prefer things presented in a more subdued manner. Entire domains of human activity are affected by this new idea: business negotiations, decisions, creativity sessions, etc. that are always constrained by a set of limits and conditions. Based on this observation, we can define an artificial emotion space (AES) for our own digital human representation and employ many strategies to link it to the real person on the other side or edge of the network. As an example, our digital human might exhibit layers of emotion that coincide with or in other cases directly oppose the other user's mood, thus creating the illusion of compassion or "playing devil's advocate", respectively.

Virtual humans are a key enabling factor to achieve emotional engagement, open up meaningful communication bandwidth and thereby successful business communication in future networked systems. They unlock the not yet fully realized commercial potential by creating an emotional link between user and a digital character. Face-to-face communication with a digital interactive virtual human is therefore one of the most powerful methods for providing personalized and highly efficient information exchange. In the so called *closed-loop* model of interaction, the digital character is capable of perceiving the moods and momentary emotions of the user and of expressing their own feelings in a way to reflect empathy and understanding. In addition, the VE system may gauge the user's level of attention, fatigue or emotional state by actively *prompting* them and looking for an appropriate reaction or using built-in sensors, such as low-cost biofeedback devices.

3.7. The requirements for a VHI

A human face reveals a great deal of information to an onlooker. Besides conveying a person's identity, it can tell about mood, attentiveness, or even intentions. For humans, little effort is required to recognize and process the facial signals of another person. Therefore, to create virtual humans that "pass the test of increased scrutiny" and know the principles of facial information processing and recognition mechanisms is of critical importance. The meaning of facial signals is revealed through animated *visual cues*, which convey a multitude of information about the personality, emotions and intentions of the viewed virtual character. While information from each cue is ambiguous and incomplete, agreement across cues provides a vital working-out that helps perception and recognition of these events.

The conscious process of controling facial display is far from being perfect. Even in its neutral, relaxed and expressionless state, a living face is never rigid or motionless. Thus, adding high frequency, but low intensity, motion is one of the key essences of animating lifelike virtual humans. Micro-expressions are quick facial displays that flash across the face in a fraction of a second. They happen so fast that neither the person displaying the expression nor the audience viewing the expression is fully consciously aware of the existence of the expression. Although we perceive micro-expressions only subliminally, they provide a powerful channel of visual cues that convey the internal feelings of a virtual character, or of a living human for that matter. Despite the relative "invisibility" of micro-expressions, careful psychological experiments have demonstrated that people definitely, if subconsciously, perceive and recognize these micro-expressions and use them to guide their decision processes during the course of interacting with others. Micro-expressions belong to a special type of facial muscle actions that bear special importance for creating affect in a virtual human. Another important challenge is high-quality eye motion (and blinking), which is an important key to making a virtual human appear alive. It is also one of the most difficult problems to address. Eye motion and gaze is a complex mechanism governed by many different factors that are constantly affected by the environment of the digital character. Understanding how attention mechanisms and eye motion work is important in order to be able to create procedural animation routines that automatically generate realistic gaze behavior.

The link between emotional processing and facial display have been repeatedly demonstrated and verified by many studies during the past decade. Experiments have shown high correlations between facial muscle activity and emotional state. In particular, strong facial expressions can intensify the experience of other emotions. People perceive emotions most frequently, but not exclusively, through their visual system. Thus, micro-expressions, when properly animated, can be used to provide subliminal cues to the viewer and therefore they form a secondary language of communication which links users to virtual humans via non-verbal behaviors.

To create the necessary emotional engagement that is required to actively modulate another person's emotional and motivational state, a virtual human needs to possess a high level of control of its *body language*. Meta-communication is implemented in the form of subtle non-verbal signals that encourage or oppose the other user's actions in a given situation, and thus guide the other user to associate positive feelings with the piece of knowledge received. These layers of communication occur at many different subliminal levels and support verbal content by means of subtle facial animation, body movements and hand gestures. The final result is an original process that starts by using photo-real digital humans and their synthetic 3D environments and engages the user in a series of emotional responses, which in turn opens the channel to "engrave" knowledge and/or help modify behavior.

In the near future, advanced VE, such as the VHI, will serve as the primary medium to propagate innovation via the accumulation of encounters in a digital domain. Data, information and knowledge are transformed into knowledge with the help of emotions that encourage local problem-solving which can be shared instantly and efficiently. This form of local problem-solving compels us to be in close and constant contact with the field. By bringing local responses, it is also a way to sensitively listen and to learn; but the real reward in innovating lies in growing deeper models for business operations.

3.8. Bridging the digital divide: should not we replace the ill-fated WIMP interface?

Modern innovation should invest in people: intelligence, knowledge and wisdom, and not only in infrastructures, such as computers, routers, optical cables or mobile broadcast stations. Communicative intelligence is intentional. It is the representation of the will, desire and intention of people using the networks from their periphery. These intentions are to be found at cognitive level and are often undervalued and underestimated in the course of developing and deploying virtual environments. The systems that we created in the past limited our ability to express ourselves. It is with those encumbering systems that we grew an intrinsic digital divide instead of bridging the gap between knowledge and business; and yet intelligence is lurking right there at the periphery of these systems. Computing that is based on intermediary systems muddy the networks, blur business opportunities, and focus may be quickly lost focus if one's back is turned to real value.

Thinking, talking, communicating and sharing the basic experiences of others should not require technological tools that hide the bigger picture; instead, they should spur innovation and help users to encounter new knowledge. A typical counter-example is the free of charge Internet browsers, sometimes sold as a built-in feature of an operating system. Browsers have made us believe that fantastic illusion: that a browser window *is* the Web itself! Hence, the Web has become a perceived software within an operating system! Of course, the Web is not simply a product of micro-informatics or the product of a single company; it is a medium constantly formed by a community of intentions and inventions. Whatever the form of the next generation interface, we should strive not to make the same mistake as it reduces our collective business opportunities. The goal of innovation must be to create an open-architecture, perhaps open-source layer, that fully benefits from the value added by each individual.

The VHI may offer one such avenue. When creating emotions we actually improve learning. Emotional intelligence is not a dull process at all, but rather a highly communicative experience. It offers a new teaching and interaction paradigm that departs from the old "learning by failure" model by implementing a personalized and attentive mentoring attitude. Cognitive-assisted applications – that build internal behavior models of their users and become ambiently intelligent by detecting characteristic behavior patterns to improve user experience – stress the importance of affect and emotions. For millions of years, the fact that we had emotions was our evolutionary advantage over other species. Today and tomorrow, emotions mean knowledge – knowledge with vast business value. While the virtual humans with ever-increasing fidelity enable our networks with means to provide true emotional modulation – a sort of modem of the communicating human brain – VE becomes more transparent and what was once the end (virtual reality artifacts) becomes the means to achieving new business models.

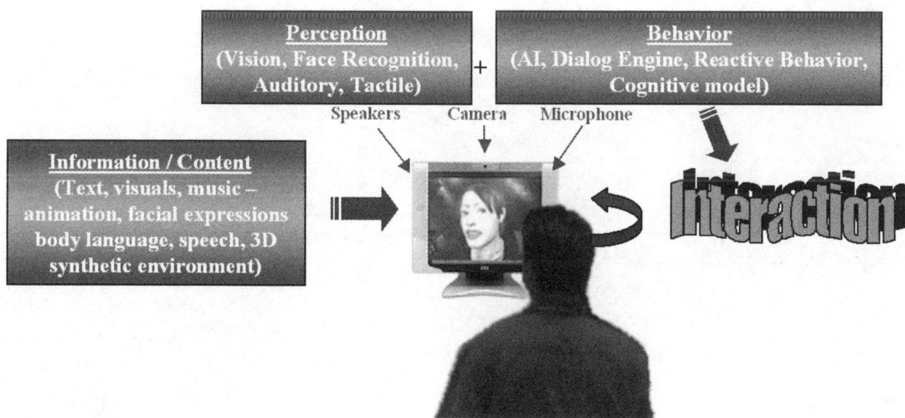

Figure 3.5. *E-hostess application for information services and entertainment*

By using these new technologies our collective memory will grow and will be continuously updated. At the same time it will also capture any marginal phenomenon that may emerge (e.g., model demand generation, new product acceptance in real time). The dynamics involved are inherently tackled; causal relationships are far transcended as a primitive model; synchronicity patterns are revealed almost exclusively by the networks used. How these networks eventually disseminate information is key to leveraging their effects and fully utilizing the potential lying within them. To promote today's innovation on demand, we need to become intimate with these novel technological opportunities and become capable of measuring the effects of such virtual phenomena. The impact on society, as well as the mental image of the self in an individual, must also be modeled and considered carefully.

With the help of my virtual self, an avatar with a unique economic identifier (perhaps a bank account?), I now may have a *virtual double* at hand. It is an "eMe" and I do not even have the burden of organizing it. Tomorrow, my eMe is my unique IP identifier. The well-known "information at your fingertips' motto of the early-1990s has become the "real-time and dynamic consciousness of your uniqueness everywhere anytime, ubiquitously through the network".

The VHI represents the new wave of thinking about teaching, and engaging in commerce and business. It is *"The Face of a New Generation"* and we predict that the on-demand economy will need this power backbone within a few years.

Chapter 4

Value Management's Creative-Destruction via Digitalized Innovation: The Winning Plan

4.1. Introduction

Concepts of innovation and value management (VM) could be regarded as a tautology. For us VM is management directed towards, and in response to, notions of "value" and this is more important than a strict adherence to a prescribed methodology. A practical distinction will be discussed that shows how some see values driving an innovation management agenda. We will call such an approach "management by value" (MbV). In opposition we see the dominant attitude that is focused on the value-outcome; this we will call "management of value" (MoV). Whereas the MoV is a weak form of "the ends justify the means", the alternative argues that "the means always lead to the sustainability of the ends". It is obvious for the specialists in this discipline that practice is located in the field of innovation ([WOO 04], [YAN 04]). It is not certain, however, that this "obviousness" is always recognized and many proponents cite examples of "better communication" or "joined up thinking" as to what VM is about, and in so doing fail to acknowledge how it engages with both society and nature.

The highly divided opinion of practitioners links VM and innovation to either short- or long-term economics, but again fails to offer a comprehensive explanation as to why this may be the case and it certainly fails to link the outputs of VM to an consistent theory of value (e.g., Ricardo's theory of rents or Marx's theory of labor value or Adam Smith's theory of the Invisible Hand). Because a theory of value has

Chapter written by Jean MICHEL and Roy WOODHEAD.

not been adequately defined in this field we anticipate that many will be forced to contemplate the implication of the distinction between MoV and MbV to the relationships between established techniques, practice and the results that flow from such interventions.

When considering the pedagogy in VM, it is necessary to regularly point out good principles, the ones that are fundamental, so as to ground VM in "how shall we believe?" and place education, research and enquiry at its core. The prime object of this chapter is to give access to key elements that characterize this MbV view of VM and to make explicit some of the potentialities it brings, with particular relevance to processes of innovation, which some actors seek to facilitate. To distinguish this difference we will refer to MbV rather than VM, which we believe is largely practiced as MoV (management of value). The fact that we put forward such a distinction means we can now move away from defending training programs and move towards a richer pedagogy. We will also argue for a shift away from the strict adherence to a prescriptive methodology. We will cite Schumpeter's notion of "creative destruction" and the digital revolution as a cause that both end a "one recipe facilitated workshop" logic and point to a need to retain the key functions born out of value analysis (VA) and value engineering (VE) and bred into VM. We argue that we should not perform them as before, but from within a richer framework built around "research" and "participatory inclusion". We shall challenge VM and its narrow view of what constitutes "value", and widen it in a "best fit" and "ethical" notion that can only be articulated through sensitive and empathetic research-based enquiry. To distinguish such an approach, we have used the notion of "valorique" which is already in use in this field.

4.2. The straightjacket of selling training and certification agenda

The training agenda in this field of VM has made many recoveries during the last 30 to 40 years. Generations of methodologists, convinced of the capability of innovation through an analysis of value and function, sought to convince owners of companies and financial comptrollers to invest in this field, and to implement such practices in support of progress and competitiveness. The "cost cutting" agenda has been given too much emphasis at the expense of a true search for value. One could naturally question our interest in putting forward this work which suggests "cost cutting" is not the only view of value available.

Several reasons cause us to return to history and to seek to deepen our understanding of it. It is, on the one hand, a fact that innovation, like the tides that come and go, is cyclical and therefore systemic. It is marked by periods when one "core technology" dominates strategic agenda which are then followed by other periods when such "core technologies" are eclipsed, pushed aside and demoted. The steam trains gave way to electric trains and the ease of traveling in cities has been challenged by the success of cars and the resultant traffic congestion. Similarly, the heavy sack of letters that the postman carried is now reduced by the phenomena of email. The way society functions is transformed by technology. Today a movement appears to be revolutionizing innovation ([PER 01a], [EUR 99]) and the difference will enable a durable or sustainable VM if managed adequately. This is our polite way of warning that we are calling for VM to adapt. It is important to recall how approaches to value, function and innovation still remain potentially rich sources of an innovation-capability.

In addition, it is interesting to consider the fact that these same approaches to liberate greater "value" have been known for many years and that a change that is not seen as having a "pure form" is the basis for change resistance which exposes the true character of the field's ambition. It seems as though we can only sing but a few songs and repeat them over and over, as well as trying to get the audience to call out for the songs to be sung. One operational model (the Job Plan) passed from VA through VE and on to VM, and is now concerned with the practices of creativity and consolidation within the management of organizations, rather than within its processes and products as was the case previously. Instead it became a consultancy tool that lost sight of an integrated role within the organization's strategic planning and this deserves to be examined so that a new potential can be created.

Finally, this text aims to bring something original that will enable the transformation of organizations. It is the realization and recognition of the extraordinary development of digital technologies and better information, communication, and new working practices that impacts on organizational design and social networks. It is time for VM to be directly confronted by the digital revolution. It is necessary to look further into possible synergies that may exist, or could be developed, between a more informed view of VM and its embodiment in the digital environment. We would see more integration and support for innovation by combining an intention that links a more considerate method of VM to a relationship between innovation, function, and value, and opportunities emerging from the growing digital infrastructure.

4.3. What exactly does innovation mean?

Before focusing on the relationships between VA, VM and innovation, we need to examine what is generally understood by "innovation" and to highlight certain characteristics of the processes within innovation on which a new approach to VM could act. In its broadest sense, innovation can indicate any change introduced knowingly into a solution by an agent for the goal of greater effectiveness and better use of scarce resources.

Similar definitions, to which many specialists adhere [PER 01a], emphasize the idea that innovation is the insertion, diffusion and deployment of an "invention" into the fabric of socio-economic contexts. For example, when Henry Ford enabled mass production, the invention of the "Model T" triggered new ways of assembling cars, the need for road building and car parks, and so on; our way of living was thus transformed. In this text "innovation" is thus far larger than "invention". This distinction highlights several interesting features:

1. The voluntary introduction of a change, innovation, and intervention that alters things for the better.

2. Determination or will of an actor or authority who "innovates" and the searching for a more efficient use of the resources.

3. The knock-on effects of an invention being woven into the fabric of society (i.e., the innovator's role).

4. The role that major innovations play in shaping society, our priorities and values, and how we form our own identities and beliefs.

VM declares its outward gaze to be fixed on principles of innovation and increased value, but the distinction between invention and innovation is that optimum solutions, or elegant solutions, are answers to problems that better satisfy varied needs. It is from this "diverse range of needs" perspective that the work of VM is more often than not practiced as MoV and consensus seeking and sense-making are carried out with a limited number of "internal people". One can read various articles about VM which use words such as "consensus" and "customer involvement", but these are limited by the prescriptive need to run VM studies in the form of workshops which often prevent widespread involvement and so minimize the actuality of what the words mean in practice.

Schumpeter [SCH 43] brings forth the concept of "creative destruction" whereby an old core technology is pushed aside and firms unable to adapt wither on the vine like a grape missed by the pickers. The digital phenomenon presents the same "creative destruction" to VM as the old training agenda and workshop technologies face the cyber-challenge. Schumpeter distinguished five characteristic situations of innovation:

1. The manufacture of new products.

2. The introduction of new methods of production.

3. The realization of new ways to organize business.

4. The opening of new markets.

5. The capture of new sources of supply.

For Schumpeter, all these types of innovation are more concrete through "the execution of new combinations" introduced by entrepreneurial and dynamic heads of companies. These senior managers are agents who are responsible for their firms and indirectly for society as economic development unfolds or recedes. Just as the successful companies of the past gave way to today's Microsoft and McDonald's, they too will be pushed aside as they are trapped inside their own outdated world views and rigid operational procedures. VM is identified with characterizations of innovation but its success is the cause of a promotion of its exacting methods and its standardization so as to enable training programs. VM enables an entrepreneurial spirit in organizations and seeks to rearrange the factors, or means, of production more efficiently and more effectively. Yet it too has not embraced new core technologies and so faces a Schumpeterian destruction unless it can apply innovation to itself.

It is still usual to distinguish two types of innovations according to whether one seeks to respond to the market and customer expectations (i.e., market pull). The alternative is where scientific and technical research enables new inventions to become innovation (i.e., technology push). Technology transfer is where an existing technology is applied in another field and offers enterprise revenue from licenses. Again VM has failed to capitalize on enabling this to happen more systematically. Innovating achievements are not always founded on new scientific knowledge but can be of socio-economic interest and as such are often in response to the realization that customers elsewhere would benefit; and so their market pulls the invention towards their innovation. The second type of innovation is based on the discovery or creation of scientific facts which were not known before. Here the properties of things are explored to see how they might be useful.

In the two types of innovation "the inventor intervenes into innovation as someone recognizes both a need and a relevant technique". VM is at the heart of this double process of invention and innovation by virtue of the importance that it places on the satisfaction of needs and the search for the best economic solutions (and thus the creation of greater economic value); that is, VM offers a means to meld all the potentialities, including practical know-how and scientific know-how, and is an active mobilization of the data-information, knowledge and competence necessary to achieve "capability".

4.4. Value management: a long history

It is not our intention to repeat the chronological history of VM. Many such narratives have already been published in the specialized writings of societies such as SAVE International (Society of American Value Engineering International), AFAV (French Association for the Analysis of Value), and others. We will limit ourselves to a few key points as we head towards the realization that a

Schumpeterian [SCH 61] moment is upon this field as its "craft based" culture must give way to the needs of a digital age, or at least adapt to accommodate it.

It is generally considered that Larry D. Miles was the inventor of VA just after the Second World War. VA was an intelligent and effective technique to reduce costs by way of making explicit something's function. While working in "purchasing" within General Electric in the USA, Miles observed and understood the formation of the cost of components within industrial products as money spent to achieve the performance of functions ([MIL 61], [MIL 89]). Miles's brilliant idea was based on the realization that many of the causes of cost (i.e., particular design solutions) were not focused on satisfying customer requirements and lacked "systematic" ingenuity. In fact, sometimes the customer was almost inconsequential to an arrogant design culture, the "experts" of which inadvertently failed to listen to customers. Following World War II, demand far outstripped supply and thus the commercial power of customers was weak. However, this trend changed, but the design culture did not adapt as quickly. The question or key focus of VA was to discover what a product, component or design solution did that made it valuable and useful in some way to customers. To identify functions, Miles asked of a component "What does it do?". This interrogation method helped to draw the design team and other disciplines into an in-depth questioning of the design and the commercialization strategy of the company. The firm's "inventors" were connected to others so that "innovation" was possible.

While recognizing Miles as the father of today's VM, it should however be acknowledged that the "function analysis" element (the method used to question "what does it do?") has much more remote origins. Thus one finds this thought within, for example, the work of the eminent French architect and architectural theorist, Viollet-le-Duc, one century before Miles. Even Aristotle's concept of "teleology" has the same "purpose-function-goal" logic and still exists in the study of biology. However Miles was the first to formulate a technique of naming functions with an active verb and a measurable noun. Even this "verb-noun" technique caused, and causes, confusion between what is a function and what is a process and, therefore, a deeper level of intellectual enquiry is needed. Miles acknowledged this and called for intense concentration when trying to name the function of a thing. Miles was arguing for a technique that focused a rigorous intellectual investment.

VA was developed in American industry during the 1950s and thus emerged from within an American culture that saw itself as entrepreneurial. VA was used by captains of industry as well as by government agencies. The public sector's need for probity and accountability also imposed expectations on the field as the need for such devices as contractual clauses started to shape and define VA practice which played a role in the emergence of a variation that became known as value engineering. These "requests" later became codified into expectations such as "a VA workshop will take five days" and since then have become core rigidities that now hamper their own adaptation. In fact what has happened is that some practitioners, who need to get work from clients who are unwilling to commit key personnel to be out of the office for five days, have adapted their approaches, and sometimes furtively. This has placed some professional societies in a policing role as they try to ensure that a consistent and common mode of practice is played out by its members.

VA gained popularity from the 1960s to the early 1970s and the terms VA and VE were used interchangeably. A distinction did exist between the two terms as VA looked at components that already existed and VE looked within the design stages. They spread to a number of industrialized countries and were applied in large defense, aeronautics, automotive, engineering, telecommunications industries, then in various supply chains which included SMEs, and finally in tertiary sectors. In the late 1980s, they began to be applied in the European construction industries as well and it is from that context that VM emerged in the UK as the larger discipline of construction management tried to distinguish itself from civil engineering and the kind of design work civil engineers undertook. Trade associations gathered experts practicing these approaches to VA (SAVE in the USA, (IVM) in the UK, (SJVE) in Japan, AFAV in France, etc.). The need to clearly define VA and VE as products led to the acceleration of codification and rigidity as the call began for British, French, European and ISO standards. The work of achieving standardization was then undertaken which made it possible to recognize VA and VE as effective means to competitiveness and innovation [AFN 98]. Professional certifications were also developed and the formation of training programs was systematically developed. As stated above, in the late 1980s and early 1990s, VA had already given way to VE which then gave way to VM. All such developments can be seen as representing a consultancy view of design, invention, and innovation processes being widened with a view to engagements and commissions.

The 1980s saw a type of "popularization" of VA and the start of its decline in popularity as its critics "dumbed it down" and linked it to "basic" technology ([EUR 99]). Some people could be heard saying things such as "value engineering is only useful at the technical stage of design" or "VA is only concerned with widgets" and in so doing they lost sight of the relationship between these methods and the underlying functions for which they were methods.

The basis of VA is regularly taught in many higher educational establishments in France and also in the introductory technology courses in secondary education, but this has been much less the case elsewhere in Europe. VA, VE and VM must now face competition (seemingly at least) from new philosophies of action and new methods such as quality management, project management, risk management (TRIZ), etc. VA, VE and VM can no longer evolve as businesses move from national to global; the field has to face its own innovation as the digital age bites. It is necessary for this whole field (VA, VE and VM) to be repositioned, redefined, and integrated into other modes of innovation and competitiveness ([YAN 04]).

It is interesting to underline the very particular contribution, in the early 1990s, of the French experts of VA who specified very rigorous modes of function analysis and how to make a CdCF – *Cahier des Charges Fonctionnel* or Functional Schedule of Conditions – a tool at the heart of "the expression of need" and thus the need for both invention and innovation ([AFA 97]). This formalization, a product of a French "rationalist" culture along with the weight of government expectations of auditing, and dominated by large professional engineering bodies, was not replicated by key economic protagonists from other countries in Europe. There were many parallels between France and the USA, but rather surprisingly not with the UK which had been heavily influenced by less rational approaches that stemmed from a school of VE that operated with a technique named "Customer FAST" and from two academics, John Kelly and Steve Male, who ran an MSc module in VE at Heriot-Watt University in Edinburgh. Through these courses a small number of American consultants were brought in to run training courses and their interpretations of what was good practice influenced norms in the UK.

The word "customer" caused much debate for it often resulted in an assumption of customer preferences without "actual" customers being involved. Given that the context was mainly construction projects, the distinction between customers, sponsors, consumers and end users demanded a participatory approach to VM, but the logic of people in workshops and the attitude of that industry to more often than not seek cost minimization meant that such a capability was invariably difficult to achieve. Not even an approach named "soft VM" which gained popularity in the UK, Hong Kong and Australia made this distinction between MbV and MoV explicit, but some people, for example Roy Barton from Australia, did promote "learning" and VM as an educative process which could claim to have had an impact on our thinking today. A highly-regarded academic in the UK, Stuart Green, borrowed from Peter Checkland's "soft system methodology" (SSM) of the early 1990s and was a key advocate of "soft VM" which characterized value management as "hard VM" and inappropriate for the social problems of management. Checkland and Scholes ([CHE 99]) have recently reviewed SSM and the differences between a systems view and a "functional view of systems" is clearly missing in their perspective; thus Green's use of the SSM schematics was consistent but also missed the point of VE's call for functionality to be made explicit. Green led UK practice toward general views and techniques of group decision support and smart VM. VE was no longer "fashionable" in the UK and few continued to develop function analysis in order to continue the project which had started with VA. As a consequence, VM in the UK has had a difficult time defining itself as its underlying principles, based on a relationship between "value", "function" and "innovation" were lost in the debates around rhetoric. Figure 4.1 depicts a view of the genealogy of this field.

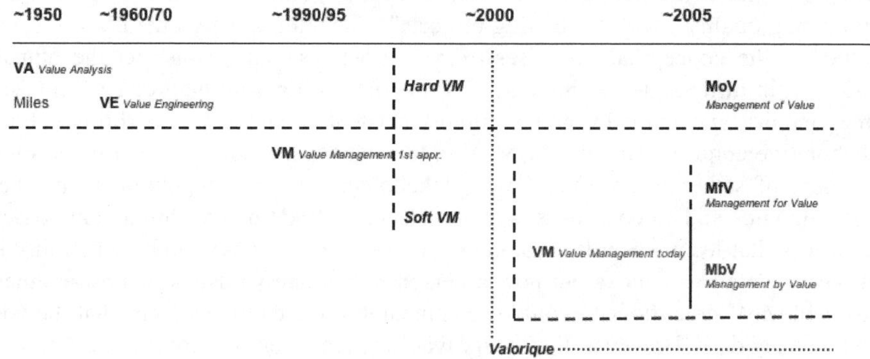

Figure 4.1. *The genealogy of the "value management" concept*

During the last decade, various paths became seriously scrambled. At the international level, VA, traditionally focused on hardware or components, gave way to VE, which focused on the design processes ([WOO 02]). The applications of VA methods left the traditional venue of large-scale industry and moved closer to assisting with decision-making in complex environments (i.e., the design process and VE) – "value information engineering" – ([MIC 96], [MIC 99a]). Many new and interesting applications of VM outside industry, for example in the tertiary sectors (social, cultural, health, etc.), were developed which also made obvious the need for "softening" the traditional (and hard) value approaches ([MIC 04], [MIC 01a]). The meetings of experts in Europe resulted in discussions that tried to widen the scope of VA and VE so that they could become more appropriate in the boardroom. The key was to challenge "cost reduction" as the only source of value, which was an obsession with some practitioners. In this transformation the emphasis was on managerial and systemic agenda and thus about VM and how to standardize this MoV logic ([BRU 01], [AFN 00], [WOO 01]). It is necessary, however, to recognize that if the desire to pass this methodology on to a managerial level was real and genuine, in practice it faced extremely different cultural and contextual realities, even within Europe.

Within Anglo-Saxon practice, in the USA and the UK in particular, a strong ambiguity of this concept of VM as MoV has existed. The French experts choose to speak about *Management par la Valeur* or MbV, rather than about MoV. This point is not trivial and has profound implications in practice. For VM and the MoV school, there is a mindset in which the notion of value exists outside the heads of people and as if it is a tangible phenomenon. Such a mindset invariably seeks to measuring value and often in the form of money. As such a narrow view of value is brought to bear, cultural value, moral value and personal values are seen as inconsequential and external to the task of "cutting costs", "reducing schedules", "optimizing quality", and "delivering projects". The alternative view of VM locates "value" in the conceptual processes of people who socially construct the human world that in turn resides within a symbiotic relationship with nature. This mindset forces inquiry and empathy and is almost existential in the way it brings about collaborative enquiry. The aim of MbV is to seek "best value" as determined by a collection of stakeholders. Often such stakeholders are in competition with other rival groupings and so consensus is about accepting trade-offs within a value-based framework that leans towards Aristotle's views of "the virtues" and how morality is not within stated yardsticks but how a person trades one yardstick against another. MbV and MoV practitioners would use comparable words to describe what they do and how they do it. However, their work would operate very differently and arrive at very different outcomes. We must not assume paternalistic roles or that we know what's good for other people; it's about doing VM with people rather than to them.

The drive to define a "European standard" in a tight timescale meant that such distinctions were never fully discussed. Furthermore, some people in the UK saw the project as being about defining a methodology that clients could specify and therefore sought to create a rigid conceptual framework of "commoditized" methods and techniques laid out in a prescriptive logic with a "product definition" type endeavor where methods such as function analysis were not seen as ways of interrogating our assumptions, but as a "unique selling point" that distinguished VM from other methods so that it would be easier for clients to procure services. MbV approaches are about "not assuming"; they are not about selling a cookbook-recipe approach to problem solving and therefore they seek to develop an approach that begins with inquiry in its truly "action-research" sense.

The last few years have seen the French and enlightened scholars from the UK and Canada join to emphasize the concept of MbV and the need for practices that are recognized as intellectually rigorous by decision-makers. It is not about "I don't understand your issues, but this is how I will solve your problem with a standardized game plan", it is about "What does value mean to the different stakeholders and how can it be best achieved, for an enduring judgment that has actually been achieved?". Paradoxically, "value" disciplines are not taught in many universities or schools of engineering across Europe, but are discussed, in part, within business schools, especially in the topic of marketing. For reasons of convenience, we want to indicate this set of disciplines and approaches to MbV the generic term of "valorique" ([MIC 01a], [SEN 01]). The purpose of this term is to allow the differences in attitude and approaches to be made explicit as a means for the value community to explain and deepen their understanding; we use it to distinguish and show how practitioners can shift from the MoV to the MbV perspective and hopefully to a combination of the two at a later stage in their professional development. We are not simply talking about a method; we are advocating a new fundamental basis for practice which is grounded in a research culture.

This approach, which we call "valorique", must still achieve results. It must acknowledge the strong concerns which have emerged and seek factual truths in order to fuel innovation and achieve competitiveness. It is about revisiting VA before the rigidities and standardization trapped it into becoming a "routine" as its rules were taught but not the underpinning axioms that made it powerful. It is about using logic and hypotheses to generate management theories that are tested for both truth and value. It is about providing a catalyst for invention and innovation within the functioning organization. The fundamental thinking within the term "valorique", as a philosophy, will develop in practice as it itself searches for its own articulation within other approaches to innovation, such as project management, quality, engineering system, knowledge management, etc. "Valorique" enables a kind of federation of doctrines and methods seeking value and economic efficiency ([MIC 99a], [LAV 01]). The pursuit of value, the recognition of function, the understanding of invention and innovation synthesize to become a grounding from which other methods then become tools. If a project is late, use project management as a context for the fundamental thinking within the term "valorique". If a new drug is needed, use "valorique" as a context of invention. If there is the need to get food to starving people, use "valorique" as the context for innovation.

4.5. Definitions and rigidity

We will always face a dilemma between the definition of a word and what it means in a specific context. If we see words as signposts to meaning then the "meaning" becomes more important than the assembly of the letters to form a word. Let us avoid slipping into vapid semantics and arguing about slogans that have no relationship with the truth. Instead, let us bring a process of rigorous thinking to bear on the search for function, innovation, and value. To fully understand the fundamental thinking within the term "valorique" one must learn research methodologies and the VA to VE and on to VM tools and techniques, for they are necessary and powerful when combined.

The weakness of the MoV toolkit is that with "standardization" comes a rigidity and hence a commoditization of such "tools". Practitioners of MoV do not engage in research, but rather steer the client's situation towards the "solution machine" that is their rigid process. The past attempts of consultants to develop a clear brand image and to make the USP (unique selling point) stand out, have resulted in a solidification of a body of theories that in the past yielded VA, VE, and later VM; "valorique" is trying to bring the underpinning theories back to the fore and thus is not so obsessed with particular toolkit perspectives and seeks to let the client's "problem" determine the choice of action-research protocol. Clients only look for commoditized methodologies when both the problem suitability and method-appropriateness are understood and this has hampered the uptake of VM by clients. Ironically, the practitioners have created their own "Catch-22" as they defend the "standards" and thus prevented any change. Furthermore, important variables can be overlooked as practitioners guide their clients to "fit within" the standard approach. Even more important to this work is when a revolutionary new technology comes along and practitioners stuck in the old ways of doing things, are unable to adapt.

The defining ground of the fundamental thinking within the term "valorique" is that we can agree that it is about enabling innovation, about changing something in order to bring or create an enduring value, or by limiting the degradation of value ([PER 01b]). This "value" for organizations is correlated with terms of "added value" (i.e. satisfaction) and the functional needs of users and other managerial aspects of control, revenue, and benefit maximization, optimization or reduction of costs, and other expenditures on social values such as skill development. To achieve this internal collaboration, implementation is invisibly linked to competences and the ability to mobilize the synergy between creative forces within the organization ([GRA 01]). "Valorique" is about using value as a means to bring stakeholders into the organization's *thinking-processes* and to achieve an advantageous "capability" that other organizations will find difficult to replicate; especially those rivals located in far-off lands. It should however be admitted that the concept of value remains one of most difficult to encapsulate and that definitions from multidisciplinary efforts fail to define value outside a particular context ([PER 01b]). We argue this because "value" is about preferences and so will always require a context in which "choice" is a central concern.

Written works abounds on the question of value and the fact that multiple meanings have been given to the word "value" reflects an "obviousness" of the various perspectives and legitimate choices of personal or collective "values". From the practical, MoV rather than MbV point of view of the observer who is searching for utility (such as an inventor, scientist, creator of new products or services) we can say that value is the differential of advantages with respect to the sacrifices (e.g. costs) or disadvantages that a given product can present, compared to another offering ([MIC 01c]). Value is a conceptual process born out of intelligence that allows us to prefer something – such as "luxury" instead of "discomfort" – and so also requires an ability to imagine future states and envisage possible outcomes. In less sophisticated animals than humans, a lack of intelligence inhibits their ability to prefer and thus survive; for example, the hunted fish that cannot distinguish the values of bait and food. We cannot "only" look to outcomes and ignore the problems of how to get to the outcomes; this is why the study of functionality is so central to our field. Value could be treated as the differential (or comparative) of advantages brought by money that is invested in a given product; but this is far from "adequate". A satisfactory theoretical definition of value must show how it is a reflection of "action" and "reality" and how things can be used to advantage. An advertisement in a newspaper tries to convey a message that guides our preferences so that we buy specific goods, products or services; here the customer must learn from the hunted fish and make wise choices or pay more money than any value yielded. Such judgments can be modeled on a multi-criteria basis ([YAN 04]), but are limited to problems of "selection" and do not help us to innovate; our role as customers in an advertisement campaign is usually to passively give some of our personal wealth in exchange for "ownership". Therefore, the concept of "property" is vital to any consideration of economics; if a thing cannot be owned then it cannot be exchanged for money.

The evaluation and the measurement of value thus remain ([FON 97], [KLI 04]) and can be approached, but not defined, with Bayesian logic (e.g. decision trees) which misses out the richness of "function" ([WOO 02]). The dynamics of value creation and stage gate processes are an attempt to structure the relationship between the context, the usefulness, the technological technique, the invention, and thus innovation. For us, value is about "preferred-usefulness" within a dynamically changing environment and is why function is at the core of our whole process of innovation. As "useful" and "preference" are always contextual, and any such notion of context is always a snapshot of a changing reality, so the notion of value must change, for they are all concepts that are inextricably linked. A chair is useful to sit upon and it is useful to prop a door open on a warm day. Our creativity is directed towards usefulness and preferences as well as understanding whether things will function as we want them to. Such a mode of inquiry determines innovation and risk-reward investment logic is generally accepted in an intuitive way, without too many questions for trivial problems, but is made formal in major projects ([WOO 02]). We are more comfortable with economic value when we define the context as "investment". However, just because we define the context does not mean others will see things in the same way. Here we return to the distinction between MbV and MoV and the approach of the fundamental thinking within the term "valorique" versus management consultancy.

For those steeped in VA, VE or VM and its core rigidities, it seems obvious that many tools and techniques can be used to go in the direction of an objective guided by the logic of MoV and claims of value creation. However, the goal of consultants is often focused on commissions, client approval, and repeat business within a logic of "quick wins" and "short-termism" rather than on research and long-term value. (This distinction is a source of sustainable competitive advantage for companies whose rivals cannot look further than the next quarter's results.) As such, they bring an attitude to collaboration that would conflict with an educative attempt at knowledge and skill transfer. It is not in the interests of the consultant to develop a capability that loses future commissions. Academics with PhDs are also researchers and transmitters of generic knowledge and endeavor to give a broader referential framework to the underpinning principles upon which are founded value approaches. It is culture and competencies that make the fundamental thinking within the term "valorique" feasible and transferable to non-academics, and thus must not be seen as a few scholars trying to muscle in with their own brand management USP. French and foreign researchers propose "valorique" for this corpus of disciplines based around research, knowledge, and know-how ([MIC 01], [SEN 01], [YAN 01]).

4.6. Potential of "valorique" in relation to the innovation

The force of the research perspective and "valorique" accord with several principles which, all taken separately are interesting, but are more powerful when they are integrated and articulated as a body. The following principles will be discussed in this section:

– problem scanning and framing: "inquiry and questioning";

– a "systemic" step with mobilization-confrontation from multiple points of view;

– a reference frame that defines "functional need" based on function analysis;

– cost intelligence and focusing on the economy of the means;

– mobilization of information, knowledge, and competences;

– project management and the rigor of value analysis; and

– explicit or implicit recourse to the practices and techniques that enable creativity.

4.6.1. *Problem scanning and framing: "inquiry and questioning"*

VA was born from the need to meet difficult challenges, such as that of substantial cost reductions, without degrading service or quality levels of the products placed in the market ([MIL 61], [MIL 89]). The value methods are of interest to innovation as they make it possible to confront critical problems and to grapple with contradictory requirements through an efficient use of a multidisciplinary team. VA is therefore a stepwise method of "problem solving" which is based on a clear identification and determination of the problems and their context, how problems can be solved in theory, and how a team can then move into implementation, and the testing of theories in action. These steps are at the heart of active pedagogies, such as "problem-based learning", used in the vocational training of engineers ([YAN 04], [KLI 04], [FON 04], [WOO 04]) and of doctors who are learning the skill of applying theoretical knowledge as they try to diagnose real ailments in real people. The key difference is the rational basis upon which the way we think is structured, rather than rushing in with the first potential solution someone suggests. There are stages of observation, analysis, diagnosis and treatment which seem to have been lost from the modern approach to MoV, as opposed to MbV and "valorique".

4.6.2. *A "systemic" step with mobilization-confrontation from multiple points of view*

At the heart of VA is an almost Aristotelian synthetic view of a joined-up reality that functions in a composite way [KAU 06]. Practice stresses the need to adopt an integrated perspective drawn from multiple interpretations and observations. The current operating modes in organizations make it possible to discover the dysfunctional effects of Taylorism, the "demarked" segmentation of trades and disciplines and a workforce alienated from the creative management processes needed to "invent", "innovate", and "invest wisely". The consequence of such "us and them" management approaches encourages conflict between departments as different interests fracture an organization into a collection of multiple and conflicting objectives. As mass production loses its advantage, the need for "unthinking" automatons as employees becomes even more ridiculous than it ever was. Today we need people with all kinds of know-how to fuel the organization's innovation capability. We argue the old styles of management, from Fayol and Taylor through Gilbreth, Bernard and Sloan and onto Ohmae and Kanter, to name but a few, have a significant responsibility for loss of competitiveness and loss of quality that could have been achieved if a "divide and rule" assumption that separates the manager from the employee had been challenged head on. It has seeped into most schools of management as the concept of "manager" is somehow socially elevated. It stems from this parceling out of fragmented commitments to "trust", "integrity", and "reliability". In some conflicts the badly managed interests subsumed values of mutual respect and shared destiny as an organization had its public vision and public mission statements as well as a myriad of hidden or unarticulated views which caused a loss of potential. It is for reasons such as this that VA has credibility for the fundamental thinking within the term "valorique" as leaders seek to cement the fissures within an organization's thinking systems which have been fractured by the organization's own policies, procedures, and internal power-games. While some managers looked to Machiavelli, they imposed their leadership through power and authority and, once hailed as Caesar, ensured that their Brutus would not repeat history. However, lack of commitment often follows and such enforced views and agenda fail to win the hearts and minds so necessary for a cohesive implementation and achievement of potential.

MbV and "valorique" give priority to this search for "best decision" in a coherent, forced and intelligent expression of particular points of view to be considered in the fullest sense of value. The passage of VA to MoV and to MbV (and the fundamental thinking within the term "valorique") accentuates a "systemic" dimension by considering multiple interests and complex social and technological relationships. "Value Management is generally concerned with maximizing the benefits of a project by satisfying the requirements of the various stakeholders involved" ([KLI 04]). Even so, it's not "MbV" but "MoV", and the language of short-termism is heard when we are told of "deliverables", "cost reduction", and "schedule improvement".

With globalization, the problem of cultural imperatives becomes more prominent. Where once we would not know how people in far-off lands were exploited, now our television sets reveal this to us on a daily basis and we are forced to confront our moral values and our economic values where "deliverables", "cost reduction", and "schedule improvement" mean children are exploited in far-off lands. Now we clearly see a major distinction between MoV and MbV. Globalization raises the issue of insisting that one view dominates the innovation agenda and promotes an enquiry into often conflicting views as is the core function of "democracy". If the goal is to maximize profit, then the global economy needs to be seen as a single entity and the trend of capital moving to low-cost economies to be seen as systemic. It is a system regulating itself as it tries to find equilibrium. The Schumpeterian view of creative destruction means we are forced to try to buck the systemic trend. This trend of a global economic system trying to regulate itself by moving wealth-creating factories from Europe and North America to emerging markets in the East ignores the social value of "work" in the towns and cities left to cope with the aftermath of "off-shoring".

4.6.3. *A reference frame that defines "functional need" based on function analysis*

The originality most cited, and appreciated, in value methods is related to a constantly asked key question, "What does it do?" in order to tease out a function. In other words, the check on needs is correlated to elements and parts to ensure they "work" toward a stated purpose. This functional frame of reference ([AFN 96]) makes it possible to engage in penetrative questioning of how things are currently done, and thus to innovate ([AFA 97]). This functional "questioning" can appear obvious, but it proves to be a powerful stimulus as it enables people with different technical and cultural languages to build a shared understanding and innovate as a team. The functional approach gives a real coherence, stability, and rigor to the reasoning and insights developed as a situation is mapped as functions by a group. Furthermore, as functions remain constant, solutions frequently change with scientific progress and technical advancement and so this functional logic enables forward-looking innovation to not only meet today's customer preferences, but also to bring new offerings by combining the latest scientific and engineering knowledge.

This "capability" is undoubtedly a good way to release creative potentialities and stimulate major innovation because it introduces a creative challenge which the whole team can take up. This challenge is: "how can we satisfy needs without imposing any notion of a pre-conceived solution that would undermine the search for alternative solutions?"

4.6.4. *Cost intelligence and focusing on the economy of the means*

A crucial question posed at the commencement of VA was how to substantially reduce the costs of industrial products. The answers to this question, which recurs for managers responsible for economic performance, can be numerous. If we see management as a game, then "fudging" is a possibility and short-term tactics, such as dumping annual sales stock, or outsourcing, or off-shoring, or downsizing or other steps that hide the "true" inefficiencies of existing ways of organizing business. VA and VM can bring an original answer that is both effective and efficient to this question, and again generates innovation and stimulates practical creativity; but only MbV can do so in a way that builds social relationships between the local firm and local customers and offer a new way of designing "supply and demand".

VA rests on what one can call a true control of costs (and all factors of expenditure, such as energy consumption, time, space, weight, etc.). This knowledge of costs is used to articulate cost factors necessary to generate value for customers, users or interested parties, and those cost factors which do not lead to the satisfaction of the needs therefore degrade and destroy value. That is, if the customer's ability to prefer is degraded because the producer believes poorer offerings will increase profit, then the difference between MoV and MbV shows "value" in terms of the true relationship (i.e. respect) between the firm and customers. It also relates to the durability of the "value" as one leads to short-term profitability at the expense of long-term trust. VA results in the basis of functional accountancy which makes it possible to connect the cost of solutions to the functions of products. These are correlated in a hierarchical basis, according to the "needs to be satisfied" and then take corrective actions in the form of new solutions that reduce the "useless costs" and augment "desirable benefits" ([WOO 02]).

This intelligence, or costs control, finds a natural location in design phases as an objective cost model (e.g. design to cost), and is now extended into the design phases design through more informed objectives (*Conception à Objectifs Désignés* or CCO) that reflect more pluralist design briefs (e.g. the ratio of benefits to increased costs). In this "design to cost" (DTC) approach, the objective of cost becomes the ambition to be achieved, which forces the innovator to seek more "new" solutions. It is known, with respect to artistic creation and design excellence, that the imposition of a strong cost "obsession" directs innovation away from solutions that focus on satisfying customer needs and attracting revenue and toward meeting minimum acceptance criteria.

4.6.5. *The mobilization of information, knowledge and competences*

VA is generally regarded as a demanding method, often difficult to implement by those not trained in its processes. It helps by framing problems that require a collective "sense making" to make innovating breakthroughs possible. The first proponents of VA stressed the importance of collecting the information available on the topics to be addressed. Thus, the mobilization of intelligence, knowledge and competence through the actors in the company or organization was quickly translated into a form of methodological instructions. They recommended the setting-up of a multidisciplinary working group and taking into account a scheme of work (i.e. the job plan) and in every VA study, a key stage known as the "information phase" is included. Today everyone agrees and recognizes the importance of these original schemata (although poorly articulated) and the capitalization of information and knowledge associated with VA and which is still present in VM today. MbV and "valorique" are grounded in a philosophy of action based on the interpretation and commitment of people and the alignment of competences in the resolution of their own problems ([GRA 01], [MIC 01a]). The functions of a "job plan" become more important than the scripting of it.

One can, without any doubt, affirm today that VA anticipated and aligned itself with modern steps of design and innovation, and that it is also well suited to the practice of knowledge management ([MIC 01b], [MIC 03b]). The methodological considerations of the last 50 years, the VA working group (i.e. VA team), and the mobilization of information, remain central to VM and "valorique", but can also be widened and extended to a more participatory "current" vision that seeks to reduce the "division of knowledge" (i.e. a product of fractured organizational designs) and increase technological competencies in the emergence of new information and knowledge-management capabilities. It is toward a certain form of 'hybridization' of the steps that VA and VM can combine themselves with the practices of previous years in new forms of collaboration in virtual mode (electronic forums, intranet sites, "real-time" virtual workshops, communities of practice, work or training) that are at the heart of this Schumpeterian "creative destruction" of VM that we argue makes the need for MbV and "real valorique". It is our view that by mobilizing, in an effective way, information, knowledge and competences, one can enable innovation as an "obviousness" and "acceleration of progress" empowered by the Internet and the generalization of the digital exchanges of data. We are able to construct a more informed basis upon which to interpret and understand reality.

4.6.6. *Project management and the rigor of VA*

In order to achieve the efficient and effective use of knowledge stored in the heads of key personnel involved in design, the founding principles of VA stand on wider inclusion within structured decision-making methods sequenced with the needs of real work. Thus, the current standards insist on early introduction of VA in new project or product development, whereby VA clearly defines various phases through a rigorous scheme of work (i.e. job plan) ([YAN 04]). In addition to this scheme of work, one can assert a clear determination of the various actors concerned (partner, decision-maker, working group VA, stimulating VA) and the whole mechanism of the validation of choices and decision-making. Without explicitly acknowledging it, the standards for VA introduced good practices for project management in the 1960s (that other methodologies may make the same claim should be seen as contextual and that actors of the day were responding to the issues of that moment). Today, VM and project management are closely related, to the point that we see corresponding trade associations, such as AFAV and AFITEP in France, converging to form an integrated methodology (e.g. project management, risk management and VM) that is fully aligned to ensure maximum economic efficiency. However, if one refers to our definition of MbV and the fundamental thinking within the term "valorique", one could say here that this "methodological rigor" for project control is not necessarily favorable for invention or innovation (its goal is to limit "target misses" that stretch such goals or even search for revolutionary innovation).

4.6.7. *The explicit or implicit recourse to the practices and techniques that enable creativity*

VA explicitly mentions a phase known as the "creative phase" or "speculation phase"; the VA standards mention this in respect of "one" technique, namely "brainstorming". The "analysis" within VA was often criticized because it emphasized the "analytical" too much. However, Carlos Fallon captured the essence of this debate when he joined up the need for analysts to break things down into parts to understand them and that their recombination was vital for them to function as a "thing" ([FAL 70]). It goes without saying that the analysis phase underpins the need for comprehension of problems that arise and constraints that are to be faced. However, for "analysis" it is also necessary to be able to imagine and name another possibility, to invent, and to find solutions. There is a necessary tension between orderly thinking and chaotic or lateral thinking that probably has a greater effect on how humans think than we assume. It is important to underline the fact that VA specialists understood this need. They knew that to couple an analytical step consisting of rigor with other creative approaches (brainstorming, synthetics, etc.) required different modes of thinking. Paradoxically, the stronger the analytical phase, the more powerful the creativity stage as it focused on innovation "pinch points" that, if improved, yielded significant benefits. This principle is also at the heart of MbV and "valorique".

In the previous section we discussed what we believe are fundamental principles established in traditional VA and in VM. We have also pointed out how attitudes based around a distinction between MoV and MbV are at the heart of our call for a new approach. We have named this "valorique" so that the presence of a distinction is firmly and irrefutably placed before this community of value innovators. The symbol of "valorique" is thus a tool to force a deeper inquiry into why and how value practice is as it is and also how it should develop in the future. We will explore this distinction in more detail in the next section.

4.7. Digital technology, networking and an ability to innovate differently

If we step back and see the Internet causing a shift in the way business is conducted, we can map, or observe, the specific contributions to "valorique" that information technologies and communication can make. To wonder about the complementarities or synergies that can develop between "valorique" and the "digital revolution" is to see VA and VM as solutions to some deeper function, and to see that new technologies enable us to contemplate and find new solutions. For us the shift is away from commoditized methodologies and the establishment of core rigidities towards a contextually designed approach based on underpinning principles rather than dogmatic adherence to prescribed methods, tools and techniques. Now VM and the fundamental thinking within the term "valorique" are about a research context and one that seeks a link between hypotheses and increased value. In a certain way, each one of these two terms voluntarily reduces to simple expressions, philosophies, concepts, socio-economic methods and organizations which influence our ways of thinking, acting, working and designing new and innovative products. It seems logical to test whether the VM and "valorique" approaches can lead to greater effectiveness and innovation in a mutually compatible way; can they co-evolve or must they fight each other for "market share"? We wonder if it is possible to distill their "essences" and thereby form a new combination that acknowledges the Schumpeterian "creative destruction" caused by the Internet and the digital revolution. We believe it is and will be so.

4.7.1. *The "valorique" culture*

The value approaches have changed our ways of thinking, and in particular with regard to the design of products and processes, and have been used successfully. We look back on past achievements with great respect and a sense of pride for having been in this unfolding history. Even if the canonical steps of VA (those that the various standards recommend) were not always applied everywhere with the same levels of competence, intensity, exhaustiveness, tenacity and regularity, it is important to underline several essential facts.

During the last 30 to 40 years, the functional approaches have spread into many spheres of practice and have become a methodological reference frame that unfortunately also has become a core rigidity that is difficult to circumvent. Thus, in the context of routine problems, functional approaches are found in procedures and are applicable to topics such as quality management and risk management. It also became the norm to view "value" in terms of some obligatory result (objectives to be reached, "outputs") and either, as was traditionally done, in terms of means ("inputs") as seen in Michael Porter's "value chain" logic which is simply a macro-economic approach to the determination of GDP (gross domestic product) translated into a micro-economic context (but where are the shareholders and institutional investors in his famous "Five Forces"?). With regards to consultations in the public sector, VM is often recommended and offers candidates a basis of a functional schedule of conditions (CdCF – *Cahier des Charges Fonctionnel* or function-based performance specifications) to help define performance expectations. Even in fields such as "software development" one comes to observe functional inquiries. The current orientations and cultural expectations of "certification of competences" or "accreditation of programs" implies a need to prove that value has been demonstrated, but not necessarily "achieved" as "politics" allows the subjective to reign over the objective.

During the previous three to four decades, one could note the regular questioning of design modes in product development and market launch. The design methods must satisfy the needs of customers with increasingly complex requirements and so, inevitably, also become complex themselves; but they must also enable value, not only to customers, but also to the various interested parties concerned with profitability. It is quite natural that one comes from that position to today to regard MbV, rather than MoV, as one of the key steps needed to face new expectations of competitiveness and "sustainable development" at a time when the objective of value creation has become a true motive of economic ambitions. Note that here we are talking of MoV which we have already argued promotes short-term gain over long-term commercial sustainability.

The use of intelligence (or more intelligence) in the design of products, while being constrained to the VA techniques (multidisciplinary working group, rigid scheme of work, mobilization of information from within the organization and its projects), was already established in the 1960s-1980s and to go towards what is possible today, namely the concerted design effort based on a solid knowledge management and mobilization of competences, requires the weakening of the rigidity of compliance with past mental-models.

This change is necessary to lead a company through the industrial era to the post-industrial era and is characterized by a necessary economic evolution toward the needs of service industries. The service industries are themselves based on intangible values, such as service quality and customer loyalty, and the need for the mobilization of knowledge and competences from a geographically dispersed multidisciplinary team that reflects the cultural complications of the global economy.

4.7.2. *The digital revolution*

Value approaches were born approximately 50 years ago, and they gradually developed throughout the second half of the 20th century. The "digital revolution" is a more recent phenomenon and is characterized by a very fast evolution and impact on the way new working methods have developed. This is a significant technological revolution that changes many aspects of the way societies work.

The Internet constitutes the most visible and spectacular part of this digital revolution, but it is necessary to also take into account some of the other digital technologies and their applications: CD-ROM, DVD, cell phones, virtual reality, blogs and wikis, and networks with or without wire.

It is known today that this digital revolution is as significant as that related to the arrival of printing works. It is a core technological advancement that has changed society. One can see that it greatly modifies the modes of thought, organization and work, and that it has had an unquestionable impact on the professions and the trades ([MIC 00a], [MIC 00b], [MIC 01c]). It is also known that it introduces new prospects regarding "action" and "location". It has transformed the relational fabric between agents in a spectacular way. Where once meetings required people to be in the same room, it is not always the case today. Digital technology is a transformative phenomenon, such as the steam engine and the Industrial Revolution, that changes the way we organize our work methods on a global scale. It technologically enables the concept of "globalization" with all the positive and negative effects that flow from a new mode of communication and shaping of ideas. The digital revolution is forcing us to change the way we develop our beliefs, and that changes our views and priorities and inner values which subsequently steer our actions and behaviors.

It is incontestable that the Internet and digital associated technologies stimulate operations in networks and make it possible to create new assemblies of competences according to the most varied geography and temporalities, often conflicting with the older and more traditional modes of organization. Whilst this chapter talks of a particular field, the subject is far larger than the topic we are concentrating on. The revolution has not finished astonishing us and involves us in a high speed, unbounded exploration of new continents and sometimes brings a feeling of faintness or fear (the images of the sorcerer's apprentice easily come to mind). The words "there has been a tsunami" are made much more real as technology widens our awareness of distant places and even shapes our response.

4.7.3. *Two innovating processes of different natures*

Again using the distinction made at the beginning of this chapter between two types of innovation, one could characterize and oppose the two processes of "transformational-innovation of valorique" and the "digital revolution" that seems unmanaged, but driven by attitudes of many with a MoV culture.

The value approaches were born from the need to rationalize, to better satisfy needs, to reduce or control costs. In their time they were an efficient and effective solution, but that time has changed and continues to change. They stress adherence to methods, rules of action and organization as conceived long ago. They aim at a goal, a finality, while endeavoring not to lock actors in "given" solutions. They are on the side of innovation that is prescriptively structured, which inevitably makes its generalization and subsequent adaptation more difficult. The way forward is to see the rigidity as a solution to underlying functions and in so doing this enables us to loosen the grip of past preferences and liberate new modes of value-practice.

The digital revolution challenges adherence to a methodology born in the industrial age. It is more innovation "push", innovation induced by scientific, technical developments that are intrinsic, privileging a logic set in the development of upstream supply and an assumed anticipation of "need". It results in the marketing of new solutions increasingly more strange or spectacular than the ones many dreamers, engineers or commercial agents had imagined (e.g. toilets that conduct medical tests as they are used). The significance and impact of the digital sphere is trivialized as its technology is embedded in more visible products such as telephones, television, heating systems and so on. That is, we take for granted many of its applications and effects on society. By taking as an example the success of the sending messages via SMS or MMS, we understand this logic of innovation made possible by a new technology and its flourishing uses, each of which in turn ends up as a series of needs as one invention begets another. One could describe this digital revolution of "uncoordinated" innovation, based on values of spontaneity and lack of a controlled collective reactivity (with the image of P2P practices, the creation of websites, and also the avalanche of spam and pornography).

4.7.4. *The digital arrival of "valorique"*

It becomes obvious today that innovation by way of the digital revolution would benefit from being grounded in the robustness and the fertility of value approaches. The application of digital technologies brings with it a fear of exorbitant costs for time spent in endless email communications, for time spent dealing with e-messages of all kinds, and for non-accountable time spent browsing Web pages (Internet or Intranet), as well as for the cost of "derived products" (e.g. the exponential growth of printouts and the cost of paper). The obvious advantages of the digital age stand opposed to the number of dysfunctions and new problems that need to be regulated (e.g. informational pollution, digital interference). The need for electric power and the bursting of financial bubbles relating to the "new economy" also worry investors. One could say that it became necessary to move gradually to a new era of rationalization in the digital environment, but it is the digital revolution itself that dictates the pace of change. The real tension is between agendas flowing from MoV and MbV and this is how we can turn the tables on a technology that is moving with no one at the steering wheel.

As an example, it is possible to state what should be urgently undertaken on the intranets of many companies so that the intranets become better tools of collaboration. A total re-design of these intranets could be unfolded, designed, decided, engaged, and rationally implemented so that a new more powerful way to enable innovation follows. Such a capability requires a new form of MoV and here we argue for MbV and "valorique".

Also, let us consider the products that did not reach their final stage of maturity and which are very often presented in the form of accumulations of gadgets (e.g. the constant reinvention of microchips that are even smaller and more powerful than before). It is necessary to imagine a day when an environmentally-respectful system of the basic needs, such as masts, is achieved. People want the benefits of progress but only if such benefits are delivered in undetectable ways and cause no loss of personal value.

A priori, it should not be too difficult to take into account the requirements of the fundamental thinking within the term "valorique" by economic actors or agents implied in digital technologies and their applications. There are several reasons for that, which include the following:

– The market is global, total, and competition is extremely sharp, and therefore only solutions proving to be of most value and best suited to needs will end up in a sustainable implementation.

– Products or applications that are concerned with the provision of contents (Web pages, platforms intranet, DVD) ensure an ongoing search for more simplified and effective solutions by the very fact that they relate to mass markets with end-users.

– The actors in the digital revolution are themselves the first promoters of collaborative steps. These people, who have know-how and are involved in various knowledge exchanges of information through operations in networks, also argue for open solutions and an open source code.

It is advisable to encourage professionals of all kinds, involved closely or at a distance in the applications of digital technologies, to adapt VM and valorique with respect to the underpinning principles of functionality and function analysis. The design must commence with the articulation of values from stakeholders that are translated into objectives and performance specifications. It is highly desirable, as young professionals direct themselves towards the digital trades (i.e. software engineers), that they receive educational exposure to the value approaches and are encouraged to combine digital capabilities and the fundamental thinking within "valorique".

4.8. VM and digital networks

Born well before the digital era, VA did not and could not have known how to integrate all the contributions of this last revolution. The recent passage of VA to VM can be regarded as evidence of a transformation in progress which should find further progress by reflecting on its own methodological assumptions given the changed context brought about by the digital environment. We argue for new attitudes, which we have named MbV and "valorique", so as to stimulate debate within the value communities.

It would be particularly advisable to reinterpret the fundamentals of value practice and to do so in light of possibilities offered by the digital era and the networks, to exceed the culture of complying with rigid methodological rules laid down in this field over the last 70 to 80 years, and to promote more inventive methods of implementation of collective intelligence for a management of innovation and value creation. Digital technologies allow information and substantial savings of time in the processes of design and decision-making and at the same time they allow a widening and greater level of user participation in design, which is unprecedented in the fields of innovation. This is possible because the relationship between function and solution is so central to this body of practice.

The traditional multidisciplinary working group, often considered as a key to the success of VA studies, can be substituted or added to by other devices or organizations via virtual workshops for design and for collaborative networks of expertise, and so can solicit a larger number of competences without it being necessary to physically meet in the traditional forum of workshops.

We can now reconsider VA and move towards the practices of "valorique", calling upon digital technologies systematically, as well as for functional articulation and simulation against objective expressions of needs, as was and could still be comparable to the information stage of VA (databases of cost, knowledge bases, technological trees of alternatives). Even more capabilities are possible for the fundamental thinking within the term "valorique" as the inventive phases (e.g. automatic generation of solutions, engineering software, and CAD (computer assisted design with widely dispersed teams)) can be augmented by the digital revolution.

The traceability and the capitalization of VA studies can also benefit from tracking and progress-chasing because of digital techniques. The constitution of platforms (Groupware, intranet, blog, wiki, etc.) dedicated to or accompanied by developments in VA or VM appears "easily and usefully" possible today. Just as operations within virtual networks could be very inventive, one could conceive, in MbV and "valorique", of the decision-making process starting from the confrontations of multiple points of view with a greater interactivity between the various interested parties made possible by the new digital environment. The risks of new product development could be reduced by integrating market research in the field with the concept stages of a new initiative. We can have customers as active members of the design team.

VM and "valorique" are not a fashion or a fad of the moment. The fundamental thinking within the term "valorique" is not a new discipline or technique with media virtues and will not become the next "dotcom" disaster. It is about applying different attitudes to, and greater emphasis on, research methods. The value approaches acquired their reputation of effectiveness and power of creativity and innovation during the last 50 years by many applications, be they intellectually rigorous or superficial, in companies or organizations. The majority of instances where VM was used focused on improving products, processes or services. All we are doing is calling for the next stage in this field's evolution to take account of the

Schumpeterian impact that digital technology has brought. We are presented with opportunities only if we are willing to relax our previous modes of practice. If we do not, then the "creative destruction" of Schumpeter will render our field obsolete and irrelevant.

MbV and "valorique" are today required to define steps, methods or techniques that constitute their bases; and in this respect, the explosion of digital technologies and the development of practices of exchange, of information-division within networks, become themselves interesting models of virtual organizations which can facilitate this actualization, and even change our attitudes from MoV to MbV. However, it is especially the intelligent and voluntarily creative coupling of "valorique" and the dynamics of information, knowledge, and networks that should support and stimulate innovation and, particularly, useful innovation that leads to a sustainable mode of consumption of natural resources.

Just as knowledge is only ever in the heads of people, the ability to inform is the key to better innovation. The traditional adherence to a rigid series of steps carried out in a prescribed weekly schedule is no longer the best way for VA or VM to deliver value. The time has arrived for a more integrated approach that unites stakeholders in facilitated information exchanges where the acts of design, development, realization, and value are unified. The time has arrived for our field to adapt or die under Schumpeter's "creative destruction".

Chapter 5

Research, Innovation and Technological Development

5.1. Introduction

A successful innovation usually has little to do with the originality of the idea behind it. What it does depend on, and crucially so, is the single-mindedness with which the business plan is executed, as countless obstacles on the road to commercialization are surmounted, by-passed or hammered flat. Life in the fast lane really is 1% inspiration and 99% perspiration.

Innovation, science, technology, research… whilst these are surely all interlinked they are sometimes mistaken for being the same thing. So what exactly do we mean by innovation? Whilst innovation often relies on scientific research and new technological developments, it is much more than the sum of these parts. Innovation is concerned with the creative and economic exploitation of research, science and technology. While some have defined innovation as a "profitable change" and others as an "economic exploitation of new ideas", "innovation means harnessing creativity to invent new or improved products, equipment or services which are successful on the market and thus add value to businesses" [VAU 00].

5.1.1. *Innovation is about taking risks and managing change*

Thus, it is clear that innovation applies to all aspects of economic activity and it must be active in every stage of economic life: markets, products, processes and

---

Chapter written by Mélissa SAADOUN and Lin YANNING.

services. Innovation embraces scientific research and development and technology, but it also includes training, marketing, design and quality, finance, logistics as well as the business management methods that join these various functions together efficiently. In this way, European support for innovation will continue to recognize the wide-ranging applicability of innovation and seek to support projects across all areas where economic benefits can be found.

What do we mean by technological innovation? The OECD (Organisation for Economic Co-operation and Development) defines technological innovation in the Oslo Manual from 1995: "technological product and process (TPP) innovations comprise technologically implemented new products and processes and significant technological improvements in products and processes. A TPP innovation has been implemented if it has been introduced on the market (product innovation) or used within a production process (process innovation). TPP innovations involve a series of scientific, technological, organizational, financial and commercial activities. A TPP innovating firm is a firm which has implemented technologically new or significantly technologically improved products or processes during the period under review".

5.1.2. *The importance of innovation in the economy*

There is ample sound economic research that demonstrates beyond question the central importance of research and technological development and innovation (RTD&I) in encouraging economic development. This research shows how high levels of RTD&I translate into:

- higher rates of economic growth;
- growth in income, output and productivity;
- increased exports and trade;
- greater business margins;
- sharpened international competitiveness.

5.2. Science, technology and innovation: building regional capacities

The long-term driving force of modern economic growth has been science-based technological advances. Without modern technologies, the world would still be where it was centuries ago, with people at the edge of survival. Technologies allow human society to fight disease, to raise crop production, to mobilize new sources of energy, to disseminate information, to transport people and goods with greater speed and safety, to limit family size, and much more. Yet these technologies are not free. They are themselves the fruits of enormous social investments in education, scientific discovery and targeted technological development in order to strengthen the national systems of innovation.

Every First World country makes special public investments in higher education and in scientific and technological capacities. Poor countries have largely been spectators, or at best users, of the technological advances produced in the wealthy parts of the world. They lack large scientific communities and their scientists are chronically underfunded, with the best and brightest moving abroad to find colleagues and support for their scientific research.

Enterprises transform scientific and technological knowledge into goods and services, but governments play an important role in promoting the application of science and technology. They need to act in the four areas described in the next sections. However, national efforts alone are not sufficient. The development of an economy requires a special global effort aimed at building scientific and technological capacities in the poorest countries and directing research and development toward specific challenges these countries are facing.

5.2.1. *Promoting business opportunities in science and technology*

Developing countries should use today's technologies to help create new business opportunities. Most developing countries still distinguish between industrial policies that emphasize building manufacturing capabilities and those that support research and development (R&D) to generate new knowledge. Adopting a "fast follower innovation strategy", aimed at making full commercial use of existing technologies, would combine these two approaches while building a foundation for future R&D. In promoting business opportunities, countries should focus on platform technologies that have broad applications or impacts in the economy, such as information and communication technology, biotechnology, nanotechnology, etc. In addition, governments should adopt policies and invest in infrastructure that stimulates small and medium-size businesses, improves access to credit and other forms of capital, increases participation in international trade, and promotes the integration of regional markets. Attracting direct foreign investment can diffuse tacit knowledge and help enterprises learn about the world's technological frontiers.

5.2.2. *Promoting infrastructure development as a technology learning process*

Infrastructure projects can also be a valuable part of a nation's technological learning process. Every stage of an infrastructure project, from planning and design to construction and operation, involves the application of a wide range of technologies and requires deep understanding and capabilities from the many engineers, managers, and government officials. Policymakers need to recognize this dynamic role of infrastructure development in economic growth and take the initiative to acquire available technical knowledge from the international and local construction and engineering firms they use for such projects.

5.2.3. *Expanding access to science and technology education and research*

Enhancing science and technology education has been one of the most critical sources of economic transformation. To build science, technology and innovation capabilities, developing countries need to expand access to higher education, but more than simply offering more places, universities need to become more entrepreneurial and oriented toward key development challenges. They can participate in technology parks and business incubator facilities. They can introduce entrepreneurial training and internships to their curricula and they can encourage students to take their research from the university to firms. Most universities will need to make some changes in order to take on these new roles. Governments should also expand and set up research centers focused on specific needs, such as agriculture or public health.

5.2.4. *Improving science and technology advice*

Governments must incorporate science and technology advice in their decisions for scientific and technological investments. They need first to set up an advisory structure, usually with a science advisor. The structure should have its own operating budget and a separate budget for funding policy research. Countries also need to strengthen the capacity of scientific and technical academies in order to participate in advisory activities in cooperation with other institutions, especially judicial academies.

5.3. Technology and global science for a better development

Many developing countries need new technologies to address specific needs. That is why the international science community, led by national research laboratories, universities, and national academies of science, must play a critical role in developing the global public goods in order to overcome these constraints. It must bring to bear its tremendous research capabilities to help solve the tough problems facing developing countries, particularly in the tropics. Global research into areas critical to developing countries, despite several efforts, remains under funded. The annual operating budget of $400 million set aside for the worldwide network of 15 tropical agricultural research centers known as the Consultative Group on

International Agricultural Research (CGIAR) is small in comparison with the combined research and development (R&D) budgets of the world's six largest agro-biotechnologies companies, estimated at roughly $3 billion a year [EVE 03].

The CGIAR specifically focuses on increasing the agricultural productivity of the poorest rural farmers in the tropics. It has had outstanding success in helping to achieve major gains in food security in many parts of the tropical world. However, the low budgets of the CGIAR system and national agricultural research centers has remained unchanged despite considerable evidence of the high social rates of return from R&D on tropical food production.

Likewise, health R&D is limited when it comes to diseases affecting the poor, with only 10% of the global funding used for research into 90% of the world's health problems (Global Forum for Health Research 2002). The WHO's Commission on Macroeconomics and Health recommends that the annual funding for R&D on global health public goods should be increased to $3 billion by 2007 and $4 billion by 2015, compared with roughly $300 million received annually today [WHO 01].

Two reasons account for the indifference of global science toward the needs of poor countries. First, public investments in research targeted at the needs of the tropics or other developing regions are insufficient due to the resource constraints in these countries. Second, while private markets in developed countries can produce development-stage science and, to a lesser extent, research-stage science, this is not the case in poor countries. No adequate incentives exist for private research to focus on tropical diseases or subsistence and small scale agriculture, since the poor would be unable to pay for the new medicines, improved plant varieties, or farming techniques. There is simply no commercially attractive market for such products. However, private research could be mobilized through two tested coordinating mechanisms:

– Prizes have been used frequently to spur innovation. An impressive example is the Ansari X Prize, recently awarded for the first commercial flight into space. Similar prizes should be offered for well defined problems, such as developing a new type of vaccine or an improved crop variety [MAS 02].

– Direct funding of private research has been used successfully by several private foundations, such as the Rockefeller Foundation and the Bill and Melinda Gates Foundation, in order to promote development-stage research in public health and agriculture.

In addition to mobilizing private research, international donors and foundations need to support more public research focused on the specific challenges facing developing countries. At least $1 billion is needed for research toward improved energy technologies. An essential priority to help African economic development is to mobilize science and technology. Tropical Sub-Saharan Africa produces roughly 1/20 of the average patents per capita in the rest of the developing world. And it has only 18 scientists and engineers per 1 million of the population compared with 69 in South Asia, 76 in the Middle East, 273 in Latin America and 903 in East Asia [WOR 04]. We stress the need for increased investments in science, higher education and R&D targeted at Africa's specific ecological challenges (energy, construction, etc.).

5.3.1. *Structural funds to support research and innovation*

Structural funds support for research, technological development and innovation (RTDI) now amounts to €10.5 billion in the form of grants. 97% of this support is made through the European Regional Development Fund (ERDF). Around 8% of the total ERDF resources are invested into research and innovation. Structural funds support for RTDI falls into four types of activities:

– research projects based in universities and research institutes receive about 26% of the total RTDI investment (€2.7 billion);

– research and innovation infrastructure (public facilities, but also technology transfer centers and incubators) receives slightly over 25% of the total, amounting to €2.8 billion;

– innovation and technology transfer, the setting up of networks and partnerships between businesses and/or research centers receives about 37% of the total (€3.6 billion);

– training for researchers (co-financed by the European social fund) receives about 3% of the total (around €350 million).

5.3.2. *Technology in today's global setting*

The countries' achievements in creating and diffusing technologies and building human skills to master new innovations can be gauged in three areas: technology creation (measured by patent and royalty receipts), the diffusion of new technologies (measured by Internet use and exports of medium and high-tech goods) and old technologies (such as telephony and electricity), and human skills (measured by mean years of schooling and the gross tertiary science enrolment ratio). A host of success stories has been analyzed and widely advertised, but the global rules governing market exchange and intellectual property rights have changed, causing developing countries today to face constraints (as well as opportunities) that their predecessors did not.

The rise of globalization (involving greater mobility, connectivity and interdependence) is changing the rules that govern innovation. Technological change is thus occurring in a vastly different global structural environment today. One manifestation of this change is the existence of globalized production networks that are dependent on geographically dispersed cost and logistical differences. These networks represent a big shift from a few decades ago, when foreign direct investment took different forms. Another factor affecting the global structural environment is the changed geopolitical climate, which has provided certain countries with preferential access to the USA and other advanced markets for new technologies, access to developing markets and significant amounts of development assistance. A third factor is the changed intellectual property regime, which played such a critical role in the early development of certain industries in advanced and developing countries. A fourth factor is the fact that revolutions in ICT and biotechnology have created new opportunities and put new pressures on skill sets and organizational practices within enterprises, universities and other R&D and manufacturing sites.

5.3.3. *Technological capabilities*

A country's induction into privileged circles of trade negotiation, economic treaties, and preferential status depends partly on its technological capability. Countries with rapid economic growth rates attract foreign attention because they represent new markets for goods and services from leading industrial powers and are considered important players with regional political power. China and India gained entry into select economic and political clubs as a result of their economic growth and advanced technological capabilities. There is no substitute innovation for scientific and technological bases, which undergird everything from agricultural self-sufficiency to public health coverage to lucrative licensing options for indigenous technology advances.

But as the case of China demonstrated, the first priority for developing countries is to build indigenous scientific and technological capacity, including research infrastructure, as part of the national planning strategies. It is through the existence of such capacity that developing countries will be able to manage technology acquisition, absorption and diffusion activities relevant to development. In other words, their capacity to utilize imported technology will depend largely on the existence of an indigenous technological competence and the learning strategies they put in place [XIE 04].

5.3.3.1. *R&D and innovation in China*

R&D in China is currently undergoing an extraordinary evolution, so much so that European research is likely to be exceeded very quickly. Globalization partly explains this important rise, although it is not a new phenomenon. In the fields of research upstream, exchanges with other countries have existed for a very long time. However, they have clearly accelerated in recent years in industrial research.

In addition, today, Chinese R&D profits from a world framework favorable to its development. Knowledge causes employment and financial flow. China, like other emergent countries, has tried to conquer this new market by creating poles of innovation on which it is a leader.

The offshore investment in R&D was stimulated in China by several positive evolutions: the abolition of the concept of distance thanks to information and communication technologies; the availability of intellectual added-value and factors of support for technological innovation, in universities in particular; the innovation in services and supports, as well as in manufacture; finally, the standardization of the rules of the international market.

Companies have various motivations for settling in China: the conquest of a new market, which is of a potentially large and promising scale; access to a network of inexpensive and well-formed resources; the development of a total network of R&D, a source of powerful innovation, and the proximity of a clustered innovation. China thus constitutes a major actor of offshore activities and a choice destination for international investments. Moreover, in 2005, it represented the preferred destination of investors and widened the gap between itself and most other emergent economies for the majority of relevant factors concerning investment. R&D offshore investments, in strong growth, move mainly towards China and India.

China currently knows an accelerated growth. Its economy is advanced more and more on the technological level since 19% of exports (and 3% of total production) are classified as "high technology", compared to 3.1% of exports in South Korea. However, China remains dependent on imports of high technology. The essence of the effort of R&D is focused on technological development. It represents $12.6 billion on the whole, distributed as follows: 5% for fundamental research; 17% for industrial research; and 78% for technological development. Scientific relations with the USA, Japan, Taiwan and Singapore dominate. If the "brain drain" towards the USA is important, partnerships with Europe are limited. Chinese expenditure of R&D accounts for 1% of the GDP, that is to say an important level; this has been rising since 1991.

Governmental authorities apply, according to fields, the policies of correction or "frog jump". These policies aim, as a whole, to benefit from foreign assistance within the framework of the process of modernization, but also to achieve a qualitative jump, while being based on national forces, so as to pass from imitation to innovation. It is thus a question of attracting many industrial partners and of causing competition between foreign investors, as is the case in particular in the sector of the nuclear power.

Most of the companies which installed themselves in China already crossed three phases: the sale and marketing; manufacture and production; design of products and localization. They start today the fourth phase, that of the installation of R&D laboratories, under very good conditions since the base of the activity is already consolidated. The R&D investment of the multinationals in China is today with the rise. The number of the R&D implementations is indeed in strong growth, but it is difficult to estimate. Essentially, these entities are concentrated in Beijing, Shanghai and Shenzhen, but certain companies have several R&D centers, which constitute amongst them true networks. American, European and Japanese companies dominate, as well as in the sector of ICTs.

Now China's industry is at a critical juncture, where its sustainable development will be determined by its indigenous technological capability. However, many Chinese firms do not have the incentive to develop their own technologies, because technology imports bring them quick and short-term benefits. Nor do they have the resources to innovate. For example, the R&D spending by Chinese enterprises was less than half of the nation's total until 2000, when the indicator topped 60% for the first time. Corporations also lack qualified scientists and engineers to engage in R&D activities. In 2000, 71% of Chinese enterprises did not have independent R&D units. Consequently and inevitably, without independent intellectual property rights in critical technologies, most of these firms have seen thin profit margin and low added value of their products.

Located at the bottom of the value chain, Chinese companies have also been seriously impacted by the changes in the upper stream, such as standards, specifications, designs, etc. Challenges also come from multinational corporations as they expand their operations in China. Since 1994, 28 leading multinational corporations have opened 32 wholly owned R&D laboratories and technology development centers, and another 16 are considering following suit. They have also sought cooperation with local universities or research institutes to undertake joint initiatives or projects.

Which lessons can be drawn from these evolutions for the future? On the one hand, China becomes a frightening competitor in the high technologies sector. In addition, it took suitable measurements to be placed in the international competition and to make its territory gravitational, by the selection of particular technologies where the R&D come to supplement a chain of carefully built values. However, the construction of competitive Chinese firms will require the formation of competences of design and marketing. It also remains to build a national system of R&D and of powerful innovation, starting from old structures, but certain universities have reached a world level right now. None of these obstacles are thus insurmountable: the vastness of the domestic market of China means it cannot be ignored by multinational corporations and it will quickly form an integral part of the saving of knowledge and its networks. The industrialized countries will have to thus adjust their policies and structures of R&D to ensure itself, as far as possible, of a competitive added-value.

5.3.4. *Infrastructure and technological innovation*

The development of new innovations and technology also contributed to the infrastructure development. Infrastructure development provides a foundation for technological learning because infrastructure uses a wide range of technologies and complex institutional arrangements. Governments traditionally view infrastructure projects from a static perspective without considering them as part of a technological learning process, even though they do recognize the fundamental role of infrastructure. Governments need to recognize the dynamic role of infrastructure development and take a more active role in acquiring knowledge about it through collaboration between local and foreign construction and engineering firms. Building railways, airports, roads and telecommunications networks, for example, could be structured to promote technological, organizational and institutional learning.

Infrastructure contributes to technological development in almost all sectors of the economy: it serves as the foundation of technological development as its establishment represents a technological and institutional investment. The infrastructure development process also provides an opportunity for technological learning.

The creation and diffusion of technology relies on the availability of infrastructure because without adequate infrastructure, technology cannot be harnessed. The advancement of information technology and its rapid diffusion in recent years could not have happened without basic telecommunications infrastructure. Many high-tech firms, such as those in the semiconductor industry, require reliable electric power and efficient logistical networks. In the manufacturing and retail sectors, efficient transportation and logistical networks make it possible for firms to adopt process and organizational innovations, such as the just-in-time approach to supply chain management.

The concepts of innovation systems and interactive relationships stress the links between firms, educational and research institutes and governments, concepts which cannot be implemented without the infrastructure that supports and facilitates the connections. Particularly in the era of globalization and knowledge-based economies, the quality and functionality of the ICT infrastructure, as well as the logistical infrastructure, is essential for the development of academic and research institutions.

While efforts to expand the use of technology in development depend on the existence of infrastructure, the development of new innovations and technology also contributes to infrastructure development. For example, the advancement in communications and data-processing technologies has fostered the development of intelligent transportation systems for more efficient traffic management and the use of geographic information systems and remote-sensing technologies makes it possible for engineers to identify groundwater resources in urban and rural areas. As infrastructure and technological innovation for development reinforce each other, the construction and maintenance of infrastructure represents a technological and institutional investment. It is clear in fact that infrastructure is a fundamental element of a comprehensive and effective science, technology and innovation policy.

5.3.5. *Research facilities as infrastructure*

Defining infrastructure in order to include technological innovation requires rethinking the strategic importance of research facilities [NIG 04]. Indeed, infrastructure projects can serve as research facilities themselves, while maintaining strong links with other research institutions [CON 03]. The management of geothermal energy facilities, for example, requires continuous *in situ* research as well as links with external research facilities. However, much of the research associated with infrastructure projects in developing countries is usually implicit.

The support to strategic technology development should be considered as part of the national infrastructure, in the same category as energy, transportation networks, and water and sanitation. A number of developing countries, such as South Africa, are starting to work toward creating networked research facilities that are accessed in a managed way. Other countries have consolidated research entities in order to create single research institutions designed to maximize synergies in human resources.

5.3.6. *Mobilizing the engineering profession*

The successful development of infrastructure services requires the full cooperation of those working in the engineering profession. Most national institutions of engineers have worldwide memberships and members in developing countries include both expatriate and local engineers. Many young engineers are the movers and shakers in these organizations and much more could be done to spread these voluntary service organizations worldwide. The United Nations and its specialized agencies should consider how they might capitalize on and reinforce these networks, particularly through their global organization, the World Federation of Engineering Organisations. In planning and implementing any project, including infrastructure projects, efforts should be made to harness the enthusiasm and drive of young professionals, many of whom are looking for an opportunity to serve the developing world.

In the current knowledge economy, a large number of young professionals in both the developed and developing world have become captains of cutting edge industries in ICT and other emerging technologies. Solidarity has always been strong among young people: knowledgeable young people in developed and developing countries alike can surely be mobilized in an organized way in order to provide help to development, following the leading example of "Médecins sans Frontières". Such a group could become a major force in harnessing science, technology and innovation for development.

5.4. Innovation and economic advance

Economic historians suggest that the prime explanation for the success of today's advanced industrial countries lies in their history of innovation along different dimensions: institutions, technology, trade, organization and the application of natural resources [MOK 02]. These factors also explain the economic transformation of developing countries that have recently become industrialized, as scientific and technological innovations come about through a process of institutional and organizational creation and modification. The defining characteristics of the West have been the institutionalization of private enterprise, continuous reductions in the cost of production, the introduction of new products and the exploitation of opportunities provided by trade and natural resources. These achievements are a tribute to the private sector and the state's ability to recognize new opportunities and the ways in which to exploit them.

Economists have recognized the critical importance of innovation and capital accumulation for growth. Empirical evidence and the modern theory of economic growth provide strong support for the thesis that long-term economic growth requires not only capital but also an understanding of innovation [JUM 92]. Innovation and technology are needed in order to set technological innovation as the basis of the development in those countries which still rely on the exploitation of natural resources.

5.4.1. *Platform technologies with wide applicability*

Even though most developing countries' governments acknowledge that science, technology and innovation are important tools for development, their policy approaches differ considerably. Most countries still distinguish between science, technology and innovation policies designed to focus on the generation of new knowledge through support for R&D and industrial policies that emphasize building manufacturing capabilities. The convergence of the two approaches would focus attention on the use of existing technologies, while building a foundation for long-term R&D activities. This approach requires that attention be paid to existing technologies, especially platform (generic) technologies that have broad applications or impact on the economy. Until recently, countries relied on investment in specific industries (textiles, automobile manufacturing and chemicals) with broad links in the productive sector in order to try and stimulate economic growth. However, their policy attention has now turned to ICT, biotechnology, nanotechnology and new materials as platform technologies, whose combined impacts will have profound implications for long-term economic transformation. Their role in meeting the development goals requires policy attention [SAA 04].

5.4.2. *Information and communication technology*

ICT has created a new way of viewing the ways in which different industrial, agricultural and service elements link together so that more than just the economic contribution of these different growth segments can be identified. These technologies challenge us to find new ways in which human efforts can enhance institutional life and sustain technological learning in developing economies so that gains in one area can be translated and multiplied as gains in learning in another. ICT can be applied so that the development goals in at least three areas will be met. Firstly, ICT plays a critical role in governance at various levels. Because of the fundamental link between technological learning and the ways societies and their industrial transformations evolve, it is important to situate technological innovation and the application of ICT at the center of governmental discussions. Secondly, ICT can have a direct impact on the efforts to improve people's lives through better information flows and communications. Thirdly, ICT can enhance economic growth and income by raising productivity, which can in turn improve governance and the quality of life.

The benefits of the new technologies are the result not only of an increase in connectivity or broader access to ICT facilities per se, but also due to the facilitation of new types of development solutions and economic opportunities that ICT deployment makes possible. When strategically deployed and integrated into the design of development interventions, ICT can stretch development resources farther by facilitating the development of cost-effective and scalable solutions.

Networking technology can be deployed in order to enable developing countries to benefit from new economic opportunities emerging from the reorganization of production and services taking place in the networked global economy. ICT will become one of the main characters in the pursuit of poverty alleviation and wealth creation in developed and developing countries alike. At the same time, as a facilitator of knowledge networking and distributed processing of information, ICT can be used to foster increased sharing of knowledge [SAA 05]. ICT does not simply differ from other development sectors and technologies because of its status as a lucrative source of revenue and taxation for business and government. As accelerators, drivers, multipliers and innovators, both established ICTs (radio, television, video, etc.) and new ICTs (mobile phones, Internet) are powerful, if not indispensable, tools in the massive scaling up and interlinks interventions and outcomes inherent in the development goals.

Even within the science, technology and innovation community itself, the seismic changes continuing to occur in computing and communications are often underestimated. For example, progress in computing is providing the foundation for innovations in industries as far a field as wireless communications and genomics. This "ripple effect" will continue to expand with the exponential growth of processing power, storage capacity and network bandwidth. The processing power available at a given price currently doubles every 18 months, storage capacity per unit area doubles every year and the volume of data that travels across a fiber optic cable doubles every 9 months. The impact of this technological progress has only just begun to be felt but we can be sure that it will be profound since the ripple effects from the Internet are still only at an embryonic stage of development. Already the fastest-growing communications medium in history, the Internet marks the beginning of the technological convergence between telephone, television, and computer.

5.4.3. *The network revolution*

In recent years the network revolution has forced a radical transformation of both developed and developing economies [SAA 00]. New network economics and dynamics have combined multiple "positive feedback mechanisms" and "network effects" with disruptive and discontinuous change. This change encompasses: rapidly decreasing technology costs with volume and innovation, vastly increased system development costs, risks and timescales, new competitive market forces, heightened user expectation, uncertain industry restructuring and financial market behavior, and standardization that is often non-proprietary. In addition, additional network benefits, such as electronic commerce, have appeared.

The International Telecommunications Union estimates that access to telephone networks in developing countries tripled between 1993 and 2002, rising from 11.6 subscribers per 100 inhabitants to 36.4. By the end of 2002 there were more mobile phones subscribers than fixed telephone lines in the world. Growth has been particularly strong in Africa, where an increasing number of countries now have more mobile phones than fixed telephones. Growth in the number of users of personal computers and the Internet has been equally impressive. By the end of 2002 there were an estimated 615 million computers in the world, up from only 120 million in 1990. In 1990 just 27 nations had direct connection to the Internet, but by the end of 2002, almost every country in the world was connected and some 600 million people worldwide were using the Internet. Growth has been most rapid in developing countries, where there were 34% of users in 2002, up from only 3% in 1992 [ITU 03].

5.5. Investing in science, technology and education

Investment in science, technology and education has been one of the most critical sources of economic transformation in the newly industrialized countries. Such investment should be part of a larger framework created in order to build capacities worldwide. The one common element of the East Asian success stories is the high level of commitment to education and economic integration within the countries. This strategy was a precursor to what have come to be known as knowledge societies [WOR 02].

The commitment of the Republic of Korea to higher education suggests that spectacular results can be achieved in a few decades. However, these experiences are not only limited to this region: in fact the impact of education on local economies is also being recorded in less developed countries, although policy approaches to education continue to generate considerable controversy in international development circles.

5.5.1. *New roles for universities*

A new view that places universities at the center for the development process is starting to emerge. This concept is also being applied at other levels of learning, such as colleges, research and technical institutes and polytechnic schools. Universities and research institutes are now deeply integrated into the productive sector as well as society at large. Universities are starting to be viewed as a valuable resource for business and industry as they can undertake entrepreneurial activities with the objective of improving regional or national economic and social performance.

In facilitating the development of business and industrial firms, universities can contribute to economic revival and high-tech growth in their surrounding regions. There are many ways in which a university can get integrated into the productive sector and into society at large. For example, it can conduct R&D for industry, it can create its own spin-off firms, it can be involved in capital formation projects, such as technology parks and business incubator facilities and it can introduce entrepreneurial training into its curricula, thus encouraging students to take research from the university to firms. It can also ensure that students become familiar with problems faced by firms, for example, through their experiences in internships. Universities should also ensure that students study the relationships between science, technology, innovation and development so that they are sensitive to societal needs. This approach is based on the strong interdependence of academia, industry and government.

Industry in the developed world has benefited from the activities of research universities, particularly from their state-of-the-art laboratories, which conduct cutting-edge research for them, while universities benefit from the research funds provided by industry. However, many universities in developing countries serve merely as degree or certificate awarding institutions, providing the necessary documentation for thousands of young people to apply for jobs. Marginalized in the development process, these universities seek only to churn out graduates. Therefore, it is important to underline that universities need to be re-envisioned as potentially powerful partners in the development process.

This adjustment can be implemented in a top-down manner by changing existing norms and procedures and it can be done for all academic departments of the university or only for certain select ones deemed to be of more importance with regard to the national development goals. Imposing new standards only on certain departments would imply widely different standards for students and faculty and would therefore be likely to require a separate administrative set up for the departments with these higher standards. Moreover, the university's location would have to be appropriate for the selected disciplines. A benefit of this approach would be working with an established institution which has libraries, staff, and very likely some links with other research institutes. Technical institutes are created to serve industry and are therefore, by nature, disposed to work with firms. Without neglecting their essential and primary roles in capability building for technologists and technicians, some of these institutes could be upgraded to university status.

New universities may also be created, particularly if new fields of knowledge in which existing universities have inadequate capability have been made a national priority or if student demand has outstripped university capacity. These universities could be entirely new institutes or expansions of industry-based training institutes.

For universities to be able to contribute to science and technology-based regional development, appropriate supporting institutions will be necessary. These include both enabling policies and organizations that can increase the pathways of interaction between academia, government and industry. Specific measures include tax breaks, venture capital funding, low-interest loans, changes in intellectual property rights, higher returns on inventions, heavy investment in ICT, business incubation and technology parks and centers within or near universities. Partnerships with other institutions, at national or regional level, could also be of great benefit. Many academics in developing countries are benefiting from institutional partnerships with universities and R&D institutes abroad. Research partnerships across academic, industry and government institutions help reduce knowledge gaps, especially in small and medium-size enterprises, which often lack adequate R&D facilities.

5.5.2. *The role of ICT in education*

The role of ICT in education is limited by the absence of business models that take advantage of the emergence of a wide range of versatile devices that can be adapted to various uses. For example, satellite technology could be combined with memory and audio devices in order to create libraries of educational materials in rural areas of developing countries. What is missing is not devices but the lack of content development. Partnerships between the ICT, media and entertainment industries and actors from developing countries could help find ways to put existing technologies to educational uses. In addition to building the foundations for the participation in creative industries, such partnerships could also revolutionize education through the use of animation in the design of teaching material [LOW 03]. A good example of the use of ICTs to promote tertiary learning is the African Virtual University, created in 1997 as a pilot project of the World Bank and described more in detail in the next section.

5.5.2.1. *Example of the African Virtual University*

The African Virtual University (AVU) is reaching thousands of young Africans including many who might not have attended a traditional university (source: www.avu.org). With its headquarters in Nairobi, Kenya, the AVU represents an important approach in using ICTs for educational purposes. In its first phase (1997-1999), the AVU used the expertise and facilities of the World Bank, with additional support from vice-chancellors from universities in various African countries and in its transitional phase, the AVU established 31 learning centers in 17 African countries. More than 23,000 people were trained in journalism, business studies, computer science, languages and accounting. Enrolment of women was more than 40%, a result of the flexibility offered by distance learning. Since 2002 the AVU has expanded its activities to all African regions, offering degree and diploma programs. The program, which is focusing simultaneously on research and development of its technology delivery model, aims to disseminate the pedagogy model as well as the general technology infrastructure in partner universities. The AVU has created a network of 33 partner institutions in 18 African countries and registered more than 3,000 students in semester-long courses. It has also enrolled a large number of African women in its specialist programs and is affiliated to a global network of leading universities. It is possible that the AVU model could be adopted at national level, linking national universities and possibly helping universities offer training to neighboring pre-university schools.

5.5.3. *The role of universities in innovation*

In many developing countries, universities suffer from unclear mandates, limited funds and do not have the flexibility to meet basic needs (often dealt with by public research centers in "mission mode") or promote competitiveness (dealt with by the private sector or government training institutes). Universities often lack the resources and the demand from a sound productive economic sector that is eager to benefit from the knowledge these universities and their students could create. They suffer from a "loneliness syndrome". Reversing this syndrome is one of the challenges for development and one that cannot be fulfilled by pushing universities to change while everything else remains the same. A better approach is to channel energies within the university environment in order to carry out a combined research, teaching and application mandate, with different types of universities taking on different challenges and government and industries engaging in effective interaction with them. However, this path is not without dangers: one potential problem is that the pendulum could swing too far, so that universities become outposts for government or private sector service functions or engage only in applied research. Incentives need to be calibrated so that as universities continue to produce knowledge, they also seek to transfer that knowledge to useful applications where appropriate. Informed science, technology and innovation policy need to account for the fact that universities must continue to have local relevance while fulfilling broader mandates of education and knowledge acquisition and diffusion.

It is more important than ever for developing countries to move ahead in scientific and technological development at an advanced level. Doing so will enable them to build local capacity that can help solve the many science and engineering-related problems they face. It will also place them in a better position when it comes to taking an active part in the global knowledge economy. Universities are still vastly underutilized even though they could potentially become powerful vehicles for development in developing countries, particularly with respect to science and technology. If both universities and industry are encouraged to work actively together, universities will be able to assume new roles that could accelerate local and national development. Rendering these institutions more effectively as key development partners will require changes at several levels of university administration. It will also require deep changes in enterprises, private as well as public, so that they can become strong demanders of the universities' capabilities, helping to transform these capabilities into capacities. Governments will need to act as a careful facilitator of interactions between these two actors. If this is achieved, the "loneliness syndrome" that for so long affected universities in developing countries will be redressed, allowing them to contribute to economic growth and social development.

5.6. Conclusion

Technological innovation is one of the least studied but most critical sources of productivity growth. Indeed, economic historians are currently changing our understanding of human history by placing greater emphasis on the role of technology and the associated institutional innovations. Even though technological innovation has played a critical role in spurring growth in the industrialized countries, the lessons derived from these experiences have not been applied in developing countries, where technological change remains a marginal part of national growth strategies. Technology presents a vast array of opportunities for improving the human condition, but many challenges lie ahead in harnessing its power. The infrastructure for development in general and technological innovation in particular must be developed. Also, competence in technical fields needs to be increased and the environment needs to be improved so that it will foster entrepreneurship and the commercialization and wider diffusion of technologies. The capacity to participate effectively in the global trading as well as in the global knowledge system needs to be increased together with the overall environmental policy needed to promote the application of science, technology and innovation to the development goals needs to be improved.

Without adequate infrastructure, developing countries will not be able to harness the power of science, technology and innovation needed to meet the development goals. Because infrastructure uses a wide range of technologies and complex institutional arrangements, its development provides a foundation for technological learning. As infrastructure is also critical in attracting foreign direct investment, developing countries need to strengthen it and enhance their ability to develop, operate and maintain infrastructure services. If a developing country is to unlock the potential to turn science, technology and innovation into business opportunities, it needs to undertake a number of core activities. These include providing broader incentive structures to all firms while creating an institutional environment that encourages entrepreneurship, rewards innovation, fosters start-ups and sustains existing firms with injections of capital. Creating links between knowledge generation and enterprise development is one of the most important challenges that developing countries face. A range of structures can be used in order to create and sustain enterprises, from taxation regimes and market-based instruments to consumption policies and sources of change within the innovation system.

Chapter 6

Sustainable Innovation through Community Based Collaborative Environments

6.1. Introduction

Innovation is an exercise which is becoming more and more difficult as it requires putting together diverse yet varied in-depth knowledge in a number of specialized fields, necessitating a close collaboration between a number of fields of expertise, from the very outset, encompassing the whole product lifecycle. In fact, companies are focusing more and more on their core competences to rationalize and optimize their processes and resources in order to remain competitive. They are increasingly using external competences and developing business relationships with other companies which have complementary activities, and sometimes are even competitors, to work jointly in the development of new business scenarios towards the creation of innovative products and services [PUL 02]. With the contribution of new technologies, business partnerships among companies are progressively becoming collaboration networks which are real sources of value creation.

Innovation is also a difficult exercise as the potential risk of failure is high due to a number of barriers or obstacles (whether economic, legal, social, organizational or technical) which absolutely need to be overcome in order to achieve a real innovation, whatever the field may be. Nowadays, it is widely recognized that the higher the number of partners (which means diversity in competence levels), the higher the creativity and innovation potential. On the other hand, the higher the number of partners involved in the same project, the higher the costs of integration

---

Chapter written by Marc PALLOT and Kulwant S. PAWAR.

and management exponentially [PAL 00]. However, the more partners there are in an innovative project, the more the inherent risks can be shared among them.

This chapter is aimed at presenting different systematic approaches which will make it possible to overcome the phenomenon of complexity brought about by interdisciplinary and inter-company collaboration, and also minimize the level of acceptable risk. These approaches are associated with the use of information technologies and collaborative work within a network. They enable the improvement or adaptation of the development processes of innovative products and services with the specific constraints of inter-disciplinary and inter-company collaboration.

6.2. Components of collaboration

6.2.1. *Different forms of collaboration*

6.2.1.1. *Employees*

At the first level are the employees and individual workers (Figure 6.1). This level does not present any problem as almost everyone has been educated to work on an individual basis. This environment does not generate any demand for a specific form of collaboration: it is thus the simplest form of work.

6.2.1.2. *Teams*

At the second level, each individual belongs to one or several teams. In essence, these people are brought together to perform or carry out collective work, tasks or projects. In this context, we as individuals are much less prepared in terms of upbringing or education. As long as these teams are composed or formed within company boundaries, they will not face any contractual problems. As a result, it is perceived to be easier to collaborate within a company, even if the team members belong to different departments or different disciplines. On the other hand, to us as individuals, the attitude or mindset of collaboration and sharing is not something that comes naturally and some resistance may be encountered. This resistance is often related to factors such as culture, organization, remuneration or even competition, rather than to problems associated with confidentiality or security. It is thus necessary to reinforce the approach of collaboration and to help the evolution of attitudes by setting up multi-disciplinary teams for projects.

6.2.1.3. *Communities of economic interests*

The third level takes us outside the boundaries of the company as it concerns communities of economic interest, which carry out a pooling of resources or share certain operations and related risks. This is the case with purchasing centers or cooperatives. Many organizations thus pool their resources to get substantial gain. These forms of organization do not always necessitate a sharing of knowledge. However, individuals can form groups according to the work that has to be carried out within the framework of these entities.

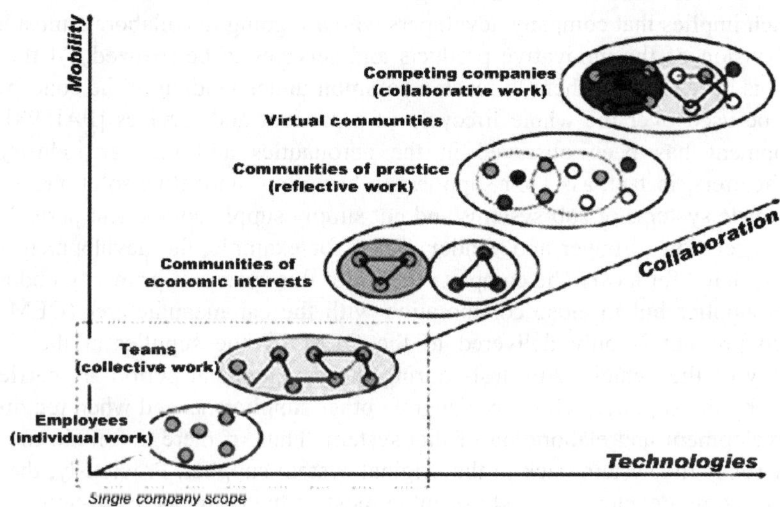

Figure 6.1. *Different forms of collaboration*

6.2.1.4. *Communities of practice*

These forms of organization are made for reflective work. In other words, the work carried out consists of developing common practices. Hence, these are groups with specific interests. The employees of companies freely subscribe to these groups, communicate with their peers and, if needed, contribute towards the elaboration of standards or norms in specific fields. It is therefore a relatively informal and open type of organization.

6.2.1.5. *Virtual or on-line communities*

Individuals, motivated by the same interests, may form groups through proposed structures on the Internet, such as forums, and exchange information or opinions. These communities do not have any physical existence as such. It is therefore a relatively informal culture. However, a community of practice can operate on the Internet and as such it becomes another on-line community where peers collaborate in contributing to the shared definition of concepts (for example, Wikipedia).

6.2.1.6. *Concurrent enterprises or the network of organizations*

This is the last level in which real collaborative work is practiced. In fact, companies form groups to be concurring to the achievement of a common objective and as a result have the need to collaborate. The employees and teams of these companies must share their knowledge to be innovative in their markets. This approach implies that company developers who are going to collaborate must have a shared vision of the innovative products and services to be realized. At the same time, it is necessary for them to reach a common understanding of the concepts that are to be used over the whole lifecycle of a product and services [PAL 98]. This development has been observed in the aeronautics and the car industry: the manufacturers, in fact, ask their suppliers to bring out innovative solutions in terms of complete systems or sub-systems and not simply supply component parts. It is no longer a relation of buyer and vendor. Take, for example, the development of the "front system" for a car. The complete design of this system is primarily undertaken by the supplier but in close collaboration with the car manufacturer (OEM). The finished product is only delivered to the OEM by the supplier at the time of assembly of the vehicle. Any tests during the development period are carried out directly by the supplier, who may integrate other suppliers as and when required for the development and elaboration of this system. Thus, if there is a fault, the OEM simply sends the system back to the original system supplier. Previously, the OEM used to design, develop, test and assemble most of the components in-house, which demanded a lot of resources and time. This type of collaborative innovation is only possible when companies are willing to exchange and share the necessary knowledge and working practices. Naturally, the contractual aspects becomes unavoidable in this type of organization, unlike collective work which does not require a contract to be drawn up between the various employees. Finally, it is observed that as one moves towards this type of organization, the requirements in technology and mobility are increasingly demanding.

6.2.2. *Different methods of work*

Collective work is about people working together as a group or a team, with two or several individuals, which does not require any specific contract until it remains inside a single company's borders.

Cooperative work brings together business partners in a specific commercial framework. In other words, the partners pool their resources together with the hope of drawing some profit from it. This framework is therefore business oriented.

Reflective work consists of building up a frame of reference of good practices which represents "reflexes" or working standards. This is associated with communities of practice or communities of knowledge.

Finally, collaborative work constitutes a collective innovation within a contractual framework of knowledge, risks, costs and benefits sharing. This type of working arrangement implies contractual complexities in terms of intellectual property rights, as the innovation is the result of the work of more than one company.

6.2.3. *Mobility*

Geographic mobility means the ability to move from one region to another to find a new job. This mobility may, of course, be related to aspects which are other than professional, such as family matters or political commitment.

Professional mobility constitutes a change of qualification or function. It may be an evolution or promotion within the company or outside of it. This type of mobility may imply geographic mobility as well.

Social mobility is the natural fact that individuals have their social position in the family, at professional or local levels evolving from time to time.

This nomadic nature is a more or less permanent type of mobility according to professional transfers to various geographic locations of clients and/or suppliers, for a generally long period of time. For example, according to the co-designing platform technique in the automotive sector, manufacturers are asking the suppliers to join them in the same geographic location to collaborate on the design of a vehicle. This leads to social and professional implications: on the one hand, the company delegating a resource sees this resource disappearing from them; on the other hand, transferred persons will be put in touch with others from the internal project team and with individuals from other suppliers, who do not necessarily have the same professional status within their respective companies. They (the employees) will be led to make comparisons whether on position, remuneration, working hours and days off work. On the whole, individuals react in two ways: either they are interested in the social aspect of the project and they communicate a lot, which may result in a loss of their own company culture; or they are uninterested in this aspect of the project and communicate relatively less with other members of the team.

Finally, ubiquity represents teleworking or distance working, or even telepresence. This type of virtual mobility makes it possible to not have to move at all while offering the possibility of working with any project team by connecting to its network. Teleworking carries a certain number of social and professional implications. Therefore, meetings and social events are organized so as to allow the persons who are going to collaborate to physically meet before the beginning of the project: this helps to reinforce the level of confidence and understanding that each team member has for others. It is thus a practical way to bring about social and professional ties so as to have a better perception of people with whom one has to collaborate.

6.2.4. *Teleworking (distance or remote working)*

The penetration rate of distance working or teleworking is 2% in France. North European countries are the most advanced in this matter, as compared to, for example, Denmark (more than 10%), Sweden (10%), Holland (more than 6%), Germany (between 5% and 6%) and the UK (6%). On the other hand, countries in southern Europe show a slight lag compared to France, where the weak penetration rate is explained by the prohibitive cost of networks, which hindered the development of teleworking. This cost is lower today thanks mainly to the low cost of Internet broadband access at high speed (ADSL).

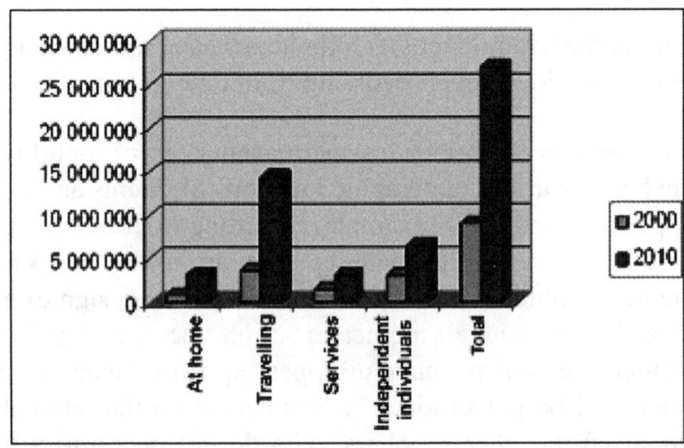

Figure 6.2a. *Evolution of teleworking in Europe (source EMERGENCE)*

Teleworking has social and professional implications. Indeed, practice of this method of work requires some organization. The USA has a high rate of penetration at more than 12% (Figure 6.2b), as Internet networks have developed there more rapidly.

The following are different types of teleworking:

– teleworking from home: this concerns, for example, translating texts, editing documents or updating websites;

– teleworking through travel: this is when individuals are traveling to their clients' geographic locations. It is possible for them to contact and to work with their company from a remote location or on the move. Salesmen who survey their geographic region, for example, can be connected to their company on a regular basis to find out the state of stocks or return orders and forecasts;

– teleservices (or distance services): this is, for example, a question of telepresence, (remote) telemonitoring, telemarketing, remote maintenance or telemedicine;

– independent individuals: there is a strong tendency to work directly from one's office at home. They only occasionally visit their clients for work, conducting studies or delivering results. The rest of the time, they collaborate with clients and other partners via the Internet by using (synchronous or asynchronous) electronic messaging systems, platforms for collaborative work, teleconferencing or video-conferencing.

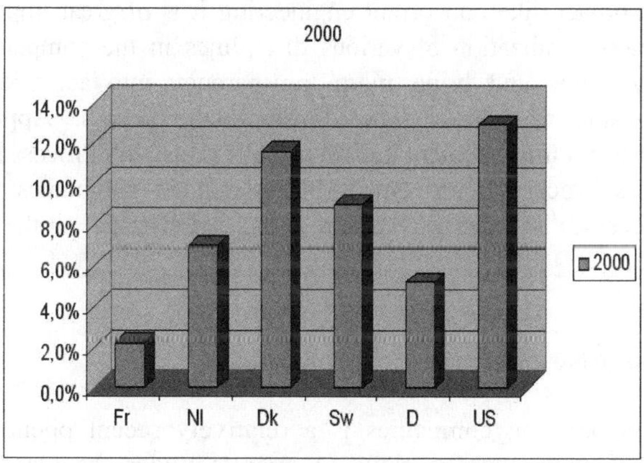

Figure 6.2b. *Penetration of teleworking (source: ECATT)*

The current average rate of penetration of teleworking in Europe is at 4%. It is likely to involve more than 25 million individuals in Europe by 2010, or three times more than at the present time (about 9 million). Moreover, there is a high co-relation with the rate of penetration of the Internet. Teleworking or distance working today involves, contrary to common belief, 80% men and 20% women. Lastly, 75% of the population is said to be interested by this method of working.

6.3. A systematic approach to collaboration

Interdisciplinary collaboration is not something new since various systematic development approaches, such as system engineering or concurrent engineering, also called simultaneous, collaborative or integrated engineering, have advocated this method of collaborative multidisciplinary work since the late 1980s; previously it was the culture of functional, vertical organization based on philosophies such as Taylorism and Fordism way of assembly line work where the activities were divided and broken down in a successive manner in a schedule which focused more on the optimization of the procedure than on the process of innovation. Nowadays, systematic approaches have already been widely deployed within large companies and have entered medium-sized companies as well as small high-tech companies. It is the increasingly global competition and the aim to become more competitive on the global market that has led companies to become interested in new types of organization and new development methods which simultaneously make it possible to speed-up the introduction of a new product on the market while reducing the costs. In fact, clients, according to their level of maturity, are becoming more and more careful by looking at overcharging related to the implementation or use of bought products which is often concealed. In order to implement an approach like concurrent engineering it is of great importance to bring down the compartmentalization of various disciplines in the company and build up multidisciplinary teams and bring more transparency into activities and decision making mechanisms to improve the development process. Approaches of re-engineering or process improvement known as BPR (Business Process Re-engineering) or BPI (Business Process Improvement) have enabled companies to review their development processes in order to consider, at the earlier stage, all the elements of the product's lifecycle.

6.4. The collaborative enterprise

Collaboration between companies is a relatively recent phenomenon since a large majority of companies continue to operate under the classic approach of purchasing, sub-contracting and supplying. This approach consists of passing an order for a sub-system or component or even for services provided by a sub-contractor or supplier through documents expressing the needs, and if necessary requirement specifications, to obtain a delivery within a timeframe, costs and quality requirements. A certain number of checkpoints/milestones are placed within the timeframe in order to verify adherence to requirements and constraints as well as the satisfaction of needs. There is no real notion of collaboration in this approach if not for the purely economic partnership according to the frequency of orders and other negotiations to bring the prices down as far as possible.

On the other hand, the notion of a company network collaborating on the same innovative project implies that partners are invited from the very beginning of the development cycle to take part in the expression of needs and requirement specifications adding their competences, specific expertise and know-how. The main goal is to maximize the creativity and innovation potential making it possible for developers to identify more alternative solutions that guarantee a higher success rate. This means that partners are no longer selected on the basis of negotiations of time, quantity, quality and costs but rather criteria such as excellence in competences, expertise and know-how. It is thus necessary that each of the companies in the network excels not only in its field but also in the implementation of an effective systematic approach, organization and optimized processing with relation to the needs of the project. This approach of collaboration between companies has a number of implications as much in matters of confidentiality, security, intellectual property and interoperability as in the evolution of attitudes (Figure 6.3).

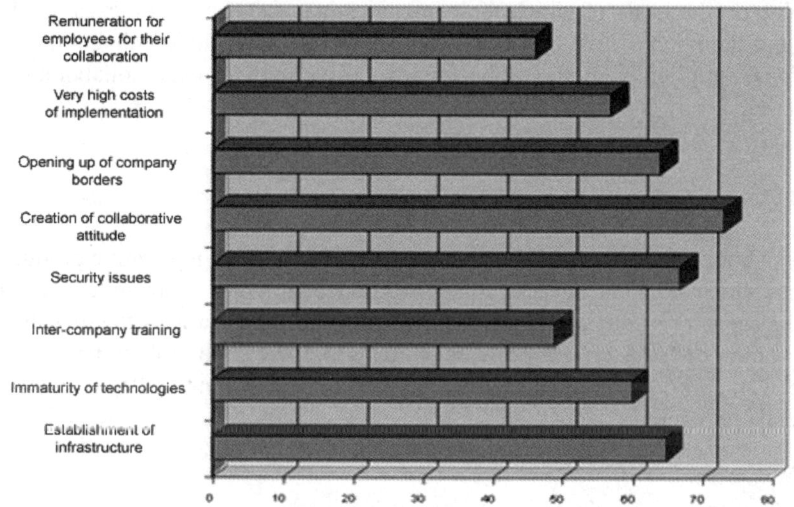

Figure 6.3. *Challenges of implementing a collaborative approach*

6.5. Innovative enterprise networks

The most important strategic reason for operating within a collaborative network is the greater creation of value through innovative capacity which is definitely higher compared to the traditional approach. To reach this stage of collaboration success, the following aspects, at least, need to be taken into account.

6.5.1. *Mixed marketing*

Companies that succeed know not only market needs today but also those of tomorrow. To rise to this challenge, it is necessary to work closely with clients and establish a very strong involvement of marketing personnel, product development services and R&D in the marketing process.

6.5.2. *Strategic coordination of partner networks*

Creative and innovative ideas alone cannot ensure the position of leader in the market. The challenge lies in entering the market quickly while keeping in mind the time required for the development of innovative products, services and processes. An important prerequisite to achieve this goal lies in coordinating the company strategy with the product strategy. There are a large number of strategic options available on the basis of which the companies may choose the quickest method to bring out their innovative products. These options can include collaborations with traditional suppliers who should become equal partners, or even collaboration with competitors.

6.5.3. *Financing innovation within a network*

Innovation requires a solid financial base. In fact, today innovative products and services which have had success in the market need not only innovative concepts but also the capacity of quick and precise evaluation of the returns of the investment. It is a question of focus on perspectives which will make it possible to bridge the gap between an approach oriented towards engineering and technical details and a purely financial approach.

6.5.4. *Company networks as incubators of innovation*

The success of a product depends on several factors. One of the most important factors is definitely the innovation team which is the incubation system in the network. This incubation system within the network should not consist only of relations with clients, partners and suppliers but also include government programs whether it is in terms of R&D, innovation or regulation.

6.5.5. *The infrastructure of collaboration*

Generating creative, innovative ideas requires a great deal of time but much more time is required to transform these ideas into products or services which will be successful in the market. The infrastructure, systematic approach and data processing tools offer project teams considerable help to overcome difficulties and complexities of the process. Choosing and implementing the most suitable infrastructure, adequate methods and appropriate tools are crucial steps in the success of an innovation. It is due mostly to the rapid evolution of new technologies that the network of innovative companies has been created.

6.6. Concurrent engineering

Of all the systematic approaches or methods known, it is without doubt Concurrent Engineering (CE) which is the most controversial name. In fact, in France, CE has often been called concurrent, simultaneous or even integrated engineering (AFNOR recommendation). In the USA, CE has had other names such as "collaborative engineering" or "cooperative engineering". In fact, the definition of CE published by the IDA (Institute for Defense Analysis) in 1988 [WIN 88] is the following: "Concurrent engineering is a systematic approach to the integrated, concurrent design of products and their related processes, including manufacturing and support. This approach is intended to allow developers from the outset to consider all elements of the product lifecycle from conception through disposal, including quality, cost, schedule and user requirements."

This approach, according to its definition, has put into focus two very innovative concepts. The first is the design of a product not only in terms of its performance independently from its capacity to be produced and maintained, but also to create a product that can be produced, used and maintained faster and at minimum cost. This means that not only the design of the processes associated with the product becomes as important, in terms of performance, as the product design itself, but the designers are also expected to analyze different aspects of its lifecycle and study their potential impact, for example, the study of the testability, fabrication, assembly and disassembly, dismantling or recycling. It constitutes the second innovative concept of concurrent engineering. It was truly a revolution at the time, when the production and maintenance processes were designed after the product design had been approved, hence being finalized without really taking into account from the very beginning the different aspects related to its lifecycle. In short, the two innovative concepts of concurrent engineering are, on the one hand, optimization in terms of performance of the couple "product and associated processes" and on the other hand the vision of "the lifecycle of the product".

This idea of the vision of the lifecycle is found in sustainable development where the entirety of potential impact on the environment, individuals and society are considered over a sufficiently long period of time.

The reasons to implement concurrent engineering are many. We can state the following: the best way to control the cost of possession and the cost of exploitation or use, as well as the costs of dismantling and recycling is when these elements are taken into consideration at the outset of the development process; quicker entry into the market as a number of iterations are removed because of integrated concurrent designing of a product and its associated processes; a better product and process quality as a result of a considerable reduction of unexpected events; a reinforced potential for innovation through multidisciplinary or interdisciplinary teams working in close collaboration from the beginning of the development process.

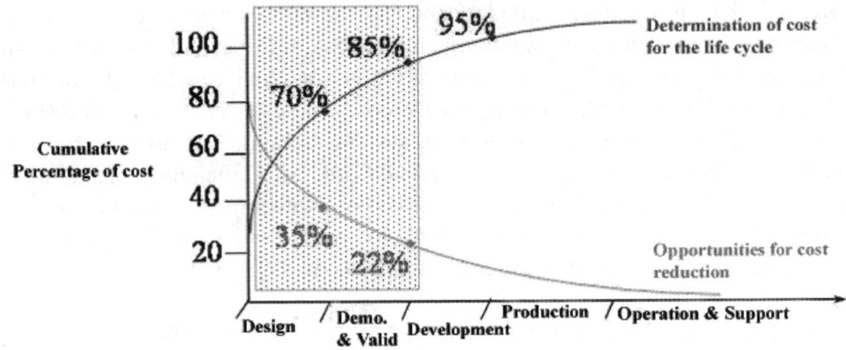

Figure 6.4. *Determination of costs of the lifecycle
and opportunities for reduction of costs*

Concurrent engineering was first implemented as an internal method in the company [PAL 96]. The organization of development often evolved towards matrix management companies of the "projects and competences" type with a host of multidisciplinary teams. For them, the process of designing and development were largely improved to increase the iterations from the beginning, including activities of analysis and study of impact of the elements in the lifecycle, in order to reduce the iterations later during the development cycle (Figure 6.4). Thanks to the rapid evolution of new technology, such as rapid prototyping, co-simulation and numeric models, it has been possible to implement concurrent activities of analysis and study of its impact on product related processes.

The necessity of leaving the internal stage and the scope of the company to integrate the expertise of partners from multidisciplinary teams has been quickly realized. To this end, techniques from the design platform were implemented, especially in the domain of automobiles, where representatives of sub-contractors of the first level met with internal teams in the same geographic location. Others in the industrial sector have also launched into distance collaborative work (teleworking), with shared virtual space and teleconferences or videoconferences to avoid meetings with experts in any particular location so as to avoid costs and constraints. A number of economic, technical and sociological issues arose in the physical design platforms as well as virtual ones (distance collaborative work).

6.7. Adaptation of the collaboration process

To resolve the complexity of collaboration on innovative projects, having a systematic approach to design integrating all competence and elements of the lifecycle is not enough. The design process and development must also be adapted to a collaborative approach and to the interoperability of various professions and of different partners. Therefore, it is necessary to review the processes and to de-compartmentalize them, verify the interaction needs, remove any activities which have no real added value, include new concurrent activities (Figure 6.5) with the objective of favoring previous iterations and to seize any opportunity for improving and implementing new approaches and technical solutions.

A process can be defined as a tree structure of activities, having at its roots the activity representing the process. The next level is made up of the entirety of activities which in turn can be made up of several activities and so forth until the activities of the lowest level are represented by the leaves of the tree. A general widespread definition arising from this idea of a process is as follows: "A process is composed of added value activities with the objective of meeting the needs of at least one client." The term "client" means either an internal or an external demand for the company.

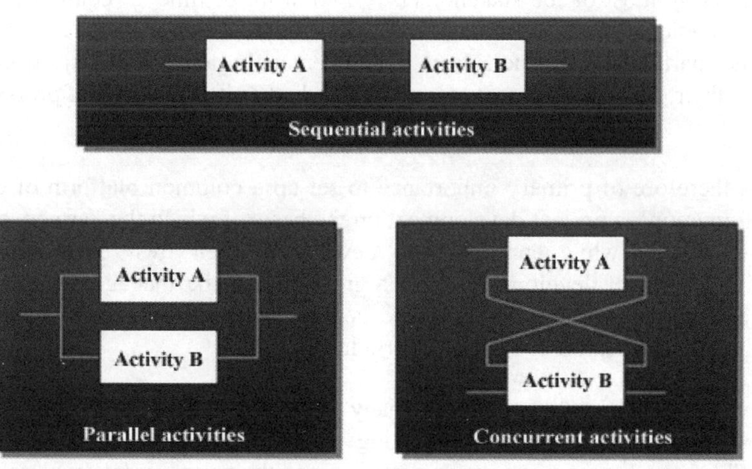

Figure 6.5. *Different combinations of activities*

The methods and tools for modeling processes, such as IDEF0 (Integrated DEFinition), are fundamental in obtaining a shared vision and common understanding of all the needed competences and stakeholders of the same project. It is possible to identify three types of combinations of activities in the analysis of the process: sequential activities, parallel and concurrent activities (Figure 6.5). These different types of combinations of activities are used to define the process according to the objectives and limitations from the operational or organizational point of view.

6.8. Management of a collaborative project

The management of a collaborative project is essential as it has been proved that, even when a project is well prepared and planned, many elements change during its course. Thus, very few projects manage to follow their initial aim and many others end up generating higher costs and substantial delays. The changes often create an impact on several partners and the interactions which result from them are either simple and obvious or more subtle and therefore uncertain. These interactions make it necessary to manage, in a very short timeframe, some trade-off between the objectives and expected performances which are the responsibility of different teams. It is therefore vital at all times to have precise information on the level of progress of a project and adhering to the many objectives so as to take the most appropriate decisions. Whether it is in the framework of a project executed within a company in the same geographic location or a project which involves several partners, it is always relatively complex to share information which can be understood by most of the stakeholders. Often, a lot of time is required to prepare and communicate this information and again more so for it to be understood by everyone, particularly partners from a different cultural or linguistic background who use their own development environment whether it is in terms of procedures or tools.

It is therefore of primary importance to set up a common platform of concepts and information on project development on the basis of which the partners can react more effectively while continuing to develop in their own environment. The activities of project development have a great need to be managed in this way of inter-company functioning which would be only for the contractual aspect for the provision of information at a definite periodicity.

Nevertheless, operating between many companies should not slow down the trade-off process or the decision-making simply because the processes are not compatible between them, or accrue costs because the partner infrastructure does not allow good interoperability. There are thus many generic needs for conducting inter-company projects such as defining and applying a common level of organization and operation which will serve as a reference model for all the partners. To ensure a good level of interoperability between existing applications of the partners and a common platform, it is necessary to increase the level of transparency and confidence between partners, and ensure a good level of confidentiality and security.

This type of platform must be common to all the partners, to which they can easily connect their data from existing applications. This platform must bring out a common level for process, organization and infrastructure and also for services such as publication and subscription to shared documents, consolidation of data and circulation of documents.

A common frame of reference for the project development and its operation must be elaborated on the basis of the taxonomy of the field [EPI 99]. This reference, once in existence, can be customized according to specific needs of each project and can be re-used systematically for projects of the same type.

This type of platform is often designed on the basis of Web technologies in order to reduce cost, while simplifying its access and implementation [PAL 00]. The use of XML language (*eXtended Markup Language*) for the description of the design dictionary offers the opportunity to create a common frame of reference for all required designs without any limit in their definition.

This frame of reference for designs is a type of standard practice for the project development which must be approved by the partners as much in terms of objectives concerning duration, cost, performance risk and resources as in terms of structures such as the WBS (Work Breakdown Structure), OBS (Organization Breakdown Structure) and PBS (Product Breakdown Structure). This access to data by partners will allow better coordination, overall transparency of the progress of the project and will offer better interactive potential between participants, whether it is at the level of developing a compromise or decision-making [PAL 00]. This mutual transparency results in a better level of confidence between the partners.

According to a synthesis of different documents and studies on the subject, project development comprises of six major activities or processes that lead to the goal and in a way that the project may accomplish its objectives [JON 99]. These activities are circular, iterative and form a spiral through the lifecycle of a project (Figure 6.6).

Identify: from the outset, a project must be divided into parts or lots of work carrying specific objectives. It is also useful for the evaluation of potential partners in accordance with their competences. A second level of division will be made by the partners on the basis of economic criteria or innovation factors.

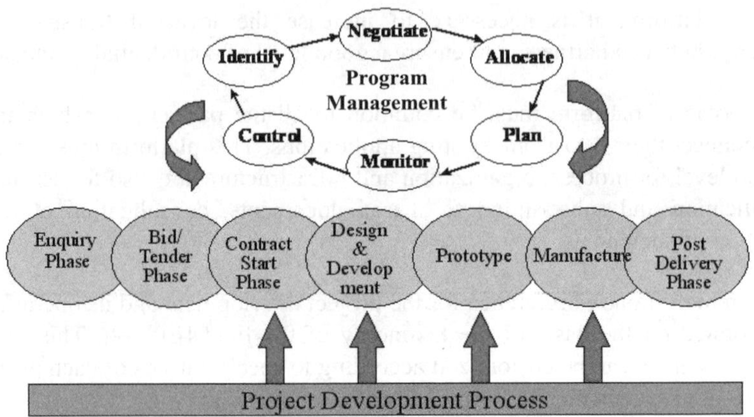

Figure 6.6. *Example of the project development process*

Negotiate: evaluating alternatives with potential partners whether it is at the beginning or during the necessary changes. In the context of a collaborative project, is not just about the sale or purchase of elements but more about the optimization through the choice of alternatives presented by the know-how of the concerned partners. This process is crucial as it allows the objectives to be discussed and their interdependence with other elements and also the impact on values that have already been consolidated.

Allocate: formalizing the allocation of resources, budgets, time limits and other technical performances for different teams responsible for different units of work. The ensemble of allocations must be consolidated so as to provide a good level of transparency to the participants.

Plan: defining the task plans and operational activities which are represented in the diagrams of Pert or Gantt. Each partner must communicate their evaluations so as to carry out an overall consolidation of all project planning.

Monitor: ensuring that the progress of a project is in accordance with the set objectives and that any eventual deviation due to changes are controlled by the participants. This process or this activity becomes very complicated when there are many partners involved as the communication of necessary data for follow-up and consolidation is often prone to several controversies.

Control: carrying out audits and reviews to check the contents of the output according to the set objectives, deciding on changes to be carried out corresponding to problems encountered and their resulting impacts. This process must intervene mainly at the level of the consortium so as to check the potential impact of any changes and to coordinate the tasks before launching any procedure with a partner, independent of the consortium, which would irrevocably delay the progress of the project.

The design dictionary for inter-company project development is based on the taxonomy of the field and on the process to be implemented. It must be approved by all the partners of the project and must present all the information which can be shared in the course of the project. This dictionary constitutes a frame of reference which can be customized according to the specifications of each type of project.

Duration, costs, technical performance, risks and resources constitute examples of the fields of interest for the development of a project. Partners must publish regularly their data on these fields from time to time. The design dictionary is a syntactic and semantic report which allows each partner to contribute his own data issuing from existing applications to a common platform so as to consolidate them at the consortium level [PAL 00].

6.9. Conclusions

Thanks to new technologies, the distance barrier has been crossed, whether at the level of shared workspace or at the level of shared understanding. With the contribution of systematic approach to development, as in concurrent engineering, the complexity barrier has also largely been removed. The management of collaborative projects and support from information technology has provided solutions to the barrier of high costs associated with the integration and management of many partners. A rapidly evolving attitude towards transparency and sharing to achieve better performance in creativity and innovation can be seen. To collaborate, companies and project teams must have a shared vision and quickly reach a common understanding of the concepts which they use throughout the lifecycle of the innovative products and services. Nowadays, communities of practice, especially their Internet or on-line version (such as Wikipedia), enable the description of shared concepts by peers. This will considerably speed up the way of reaching a common understanding among collaborative project stakeholders. What remains, among other things, are barriers of language and culture. Without doubt, the semantics of concepts will soon be an element of prime importance in the collaborative approach to ensure a better interoperability on all levels, as much for multidisciplinary teams as for the partnership networks.

Chapter 7

New Spaces for Innovation, New Challenges

7.1. Introduction

The approach to new spaces for innovation that we will develop in this chapter is based on the concept of virtual space. We will talk about new spaces for innovation precisely because they exist in great number. These spaces constitute the basis of a revolution in the exchange of ideas and methods of work. According to us, the concept of virtual space is the new common denominator of human activity, of which innovation is one of the most affected.

Next, we will deal with aspects that relate to the fundamentals upon which the process of innovation is increasingly based, fundamentals that in turn are based upon the development, the diffusion and the implementation of information technology. This can be seen right from the various Internet waves to the development of "Grid Computing" not forgetting the P2P system.

We will present the main characteristics of new spaces for innovation that have been made possible thanks to the development and implementation of information and communication technologies, Internet, P2P and Grid. It involves a new dimension of human activity that has enriched other activities or technologies that we know of and that we have been using all along. Reasoning in such terms will enable a better understanding of the potential and the opportunities that are available

---

Chapter written by Hiroshi MIZUTA, Victor SANDOVAL and Henri SAMIER.

in the field of innovation; opportunities that were perhaps previously inconceivable. The same could be resumed by the fact that we are witnessing a new generation of innovations that require a new approach and new Internet strategies so as to extract maximum benefits. The above concerns businesses and organizations, as well as industrial and service sectors in all countries.

7.2. Internet waves

The new systems are closely linked to the evolution of information and communication technologies (ICT) of computer networks and telecommunication infrastructure. The development, diffusion and implementation of information and communication technologies contribute towards the creation of new spaces in which more and more human activities take place. However, this world is dense, fast-paced, rapidly evolving and possesses its own particular dynamic of change. The Internet, in the current context, is an essential vector in the development and the utilization of these new spaces that should be given special weight in any study of generations of innovation. Its development may be considered as a succession of waves that change the way of working and communicating through computers.

The first wave is characterized by the introduction of the TCP/IP protocol and email. The network still remains the preferred tool of the research and education sectors.

The second Internet wave is characterized by the introduction of Web navigators that enable users to see data and to visually navigate or surf data on the Internet. This unleashes an unstoppable exponential growth of the network. The navigator redefines the classic concepts of space and time: one can be reached anywhere within a timeframe that is getting shorter and shorter. Over and above this, it enables interactivity that marks a new stage in the history of information technology: that of communication technology. Varied Web services enrich the capacity and the possibilities of work through the network that becomes more and more a tool used by the corporate sector [RIC 03] and individuals, "migrating" from its traditional domain, research.

The third wave of Internet technology is characterized by the "peer to peer" (P2P) technology that renders the manner of working and communicating more natural. P2P[1] makes it possible to connect heterogenous computer networks and other peripherals in an easier and more transparent manner. Rather than communicating with others in environments with centralized networks, we can now communicate and meet our colleagues in a more direct way. The hardware can now

---

1 See also www.openp2p.com.

access several services that use the new protocols and standards and receive replies from the other nodal "pairs" without requiring interpretation of the data through a centralized network. In this stage of the evolution of the Internet, there is a growth that seeks to "integrate" and "expand" the scope of the Internet within businesses and organizations and their related environments. One finds oneself in a rapidly evolving environment where one needs to determine the basic elements that characterize it. Various approaches are developed in this regard.

In this context, Laso Ballesteros proposes a simple scenario for the third wave of the Internet evolution. According to the author, utility computing, wireless technologies, interfaces and the creation of a software industry are part of this scenario in collaborative research and development. It implies an interdisciplinary domain that involves several isolated sectors of technology. This domain allows the integration of certain parts of these sectors so as to increase the new productivity as well as the advantages of creativity that are the result of the third Internet wave.

Amongst the constituents of this wave, utility computing allows individuals to access virtual resources that they may require, either in terms of calculating capacity and real time intervention with work environments, or with other individuals. Wireless technologies evolve towards Internet "objects". These wireless technologies help to create dynamic networks and the nodes of these networks allow the transmission of information from one computer post to another: the computers have a Web view of the real world. The radio frequency beacon ID, called RFID, is one of the first stages towards this objective. Today, every chip integrates wireless technologies. The arrival of operating system standards for software and hardware will encourage collaborative work environments. The development of software as a separate industrial sector on its own is another element of this third wave. This is because of the predominance of Web semantics, Web mapping and intermediary Web services that allow computer systems to communicate between themselves.

The current networks are represented in Figure 7.1. These networks act as an intermediary between users communicating between themselves, while the applications can be found in their terminating points.

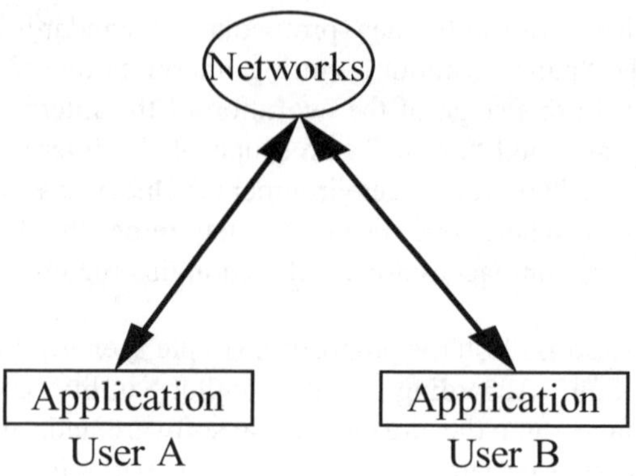

Figure 7.1. *User of a classic network*

This classic network system is progressively challenged by the new topology of applications, users and networks that are being developed. Figure 7.2 shows, for example, two users (A and B) that communicate to share the available resources in C's files, subject to communication between the two of them. This is the principle of a new stage in the evolution of information technologies that is in full expansion and that gives greater power and scope of computerized work practices to its users, independent of their physical location.

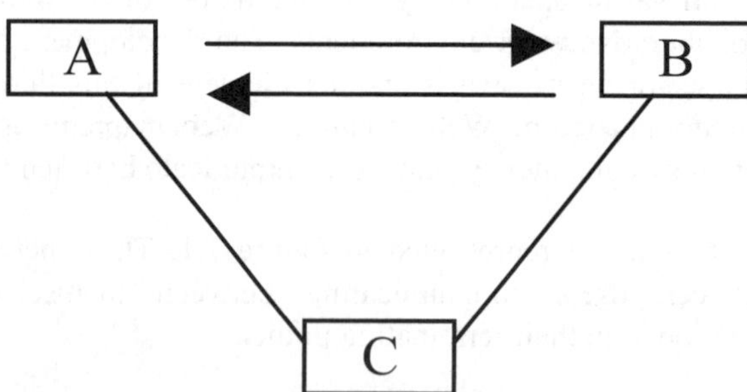

Figure 7.2. *Sharing of C's resources between A and B*

These developments are complementary and complement in a certain manner the principals of ubiquity that comes into play during the display, the diffusion and the implementation on a large scale, of the navigation of a set of pages, documents, data and content.

7.2.1. *P2P technology*

As we have seen, one of the central elements of these developments is P2P. In this section, we will briefly explain the P2P technology.

P2P technology is defined [GEA 01] as an information technology center directed towards computers that are capable of directly communicating between themselves. This allows sharing of data and system resources with a limited number of servers, managed centrally. In order for an application to be considered as a P2P application in its strictest sense:

– the application must be designed to adapt itself to the dynamic model of connectivity and to the temporary addresses of the network;

– the pairs must be able to operate with a sufficient degree of autonomy.

However, the term "P2P" is also an umbrella that covers three principal groups of technology.

The roots of distributed information technology lie in parallel computing and "clusters". The key idea is that the users can create clusters of machines connected to local networks (LAN) or the Internet in order to create a cheaper computer environment, which has the advantage of having a calculating power capable of challenging that of super-computers. Even though it must not be confused with the classic client/server model (communication between a central server and its client machines), a P2P environment is also composed of client machines. These machines work together on a problem in "pairs". In certain distributed computing environments, clients are capable of executing remote procedure calls (RPC) to another zone or resource.

Collaborative information technology allows the users located at a distance to communicate directly in real time with, for example, Instant Messaging software. Rather than sending a message through a centralized mail server and waiting for a response, the users can communicate directly with a user through their Instant Messaging application. A technical advantage of this model is that one can communicate directly with another person's machine (contrary to the classic mail sent first to a destination server and then sent to the correct local address) and one avoids the need for a centralized system. If two pairs have Internet access (through connection on request, local network, wireless network, etc.), then they can access any other. Another advantage of this technology is the detection of online presence through which it is possible to know whether the person, resource or file that one is searching for online is available. The graphic representation technology[2] makes it possible to view the network of people connected to each other.

---

2 For example, www.huminity.com.

Finally, file sharing consists of sending the information from a centralized file storage system and storing the file on one's own computer. In a centralized system, if one searches for a document, one finds it without knowing if it is the most recent version of the document on someone's computer. In the file sharing environment distribution system, one searches and one finds the document in the office of the author of the document and one knows therefore that it has been updated. This does not overload or clog the network as the data is relayed through communication routes in pairs.

P2P is an important technology for the new generation of innovation. In fact, the Internet, in its original form, introduces a platform that inducts a lot of innovation in its applications and contents. This is done in keeping with the principle of end-to-end which means that the network must remain as simple as possible and must push all intelligence and consequently all innovation towards the terminals. This vision is now possible thanks to the P2P architecture. The P2P environment is gradually becoming the environment that inspires the greatest volume of innovation related to the Internet. In the case of the third Internet wave, the applications and the data are found on the network and are not connected to any specific machine.

7.2.2. *Grid computing technology*

After the invention of the Internet and the Web, the Grid [FOS 02] [FOS 01] is becoming one of the recent innovations in cyberspace, offering a capacity and remote services that have never before been offered. The advocates of Grid swear upon its capacity claiming that it would be as though each Internet surfer had a personal super-computer and only by using computer resources not being used by computers connected to the Internet.

The need to calculate and process data increases proportionally with the data volumes created. The emergence of giga-volumes of commercial, scientific and engineering data makes "grid computing" necessary. Currently, computers can process data 700,000 times faster than the network connections in 1986, with a computer terminal transmitting a million of trillion bytes per second (an exabyte). Moreover, computers find it difficult to meet the needs of researchers and scientists. For example, a simple high-resolution brain scanner generates a file of three terabytes.

Corporations, industries, marketing, finance, education, security, and health and leisure related services are all increasingly aware of the possibilities offered by "Grid computing". For example, the Wall Street Company belonging to Charles Schwab uses the IBM computing grid to offer its clients advice in real time on their investments. The USA hopes to spend one billion dollars on providing Grid computing infrastructure to most of the industrial sectors as well as to research, education and health [DAI 03].

"Grid computing" involves, on one hand, distributed hardware infrastructure and the software that allows access to a huge calculating capacity and, on the other hand, the coordination of the sharing of resources so as to ensure the dynamic resolution of problems in virtual pluri-institutional organizations. A Grid system coordinates the resources that do not undergo centralized checks; it uses standard protocols.

The wireless telephone industry is a good example as it demonstrates the difficulty of applying rigid categories in the field of technological innovations. Even a little while ago, nobody expected this technology to expand to include Web navigation, organization, cameras and video cameras.

The Grid architecture [GEN 02] is a set of protocols that define the basic mechanisms by which the users of the virtual organization create and manage resources as well as the use of shared files. It is only an open architecture of services based on standards that facilitates interoperability and transportability. Amongst the technological layers of the Grid, the connectivity layer allows easy and safe communication. This layer defines the heart of the communication and the authentication protocols necessary for the specific functions of the network. It enables the exchange of data between the resources of the other layers. The support of this layer is designed according to the TCP/IP, IP, TCL protocols and the domain name (DNS) layers.

The resource layer defines the API and SDK protocols for safe exchanges/negotiations, administrative checks, accounts and payment of operations shared with individual resources. These two protocols define this layer: information and management.

"Grid computing" corresponds to the third Internet wave. It makes it possible to compress time by increasing the computer space through the optimal organization of calculation resources available. It provides safe and transparent access to the Information and communication technology resources such as hardware, software or any other scientific instruments. Finally, Figure 7.3 illustrates the positioning of Web services, of the semantic Web, of the classic Grid network and the third wave.

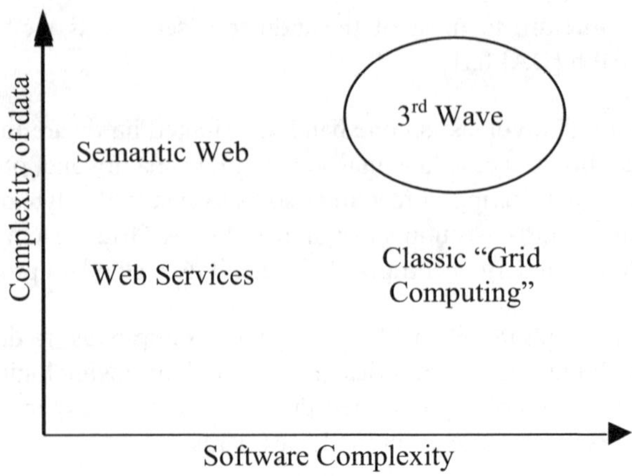

Figure 7.3. *Grid positioning and third wave*

Researchers were the first to adopt the Grid technology because of the need for greater calculating power, but the advantages of "Grid computing" go far beyond that. Each organization wishes to make better use of its resources, increase its efficiency, find solutions more rapidly and place products faster in the markets; and the Grid computing can facilitate all of the above. Security remains a concern for industry. Most managers do not think of exposing their companies or departments on the Internet if there is no security. In confidential environments within departments or an open research community, for example, where applications and data are not critical or private, security poses a lesser risk. The Global Grid Forum (GGF)[3] is working on the standardization of several existing security solutions.

An Intergrid protocol is also necessary. The Grid requires protocols (and interfaces and policies) that are not only open but are also standard. It is the standard that should make provisions for the dynamic sharing of resources with any other party. This would enable the creation of something more than a plethora of incompatible, non-interoperable balkanized distributed systems. The standards are also important as they act as levers for services and applications.

The definition of the standard "Intergrid" protocol is one of the most critical problems of the Grid community. The GGF, which has six years of experience in the field, currently provides a standard known as "Globus Toolkit". In the GGF, efforts are being made to define an Open Grid Services Architecture (OGSA) which could modernize and expand the protocols of the Globus Toolkit so as to meet new requirements including Web services. Companies such as IBM, Microsoft, Platform,

---

3 Visit www.gridforum.org.

Sun, Avaki, Entropia, and United Devices support OGSA and participate and contribute to its development.

An example of a virtual company

Grid technology is distinct from the Internet tendencies of the third wave or the peer-to-peer technology, but it allows them to benefit from the growth of virtual space (space problem) that it is supposed to process. A classic example is that of a virtual company or the Grid infrastructure that supports the sharing of resources and coordinates the use of such resources. The Grid technology is therefore a powerful tool for the development and use of virtual enterprises. Indeed, virtual organizations are by nature, distributed, dynamic, constantly changing and very flexible [SAN 95].

7.2.3. *Grid computing in Japan*

Japan searches to use a full potential of Grid. For example, in parallel programming, typical Grid applications are: parametric execution (execute the same program with different parameters using a large amount of computing resources), master-workers type of parallel program; or as typical Grid resources; a Cluster of clusters: some PC clusters are available, dynamic resources: load and status are changed from time to time. Japanese Grid is related to the world grid. There are now two GRIDs in the world: one is the science GRID which is the super-computer for scientific simulation and the other is business or commercial GRID which is the enterprise computer system for collaborating for business whilst being faster, cheaper and better.

Here is a list of most well-known Japanese Grid related projects in Japan: Information Technology Based Lab (ITBL), Super-SINET (NII), VizGrid, BioGrid, Campus Grid, National Research Grid Initiative (NAREGI), Grid Technology Research Center, Japan Virtual Observatory (JVO), (Business GRID[4], Science GRID[5]). We will now comment on some of them.

ITBL is a part of e-Japan Project (national project) with SuperSinet. e-Japan aimed to make Japan the world leading IT nation, promoting 103 projects. There are 523 users out of 30 Institutions sharing computer resources by 12 Institutions. ITBL has built an infrastructure and is now developing practical and expanding activities. It tries to join Japanese science, engineering and technology, offering places for participants themselves, and online organization in a ubiquitous society. As an

---

4 Grid Consortium Japan: www.jpgrid.org/english/index.html,
IPA: www.ipa.go.jp/index-e.html,
GTRC: www.gtrc.aist.go.jp/en/index.html,
GRID Data Farm: http://datafarm.apgrid.org/index.en.html.
5 NAREGI: www.naregi.org/index_e.html.

illustration concerning enterprise, ITBL considers this example: enterprise 1: infrastructure of computer science and technology (integration of computer environment – hardwares and softwares), advancement of telecommunications equipment, resources sharing in distributed environments; enterprise 2: ubiquitous laboratory/virtual office: anytime, anywhere, with anyone (borderless with connections between firewalls, melting pot beyond email culture); enterprise 3: the third research institution (integrated institution): colligation of interest groups.

SuperSinet concerns all optical production research networks, is a 10 Gbps photonic backbone with GbEther Bridges for peer-connections, with very low latency (3-4 ms roundtrip), and operation of photonic cross connect (OXC) for fiber/wavelength switching, 6,000 km dark fiber. It has been operational since January 2002.

VizGrid is a project started on June 2002 with a span of 5 years. This project realizes a remote collaboration environment on the Grid: volume communication environment and knowledge creating environment; spreading this technology to the academic society and industrial world, application areas for verification are medical science and nuclear fusion experiment.

NAREGI is a new R&D project funded by MEXT, a collaboration of national university labs and industry in the R&D activities (IT and Nano-science applications). The goals of this Japanese initiative are:

– to develop a Grid Software System as a prototype for future Grid Infrastructure in scientific research;

– to provide a Testbed to prove that the High-end-Grid Computing Environment (100+Tflops/s expected by 2007) can be practically used in the nano-science simulations over the SuperSinet;

– to participate in international collaboration (with the USA, Europe and the Asian Pacific area);

– to contribute to standardization activities (e.g. CGF).

NAREGI's participants are the National Institute of Informatics (NII), the Institute for Molecular Science (IMS), universities and national laboratories, research collaboration (ITBL Project), vendors (IT and chemicals/materials) and the consortium for promotion of Grid applications in industry. NAREGI software stack contains Grid VM, Grid Monitoring and Accounting, Super scheduler, Grid programming: Grid RPC and Grid MPI, Grid workflow, Grid PSE, Grid Visualization, High-performances and secure Grid networking [FRE 04].

As an example in perspective, it is possible to indicate that of MAP Project. It is a Canadian Government Project for SME manufacturing. It can be put together with

Business GRID, and will be a most innovative manufacturing system called Virtual Enterprise.

WIDE project is the international broadband project started from Japan. In this country, there is actually NAGERI which is a 40 GBIT network. And it will go the around the world soon for science[6].

7.3. Strategies of innovation

Faced with these challenges of constantly evolving technology, businesses and organizations of all types must seize opportunities that are offered. The question of strategy is first linked to information and computer network infrastructure that we have talked about briefly in previous sections. In this section, we will give some examples that will illustrate this.

Case study: e-science Grid (UK)

The e-science network connects the cities of the UK to the traditional centers of research and development and innovation like Cambridge, Oxford or London. The Grid network allows for the creation of a new scientific base that will help the country to maintain its position at the top:

a) linking of premier centers of research and training in science and engineering;

b) combining and sharing of resources that would follow: the community shares the resources of the network while working towards specific objectives;

c) new applications are developed thanks to the coordination of geographically spread out resources, for example, the distribution and collective access to data, and the distribution of computer resources of schools and universities.

The e-science Grid network in England is being emulated in several other countries: APAN[7] (Asia Pacific Advanced Network), CANARIE (Canada), Geant (Europe) or Internet 2 (US). There already exists a world forum for the Grid under the name of ANF[8] (Advanced Network Forum).

Countries like Canada develop national strategies or innovation policies taking into consideration or integrating all the concerned actors or parties.

The Canadian innovation strategy was launched on 12 February 2002, with the publication of two related documents called "Attaining excellence: investing in

---

6 Wide Project: www.wide.ad.jp/index.html, Commercial GRID: GRID today: www.gridtoday.com/gridtoday.html.

7 www.apan.net.

8 http://anf.ne.kr.

people, knowledge and possibilities" and "Knowledge, key to our future: the perfection of competence in Canada". These documents outline objectives, milestones and targets to be met in order to innovate more and to improve competency and learning in Canada[9].

The Network of Scientific Information of Quebec (RISQ) is the network of research and higher studies in Quebec. Its members are universities, middle schools/colleges, high schools (colleges of further education) and other educational establishments and centers of research, organizations promoting and diffusing culture and government organizations of Quebec. The RISQ has created its optical network with the financial support of the Ministry of Education of Quebec. The network is currently spread over more than 4,300 km of optical fibers that belongs to it. The aim of the organization is to provide the latest high tech high-speed telecommunications infrastructure, auxiliary services and access to a vast range of sources of information so as to facilitate and support the educational and research objectives of its members.

Case study: WOW project

Innovation in cyber health: better access to health services that are more affordable thanks to the WOW project.

WOW Wound Care and WOW Cardiac Care are at the heart of a revolutionary research project in which CANARIE has invested $180,000 and which receives the same amount of funds as well as help in kind from the "St. Elizabeth Health Care". This new approach makes it possible, on the one hand, to deal with the problems of patients suffering from chronic illnesses and on the other hand to obtain information, follow a training program or benefit from the necessary treatment or care. As professional health practitioners are not easily available, funds limited and the requisite expertise not always on hand, and the creation of a network of experts becomes indispensable. The development of a new approach to health care management has proved instrumental in creating better checks today that have been beneficial to health and the quality of life. The providers of health care – from rural and far away regions especially – have appreciated how such communication or contact with specialists, mentors and peers not residing in the same area has helped them to remain in touch with latest developments in the field, to multiply local facilities and to maintain their faith in their work and the care that they provide.

The WOW concept marks the convergence of three initiatives in cyber-learning and cyber-health. It has been installed in all the participating experimental sites especially the West Prince Regional Health Authority in the Anishinaabe Mino-

---

9 http://innovation.gc.ca/gol/innovation/interface.nsf/vSSGFBasic/in02424e.htm.

Ayaawin Health Authority in Manitoba and in the St. Elizabeth Health Care centers in London and North York, in Ontario.

The Web applications, baptized @YourSideMC, @YourSide ColleagueMC and @YourSide CompanionMC, offer popularization and outreach services and health management services to individuals, cyber-learning and perfection services to health care providers and evaluation services and counseling services with experts respectively. Together, these three applications constitute a "Web of Wisdom", hence the abbreviation WOW.

Case study: CANARIE

CANARIE[10], developed in Canada, is a non-profit organization whose objective is to speed up the development and exploitation of the fastest and the most efficient research networks as well as applications and services that are meant for them. By encouraging collaboration between the key sectors and by concurring on initiatives of the same kind all over the world, CANARIE helps innovation and growth so that the entire population of Canada benefits from it in social, cultural and economic terms. In 2002, the Canadian government ranked CANARIE amongst the best organizations in Canada for innovation. It is the national network for research and innovation.

7.4. Hyperspace: new dimension of innovation

Classic or ordinary space is henceforth an additional element in the innovation process. A new space is being gradually created alongside it, a new dimension which gives a reality that is much larger than what we are used to. This new dimension is part of a new space that can be simply called hyperspace. We have to get to know hyperspace in order to grasp better the possibilities and opportunities available in the world of innovation which is entering into a new generation: a new dimension created for integration into human life and activity, to be used as a tool in work, in life in the production of physical and intellectual goods and services. This is related to all the human activities and therefore innovation. There is an interface between this hyperspace and ordinary space which can be seen in a diagram of the interface Man-Machine. This interface defines a boundary between dimensions and can be called fractal boundary.

---

10 www.canarie.ca.

7.4.1. *Hyperspace laws*

Hyperspace has its own nature, its laws and principles which must be known and respected. These are distributed according to the layers of this space. In fact, this space is formed by different layers which go from the most physical (networks) to the most abstract (applications). The world in layers is already to be found in individual computers. Thus, for example, Tracy Kidder illustrates it by relating the story of the mini-computer Eclipse MV/8,000, launched by Data General [KIL 81]. In fact, the computer has physical layers, microprocessors, memories, cabling, etc., from where the circulation of electrons is organized, the electric voltage is controlled, the bits take shape and from where operations for consultation of memories, calculations and writing are carried out at high speed. To these layers must be added layers of languages so as to give instructions to this robot: microcode assembler, operating system, programming language and finally applications. Each of these layers has its own principles, nature and obeys general and specific laws. In retrospect one can see here the foundation or even the geology of the future networks and of the new dimension of the work that will be done on them.

We shall make a rapid tour of some of the laws describing hyperspace. The reader can appreciate for himself their importance and their implication for the new generations of innovations.

7.4.1.1. *Moore's Law*

Gordon Moore (1965) notes that the microprocessor, when in operation, has one tendency: each new chip has twice the power of the earlier one and the development time is about 18 to 24 months. This is valid for the size and price per unit which drop. Thus the price of the PC between 1975 and 1980 was divided by 10 and between 1980 and 1985 by 23. There is therefore an acceleration of Moore's Law. It operates in the physical layers of the hyperspace, first in the computer industry and then moves on to all the industrial activities and services. This capacity of memories allow transfer of data, CD, books, etc., directly between the creators and consumers without any intermediaries.

7.4.1.2. *Metcalfe's Law*

If there are n telephones in a network there will be n-1 values added as n-1 can call n. Very simply, the value of a network is proportional to the square of the number of subscribers. If the number of subscribers doubles, the added value is multiplied by four. The greater the number of subscribers, the greater will be the potential value added thanks to the increase in the number of communications of each one. In other words, connectivity generates added value which we call potential value of the network. Now, this is also true for a network sharing knowledge, made up of contributions or calls (a natural network doing this is the human brain). If a user enters the network, he can create new connections which would be equal to the number of users already existing. There is an increase in the possibility of sharing knowledge (a common warehouse), but each user must enter the network and put the connections to optimal use. It must be recalled that optimization of this system is very different from that of the networks of the affine world. This is a fundamental factor for the optimization of future systems, for example, classic algorithms of optimization do not totally apply. In fact in this new world, there are a very large number of feedbacks and non-linearities as opposed to the affine world. These are some of the problems faced by new generation innovation which the companies should consider.

7.4.1.3. *Growth limit or asymptote law*

Any product or service ends up saturating a given market, but this happens after a certain time. There always is a horizontal or asymptote line which marks the upper limit of this growth of technology. Beyond this lies the saturation point. The question is thus to determine the value for this line. Any growth follows a logistical type of law (rate of penetration plus rate of renewal from a given moment) affected by the network effect of diffusion of technologies. There is therefore a relative physical limit. This is particularly important in the case of the Internet and related services. They follow a model partially similar to that of the diffusion and implementation of the telephone but they depend on the diffusion from terminals and servers. These are organized according to different topologies and thus generate segments in their utilization.

7.4.1.4. *Law of value creation by the virtual community*

This is an extension of the Metcalfe's law of application layers of hyperspace. It is about the actual content of the exchanges between users on a network in the form of a virtual community. There are indications on the sources of the added value and how these are used in the creation of new knowledge taking the benefit of the network: by increasing the number of subscribers, providing the greatest facility for the broadcasting of information, establishing contacts, bringing together people (problem of distances; see below) and shortening of time (see below). But connection does not necessarily signify new added value; it is a potential which must be managed and optimized.

7.4.1.5. *Amda'hl's Law*

The calculation capacity of the machines increases faster than the input/output capacity of the system. This happens with a single machine, a set of machines in a cluster or a network of machines and it can be particularly observed in the case of internet. This signifies the need to enter data, documents and files which have the lowest load on the network.

7.4.1.6. *Dilutive nature of information*

Compared to traditional goods and services, information behaves differently because of its intrinsic nature. For example, it has an owner, but if he or she sells it, they do not necessarily lose their ownership. The user or buyer receives it and they can create or recreate new information, and neither can they prevent the other from making copies even if they respect copyright laws. One produces, sells and continues to enjoy ownership and at the same time the buyer sells and continues to be owner of his production and the buyer is transformed into the owner of the same information. This is the process of shared information, which is natural in hyperspace. The advertisers pay for an advertisement and the buyers pay for the newspaper by financing the advertisement. They help to finance the advertisers. The ownership changes in its essence as it is related to the content of the information, the cyber-owners; the capacity to copy the content of the information allows the buyers to reproduce it whilst respecting certain limits. Sharing ownership of information thus creates a richer environment to further create new data, services and goods. The network is therefore a catalyst for ownership.

7.4.2. *Hypertime or space time*

We all have 24 hours in a day in normal space but in hyperspace time has new characteristics. In fact in this space time can be sequential, parallel and concurrent at the same time. Figure 7.4 shows an example of several types of time.

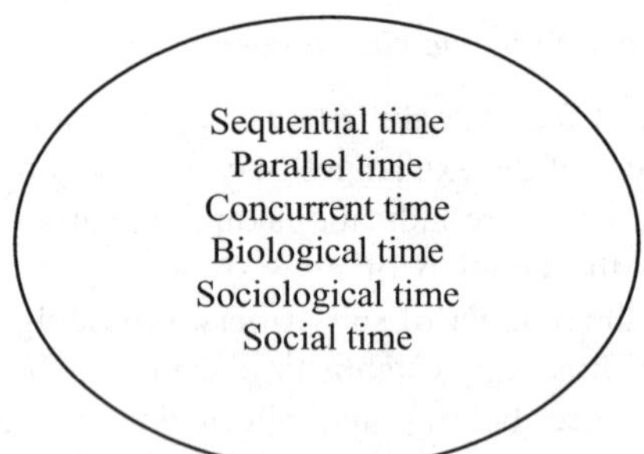

Figure 7.4. *Examples of different types of time*

It is observed that the time can be "modified" according to several forms; for example, time can be "shortened" by having parallel or concurrent time (for processors, for computers with clusters). These properties of the first computers pass to greatly parallel computers and are amplified by the computer networks.

"Shortening" time starts from the hypothesis that there exist several categories of time and that there is a possibility of reducing them by changing their sequential character thanks to the computer networks that make it possible to combine these different times. There exist different times, namely network, computer, biological, universal sociological (it is important how we perceive the time we have), etc. Let us now consider the computer network and the operators' network as having their own time and working together. This problem can quickly become complex, for example, if one considers millions of operators and computers working simultaneously and adding different perceptions of time of each person connected to the network.

The earlier characteristics form the intrinsic nature of the new time (internal). It can be called hypertime or space time. It only exists thanks to digital spaces and thus it happens that time is not so much an exogenous factor but an endogenous one. This means that we can manipulate it and control it according to our needs, for example, working together on an innovation project whatever be the latitude and its timing.

Thus, a team of engineers will have certain constraints emanating from the planning of the project but its members will be able to work on their own available time, make use of the flexibility offered by the space time which is both local and general, a sort of multiple time. The question is to know which are the elements that belong to local and which belong to general, how to use them and how to best combine them so as to achieve the maximum benefit. Solutions exist case by case but there is no general solution and we can organize the potential cases. Time must no longer be considered as computer time but as integrated time in the new dimension.

It is important to take advantage of the hyperspace to improve the quality of the time and to control it. One solution is to give more power to the engineers and make them more creative by making use of the navigation tools. This requires reorganization of working conditions and the contents taking support of the hyperlinks once these contents are put into hypertext. In order to "shorten" the time each one must make use of the hyperspace then use it to improve work.

However, time is relative to each person in hyperspace: each one can and must create and use his own time, relative by the nature of the network and the nature of the time created by each one. Hypertime is related to the computer and persons' networks and because of this, these networks must be considered and assembled. Create spaces so as to create time, do it layer by layer as per the geology of this world: thus space and time are created and any action can interact in space.

Constant competition prevails in Internet resources and time is included in it. Hyperspace is a shared space where there is competition for resources, time (shared time), ideas, etc. Competition as action for life and to do useful things by oneself or by others is the very nature of hyperspace. It is in this field that innovation now takes place.

Simply copying current reality in the new space cannot thus work: they are different realities.

Let us take the example of the Internet year. The Internet year is different from the normal year that we all know. In fact in an Internet year there are three ordinary human years of 24 hour days. But we have to eat, sleep, etc.

7.4.3. *Distance and hyperdistance*

Hyperdistance is the distance of hyperspaces. One can imagine space of the Internet as a pin head. Thus, one has the impression that the distances reduce or simply disappear if one thinks in terms of classical physical distance. However, here the physical distance is only a component of the hyperdistance. For example, there are other distances such as cultural and linguistic distances. The classic social distance is transformed as well: the traditional hierarchies tend to fade out in a network where initially all are at the same level, but the most important is the distance created by the system of links between pages of a same site or of different sites. Distances can be formed and reformed and will have to be integrated in day to day activity. Here are some types of distances that are to be considered [BER 01].

7.4.3.1. *Organic distance*

If the objects touch each other it can be said that their distance is at the organic center. If they do not touch, intermediary objects have to be added, which are deducted directly from the calculation of addresses. The "soft" distance is done by indirect addressing (distance in the addressing tree, etc.). All along the addressing tree steps are marked with "Yes" or "No" on them.

7.4.3.2. *Hamming distance [HAM, 80]*

The Hamming distance is the most functional as it corresponds to a content distance. One can have a distance of forms between two images (based on recognition of images). The distance between two objects is described by the difference between the bits describing it. If the two objects have the same length in bits then it can be done; otherwise it is more complex. According to Hamming, two strictly complementary messages have a maximal distance. Distance can also be defined as the number of bits required to describe the second message starting from the first message (and vice versa).

7.4.3.3. *Other functional distances*

The differences of meanings and the depth of the process are not indifferent to organic bases. Intermediary distance: the language difference is a form of distance. On the other hand, standardization of languages reduces the distance, but encryption (security for intermediaries between systems) introduces a new element of distance. This creates a specific factor. As the distance increases, the supplier of values must work more.

7.4.3.4. *"Sociometric" distance*

This distance is as large as the difference between communicating objects, for example, the length of the addressing. The distance measure can bring Moreno's sociometry into the picture which will make it possible to gain a definition of measures between participants in a network. More interesting are the cognitive distances; there is everything on the Web. For this, standardization must be done but logs and IP addresses can be used to measure distance.

7.5. Cyberenergy and cyberentropy

In the specific case of cyberspace, we have cyberenergy and cyberentropy. The rapid growth of the volume of information on the Web affects these two phenomena, particularly what can be qualified as "information eclipse": that is, a part of information that is hiding another part.

In this space there are many interactions between agents that populate it and the existing objects. It is possible to formalize these interactions, but one can also consider the dynamics of these objects with a model that integrates the cyberenergy and the cyberentropy.

Cyberenergy is the capacity of an object to execute an information work, that is, the capacity of a hyperspatial object to correctly pass information in space between persons who are participating in a given project. This can be reasoned out in terms of distance and in this case mistakes appear in comprehension as well as computer languages and ordinary languages. Thus, scattering of computer and ordinary languages is high if the distance between the objects is high. If we have a distributed software system, each computer of this system should at least have a common language (communication protocol between them). Thus, the cyberenergy of a hyperspatial object depends on several factors, amongst which are the size of the network and the format of the object.

Let us now take cyberentropy. This represents the tendency of an object to deteriorate with time. That means that the object tends to lose cyberenergy.

Cyberentropy tells us that an object cannot always execute its information work with the same performance level. Therefore, one must know for what reason and how to stymie this tendency.

Mathematically speaking, cyberentropy is derived from cyberenergy. It will thus have the value of the slope of the curve of cyberenergy combined with time of a given object when the latter has been left in an abandoned state. In this way, the more an object is cyberenergetic, the greater the value of the slope of the curve will be. We will have an equation as given below:

$$\textbf{Cyberentropy} = N \times dF/dt + Fx\ dN/dt$$

where N is the size of the network, and F the format of the object.

Several factors intervene in the behavior of this equation; for example, a bad choice of network. Suppose that we put the object on the Intranet: that can limit the number of potentially interested persons whom one could reach thanks to the connection and for whom the information carried by the object would be pertinent.

In any case, we have to preserve the cyberenergy from the deterioration in time. To do this we have to observe the frequency of updating of the object and choose the correct format to keep the object.

From this point we can integrate the deteriorations of the objects in the cyberenergy formula which would then become:

$$\textbf{Cyberenergy} = R(t,a,f) \times F(p(t)) \times (K(t)$$

– R: size of the network which depends on the time t, on the accessibility a and reliability f;
– F: coefficient of format of objects which depends on the share of the user's units in the total of the units of F(p(t));
– K: coefficient of deterioration related to time.

In these conditions cyberentropy will be given by the following formula:

$$\textbf{Cyberentropy} = RF \times dK(t)/dt + FK \times dR(t,a,f)/dt + RKxF(p(t))/dt$$

7.6. Conclusions

This chapter, due to its nature, has several conclusions. We start with the fact that the innovation generations are entering today into a new stage characterized by the increasing usage of the potential of software networks and the Internet (generally speaking). Innovation can only take place with relation to the past: a genius, or a voluntary or fundamental research group are no longer required to create one or several new products. This is a situation noted for the existence of new work and creation spaces which require new forms of organization, optimization and upgrading of the innovation.

In these new spaces, innovation is generated by practicing communities who have identical needs or common values such as Weblog[11] and Wiki[12].

Innovation is also created collectively by the networks; in fact companies[13] and organizations[14] organize worldwide competitions with prizes for innovation in order to gather a profusion of ideas. For example, innovation takes all the forms (continuous, in batches, momentary and in fusion form) and benefits from the contribution of scientific and technical knowledge almost in real time, whatever the geographic situation of the participants in such a process may be. The new spaces open a new type of cyber-innovation generated by these new spaces and using these new technologies as vectors of dissemination.

---

11 See http://shiva.istia.univ-angers.fr/~blogistia/agregateur.
12 Example: Autrans 2004: autrans.crao.net/index.php/AideUtilisationWiki.
13 See, for example, www.mondiinnovation.com.
14 See, for example, www.fabriquedespossibles.org.

Tooling Innovation:
Which Methods to Play and How?

Introduction

Following the exposé of foundations and new models of innovation in Part 1, we begin Part 2 on technologies, tools and methods that facilitate the putting into use of innovation processes within enterprises. Technologies bear the role of "virtualizing" certain human tasks and consequently become intelligence amplifiers. When coupled with collaborative work methods, virtual reality technologies also represent an outstanding potential for evolving innovation.

This Part introduces the methods and the tools that consultants and professionals, as well as students, favor, as they are compelled by management to implement a new organization. Ideally, methods and tools are often those designed, implemented and optimized by the enterprise itself; however, to reach such an ideal, it is preferable to benefit from numerous research activities and discoveries made in the field.

We offer a method for capitalizing innovation that aims at yielding sustainable enterprise policies: a method to determine and use stylistic trends within a target market; the creativity tool (TRIZ) which offers yet untapped possibilities; a way to develop creativity within networks; and the most recent Internet watch and survey tools from economic intelligence trends as well as various related innovation methods

Our goal is not necessarily to reveal a "meta-method" that would hypothetically chain all methods and technologies by making them work concurrently; it is to provide the reader with a global vision and a culture of innovation tools that will enable conscious choices to be made and will support his or her own methodological construct.

Chapter 8

Knowledge Management for Innovation

8.1. Introduction

Innovation is none other than the creation and application of new knowledge in order to make things productive.

How would a better approach to this knowledge favor innovation? Still often associated with the capitalization of knowledge and asset management, can "knowledge management" take on the challenge of innovation? Companies are beginning to bring together knowledge management and innovation in their applications, and numerous approaches of innovation are developing and are directed towards the communities, in order to stimulate the creation, sharing and validation of knowledge.

8.1.1. *Studies*

In December 2002, SESSI[1] published a report on the relationship in French industry between the policies of knowledge management put in place by companies and their innovative capacity.

This study identified four types of knowledge management policies: "a culture to promote the sharing of knowledge, a written policy of knowledge management,

---

Chapter written by Marc de FOUCHÉCOUR.
1 Service of the Director-General of Industry, Information Technology and Postal Services – DiGITIP.

forming partnerships for the acquisition of knowledge and a policy to motivate employees to stay with the company".

The CIS3[2] survey brought to light a strong correlation between the implementation of knowledge management policies and the capacity to innovate:

– the companies that have innovated are twice as likely to have implemented at least one knowledge management policy as the other companies (66% compared to 29%);

– the propensity to innovate of the companies that have adopted all the four policies of knowledge management is 63%, while that of the companies who have not implemented a single policy is 46% (see Figure 8.1).

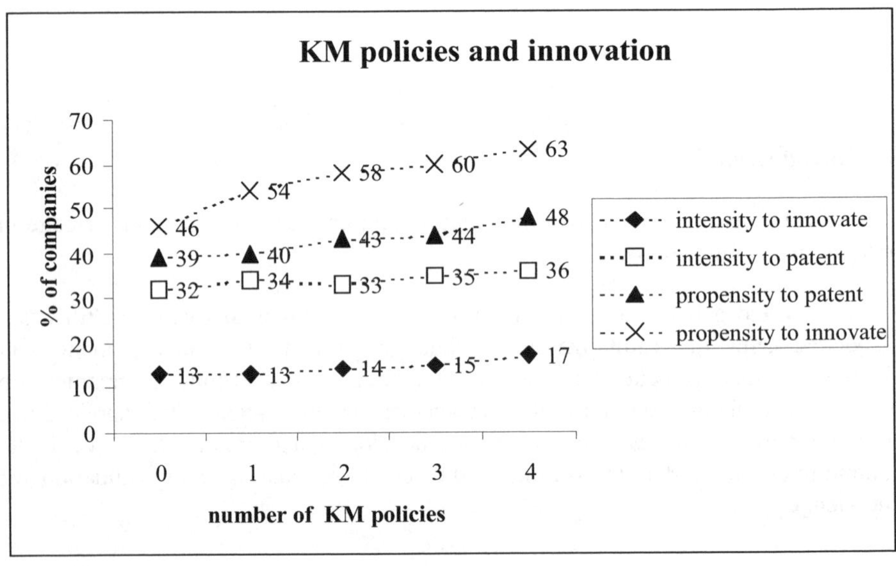

Figure 8.1. *Performance of innovation and the intensity of knowledge management*[3]

<hr>

2 Available at: http://www.industrie.gouv.fr/biblioth/docu/4pages/pdf/4p168.pdf.

3 Definitions: *the propensity to innovate* is the proportion of the turnover of companies comprised of new products or products clearly modified which were introduced between 1998 and 2000. The *intensity to innovate* is the proportion of the turnover of new products or those clearly modified in the total turnover of the companies. The *propensity to patent* is the proportion of enterprises having a turnover corresponding to the products protected by a patent in 2000. The *intensity to patent* is the proportion of the total turnover protected by a patent in the total turnover of the company (source: Sessi, CIS3 survey).

Recently the 2003 survey "Vision of Managers in the field of Knowledge Management" conducted in France by Knowings confirms "that we have entered into a new era, that of a more mature development of knowledge management, which is progressively becoming an essential management lever to help organizations adapt to the demands of the current economy: working in networks, development of immaterial capital, continuous innovation, rapidity". In particular, this report indicates that "the element most often classified as domain no. 1 with regards to returns on potential investment [is] that of conception/innovation/R&D"; moreover a third of the devices of Knowledge Management (KM) "already in place" or "in the course of being implemented" concerns this domain. In addition to this, if, as in 2002, innovation is no longer the factor that best explains the importance of KM in the eyes of managers, the economic situation is another factor.

On the other hand, the report underlines the increasing importance of the networks in knowledge diffusion and innovation.

These two studies clearly indicate a strong awakening in companies as much as in institutions that KM can be an asset for innovation, especially if it supports networks. The four categories of knowledge management policies used by SESSI do not provide an immediate answer to the question: "*which* management of *what* knowledge creates innovation?"

More recently, a project of the European KM Forum published a report on the theme "How to exploit knowledge for innovation", the outcome of surveys and workshops carried out in 2003 and 2004 in Europe. This report underlines the importance of the relationship between creativity, innovation and knowledge, all resulting from interactions between individuals ("social interaction") [COM 03]. An examination of the contributions and discussions on this theme also indicates that KM for innovation is a field of reflection that is fast-evolving and still very open. The KM initially oriented towards the collection and capitalization of knowledge is still exploring better ways of reflection on innovation[4].

8.1.2. *Objectives and plan*

The question that we shall ask in this chapter is: "How can innovation be managed *by* knowledge?" or: "how can we favor the emergence and increase the chance of success of the innovation process, and amplify its results; how can we engage a virtuous circle in which today's innovations will stimulate those of tomorrow?"

---

4 See also the website of the CIKM (Creation of Innovation through Knowledge Management) Project: www.cikm.net

We shall discuss firstly the relations that foster innovation and knowledge (section 8.2). We shall then recall certain reports to help locate the increased stakes of knowledge management and innovation (section 8.3), and we shall touch upon the "organizers of thought" that make it possible to apprehend the concepts of knowledge, their role in organizations and their management (section 8.4).

It is by following innovation and knowledge in their parallel cycles of transformation that we shall discover the processes, methods and tools of knowledge management for innovation (section 8.5). Some key factors of success (section 8.6) can be deduced naturally. The conclusions (section 8.7) are aimed at opening up reflection in evolution.

8.2. Innovation and knowledge

"If the idea at first is not absurd, then there is no hope for it" (Albert Einstein).

"That which distinguishes an innovator is his capacity to integrate a novelty into social practices, to build a new collective behavior with the help of a new idea" (Norbert Alter [ALT 03]).

To approach the relationship between knowledge and innovation is to step right into the complexity of organizations and people, and to discover a number of paradoxes where our first reflex would always be that of wanting to find solutions. Reality often leads us to accept these paradoxes, and it tries to accommodate us in them, or help us solve them, by changing the point of view or the objective. Yet, in this chapter, I shall take the side of the researchers, and will use them as energy poles, the short circuits of thought, and as invitations to reflection and to creativity. The complexity and rapidity of the world compel us to move from a binary mode of reflection – either/or – to a composed thought – and/and – where the opposites do not get terminated but get maximized, while coexisting.

Other pair of different and related notions, dualities allowing – obligating? – a multiple look, will allow us to construct a pragmatic "parametered" vision of KM for innovation, adaptable to the realities with which we are faced.

Figure 8.2 presents some dualities of innovation/knowledge landscape, which will be found throughout the chapter. The left branches rather concern knowledge and the culture while the right branches concern the organization and its processes.

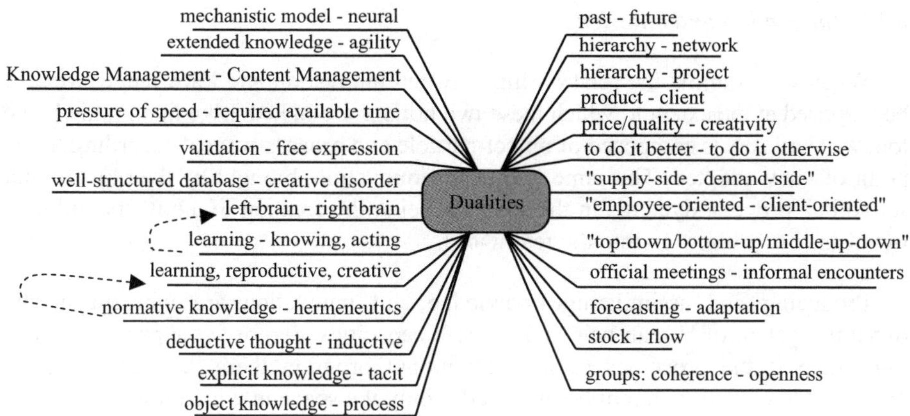

Figure 8.2. *Dualities*

8.2.1. *Some dualities*

Innovation refers to the new and the existing, to the future and the past, to the unknown and the known. Novelty is always relative; it exists only with reference to its opposite, to that which is currently available. The new knowledge defines itself with respect to established knowledge which is shared. Be it an unpublished combination of the existing knowledge, the fruit of a "destructive creation",[5] or simply born out of a change in the point of view on the object being studied, it is at the beginning of an emergence whose creation remains mysterious and in any case, difficult to express in models, only becoming innovation by its diffusion and its collective acceptance.

Innovation is, at the beginning, in its creative dimension an affair of the right brain (Einstein affirmed that he first dreamt his discoveries), of the qualitative. It soon becomes a problematic of appropriateness between the object, the service or the new procedure, and its "market", its field of application or development. We shall consider measuring the product-market distance from clients. Production needs to be optimized while their number shall be multiplied and their satisfaction and loyalty shall be increased. In short, we shall consider analysis and quantity.

Innovation is a shared novelty. It is born *singular*: unique and isolated, and only takes life by sharing done at many levels: sharing to create a favorable environment, sharing to validate and to develop it and finally, sharing by the end users of the innovation: an innovation will be successful if its users become the stakeholders, also contributing to innovation by devising new, unexpected uses for the products and, in doing so, by making such uses give rise to further innovations.

---

5 Schumpeter [SCH 61].

8.2.2. *Innovation and knowledge*

When we examine the relationship between innovation and knowledge, we will be surprised at the extent to which these two notions are linked. Like light, knowledge too, we shall see, is composed of object-particle and process-wave, "according to the point of view adopted. The same is true for innovation, brought by the object[6]" but found in a process, dynamic in the sense of being the creation of a novelty and then by means of its insertion into the real world.

Producing and assimilating knowledge and innovation are inherent in the dynamic system of human beings and these capacities are not yet apparent: we can only support the processes of innovation, not program them, as underlined by Nonaka: "Knowledge cannot be managed – only the space in which it is created can be" [NON 95]. What is this innovation made up of?

Innovation and knowledge are both, as we have seen, social phenomena: they are born in the minds of individuals but only get deployed, and affirmed collectively. Their processes are equally contagious[7] and self-procreating: knowledge begets knowledge, innovation begets innovation.

In the end, the relationship between innovation and knowledge is complex: they most certainly stimulate each other, and it is easy to comprehend, on the one hand, that innovation is a creation, a transformation and a diffusion of knowledge all at the same time, and on the other hand, that a "learning environment" is an ideal breeding ground for innovation; however, a lot of knowledge, especially that which has been rationalized and homogenous, can curb creativity. Therefore, it is not the bulk of knowledge possessed by each individual that fosters innovation, but the flux of knowledge in the form of hearing and expressing, dialogue and discussion, between different individuals.

In terms of knowledge, it is the passing from conservation to conversation, and not to conversion.

---

6 Product, service or process.

7 An advertisement of Renault appeared in April 2004: "for 100 years now, we have been creating automobile models, then one day, we invent a new model of the company".

8.3. Reports

8.3.1. *The reversal of the pyramid*

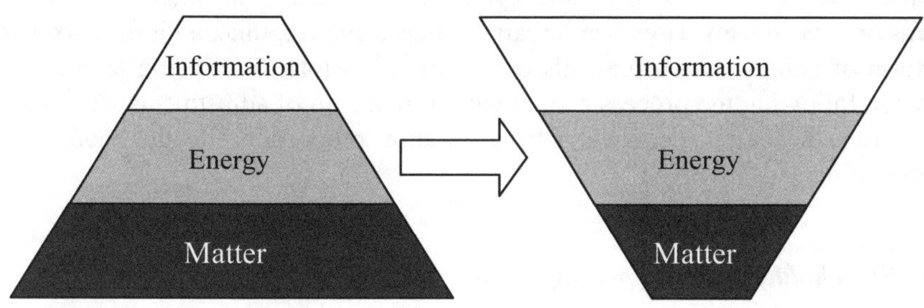

Figure 8.3. *The reversal of the pyramid*

We have passed from a materially "infinite" universe – where were the limits of the world? – in which information was rare, and expensive, to a Global Village where information is the only inexhaustible *and* renewable resource. The reversal of the pyramid engendered a necessity for innovation (less matter and energy) and provided the means (more information): more services and fewer objects, lighter objects that consumed less energy, and contained more information. Do you remember: "less petrol, more ideas"?

It is not so much about finding the information, as it is about sorting it out, and transforming it into decisions or actions. Information in all of its components, inclusive of those of knowledge, becomes a wide and complex space, enriched by exchanges and meetings, best proven by technologies today. "Knowledge is power" wrote Francis Bacon in around 1600. "Knowledge *sharing* is power" is the order of the day in a society of knowledge.

8.3.2. *Complex – collective*

The level of collective knowledge certainly increases, but that of individual knowledge and expertise diminishes, at least relatively: each of us understands less and less the increasing complexity of the world and the quantity of knowledge necessary to understand it. We are more intelligent collectively and more ignorant and helpless alone. Our new weapon is *collective* intelligence, which presupposes communication, coordination, cooperation, conversation…, co-, that is, links, exchanges, messages. Hermes, the god of travel and communication, reminds us that time has become a rare commodity; the "collective" is time-consuming.

8.3.3. *The paradox of time: compression and space*

Time is the metronome of innovation: it is necessary to compress the circuits of decision, the *"time to market"*, the cycle of renewal of the ranges, the delay of returns on investments. However, creativity needs silence, time, a bit of relaxation in the form of games, which prevents our neuronal pistons from being jammed. And the stage following the process of innovation, made up of adjustments, coordination and consensus/dissent, demands a time duration proportional to the square of the number of persons involved.

8.3.4. *Stakeholder-oriented management*

Innovation management tends more and more to consider and integrate into the process of innovation, over and above the R&D department, all the external "stakeholders", such as suppliers, partners, distributors, clients, going right up to the clients' clients and the competitors, as well as their internal departments – marketing, sales, production, services, engineering, etc. Intra and inter-organizational integrated solutions, and the co-development of new products or services in association with the client: all these approaches make it necessary for a company to find a *common* or, at least, a language which can be shared, in contrast to the language of experts of the R&D department.

8.3.5. *Matrix organization*

As stressed by Jean-François Ballay, in the matrix-like structure of the company, the employee is subject to a double constraint: the project axis demands a *result* while hierarchical axis tells him how to do it and with what *means*. None of the two logics can predict the place and time of production, the sharing or the acquisition of knowledge. "However, the exchange of knowledge does take place somewhere, but where?" [BAL 02].

8.3.6. *Methods, tools and incantations*

The theme of "Innovation and knowledge management" has given rise to a large amount of literature and symposia. Models of successful experiments were built and the list of the methods and prescriptions were published without it being always possible to distinguish them from the incantations or from the canons that promote the ideals: "Measure whatever you want to know!"; "Recruit and save talents!"; "Do not underestimate the culture of organization!"; easier said than done!

Methods are easy to find, always in list form: publications, workshops, business trips, newsletters, websites, networks and forums, projects, communities of practice, expert research tools, blogs, etc.; in which order and for whom?

One of the most nagging questions that a practitioner or a manager is faced with is how to find the connection in order to create a good combination, the right articulation of the approaches, methods and the tools that allow him to respond to a specific situation in his organization. If knowledge is many-sided, contextualized and dependent on people, this is furthermore true of the whole process concerning it, and we shall not consider the problem of retirement of experts in the same way as we would consider that of innovation, in a big company or a Small or Medium-sized Business, in the aeronautical industry or the Information Technology sector, in a period of growth or a period of stagnation, etc. True enough, there is no universal solution, but if by intuition or by chance the chosen process works today, there is little guarantee that it will work in the same way tomorrow, because the environmental changes and the process itself have brought about changes in the company and its stakeholders.

8.4. Knowledge: some "organizers"

Knowledge has been the subject of some extraordinarily complex studies, and has been explored for a very long time by scientists and philosophers. It would be out of context here to sum up the different schools of thought, the disciplines which they have produced, from philosophy to biology, touching on sociology, psychology, linguistics, ethnology, epistemology, etc., up to the recent cognitive sciences. But a slight deviation by some elaborated models in the cadre of reflection on knowledge management will provide us with guidelines and organizers of thought to facilitate a better understanding of the relationship between knowledge and innovation, and to find the connection.

There is not an article on knowledge management today that would only remind us that knowledge is incorporated, that it is a "process", that man is at the centre of the whole problematic of knowledge, and that, all that is outside this problematic is merely information or data. However, just a turn of the page or a turn of the head would be enough to show that the Internet is a *reservoir* of knowledge, that our information system contains the *basis* of knowledge, and that the number one concern of a number of companies at this time of aging population is how to *extract* the knowledge of experts. Besides, the book that you hold in your hand is none other than "explicit knowledge". Is knowledge an object or a process? In fact, it is an object *and* a process.

8.4.1. *The DIK model (Data-Information-Knowledge): knowledge as an object*

The DIK model introduced a hierarchy, a progression in immaterial objects, especially according to their degree of contextualization and human implication. The following figure is borrowed from Tim Baker, and illustrates the respective implications of technologies and man in data, information and knowledge.

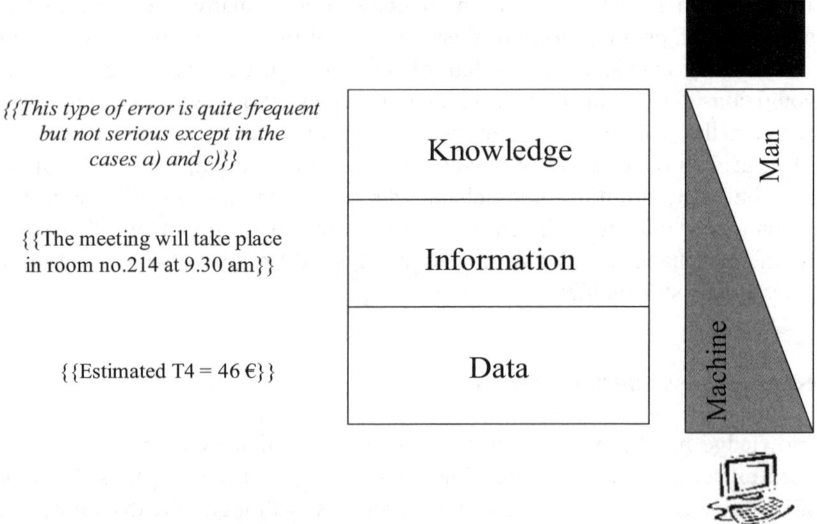

Figure 8.4. *Data-Information-Knowledge*

In the above example, data corresponds typically to the content of the cells of a spreadsheet, information recalls the context of sense, and knowledge is considered as a result of the interaction between information and a human representation system: the reader, in order to understand the sentence, must interpret notions as vague as "quite frequent", and "not very serious", related to his real life experience in the context of the sentence, and shall know or find out the cases of a) and c); he would even need to "automatically" correct spelling mistakes.

The boundaries between data, information and knowledge, as well as their definition, have given rise to many controversies and vary over time. In fact, if we ask ourselves what the future roles of man and machines in data, information and knowledge management will be, based on the projections of the recent past, we could arrive at a static conception of knowledge, a catastrophic scenario of this type:

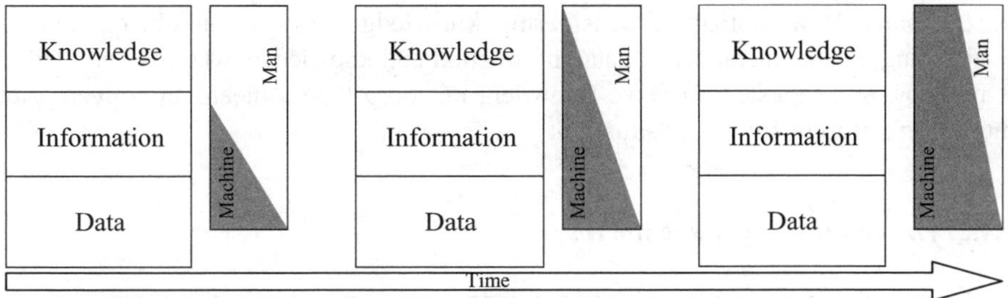

Figure 8.5. *Men and machines in the DIK model (1)*

This would leave man in an increasingly reduced position in the treatment of knowledge, and eventually, technological "progress". We could think that evolution will generally take place in the manner represented in the schematic diagram – see Figure 8.6.

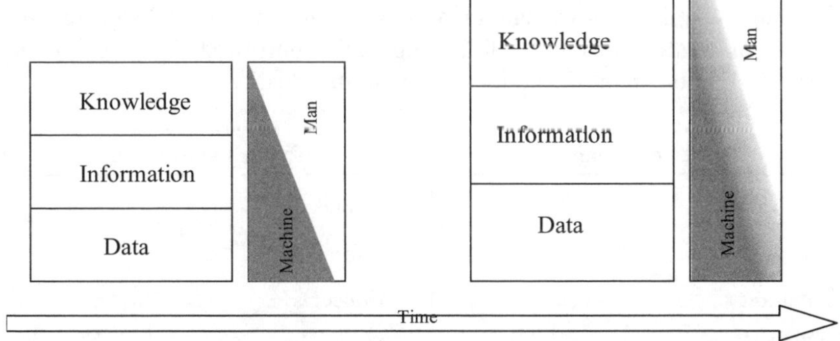

Figure 8.6. *Men and machines in the DIK model (2)*

Technological evolution plays the role of a stimulator in this hypothesis, and allows the human mind to achieve more complex levels of knowledge. That part which is considered knowledge today will become information, etc. On the other hand, the man-machine integration will sometimes make it difficult to distinguish between different roles.

The most important message of this simple model is that, in an organization, we must not confuse the two concepts: that of data treatment as in informatics, i.e., the information system – with that of knowledge, which is an integral part of human management.

Its major flaw is that of considering knowledge only as an object, and of establishing a value hierarchy of data, information and knowledge, whereas knowledge is perhaps, as suggested by Dave Snowden, no more than a means to convert data into information in any given situation.

8.4.2. *The creative spiral and the Ba*

Most of the existing methods of knowledge management have their inspiration in the works of Nonaka and Takeuchi [NON 95] and describe the model of the upcoming company, which alone is capable of responding to the challenges of competitiveness and innovation.

Their fundamental hypothesis is that knowledge exists under two modes: explicit and tacit. It is as difficult to precisely distinguish one from the other as it would be easy enough to draw up an intuitive, tacit representation. By an extreme simplification, we could associate tacit knowledge with expertise, with gesture and intuition, and, explicit knowledge to a more formalized knowledge, expressed verbally or in written form. The following table, borrowed from Jean-François Ballay [BAL 02], illustrates the notions linked to this duality.

Tacit knowledge	Explicit knowledge
Experience	Knowledge
Know-how	Information
Intuition	Concepts
Memory	Documents
Oral	Written
Socialization	Exteriorization
Informal	Formal
Subjectivity	Objectivity
Network	Hierarchy
Groups	Structures

Table 8.1. *Tacit knowledge - explicit knowledge*

The creative spiral of knowledge is generated by a process of conversion between the two modes, in a "back and forth" exchange between individual and collective knowledge.

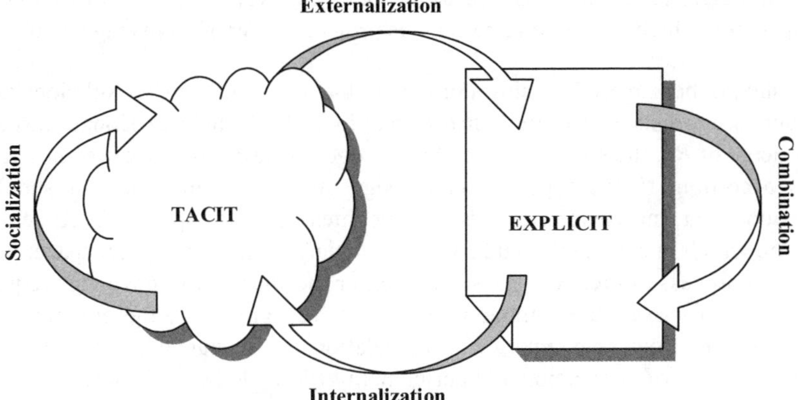

Figure 8.7. *The 4 modes of knowledge conversion according to Nonaka and Takeuchi*

Socialization: the sharing of tacit knowledge requires a direct interaction between individuals, for instance, in a learning situation where the student observes, imitates and practices, but also in the direct exchanges between pairs. The key to socialization is experience.

Externalization: the explanation of tacit knowledge (writing a report or a manual, modelization, conceptualization) entails setting aside, at least partially, the context of the initial tacit knowledge; the knowledge produced becomes much easier to duplicate and to diffuse.

Combination: this is the conversion of the existing explicit knowledge into new explicit knowledge by addition, restructuring, diffusion and confrontation.

Internalization: the "incorporation" of explicit knowledge, the outcome of the SECI loop, done individually or collectively through training programs or exercises which allow the integration of shared knowledge, in the form not only of formal documents but also spoken accounts, and the return of experience, *"best practices"*, etc.

This model earned great success, particularly in the Anglo-Saxon world, because, on the one hand, it justified the process of systematic externalization of knowledge, in order to "capitalize" on it in the so called knowledge bases, and, as such, to render them "measurable", and, on the other hand, because the developing technologies have emphasized the stage of combination that lends itself better to scientific approach. The importance of socialization and internalization was often underestimated, an attitude that led to the failure of several projects of knowledge management, which ended up remaining as projects of "content management".

It should be stressed at this point that the original theories of Nonaka and Takeuchi were distanced from their practice; in 1998, Nonaka & Konno introduced the concept of *Ba*, the space in which knowledge resides; this space assumes various forms according to the types of knowledge and their conversion. It should be understood that knowledge depends on its context, i.e. its "space", in all senses of the term, which in any case would have to be enlarged into a temporal space: putting in place a process of creation, transformation or the sharing of knowledge requires a *space*, a specific *Ba* whose importance must be recognized by the organization. In turn, this *Ba* allows the emergence of relations and engenders an autonomous dynamic system of conversion and enrichment of knowledge: it is indeed this space that can be the object of management, not knowledge (see section 8.2.2).

However, the original idea that has been in circulation ever since it was promulgated by Michel Polanyi in 1966 [POL 66] is that knowledge has two *inseparable* dimensions, tacit and explicit; the very popular quote, "*We know more than we can tell and we can know nothing without relying upon those things which we may not be able to tell*" which expresses the tacit dimension of *all* the knowledge has often been shortened to read: "*We know more than we can tell*", or distorted as: "*We know more than we can tell, and we tell more than we can write down*", to illustrate the methodologies of the externalization of knowledge, change of experience, compendium of experts and "reference works" destined to collect knowledge always considered an "object", even if it had become an "object to be processed".

8.4.3. *Knowledge as a process*

The act of knowledge, ephemeral and omnipresent, consists of three important characteristics for its facility, which are as follows:

– It requires a *voluntary* act; constraint would only engender a simulation, or an avoidance, and the intention strongly controls the quality of knowledge. On the other hand, we have a spontaneous tendency to internalize and use knowledge.

– Our knowledge capacity is far superior to all *formal codification* (oral, written, gestural) of this knowledge: "*we know more than we can tell…*". Creating a well-written formal document has its virtues, but certainly not that of accuracy or comprehensiveness.

– We know only in the moment in which we need to know: knowledge is quintessentially *contextual* and is triggered by circumstances.

8.4.4. *Cycles of innovation and of knowledge*

8.4.4.1. *Innovation – a possible cycle*

Innovation is often born out of disorder and creative desire, in a destabilized, destructured and a *chaotic* situation. The ideas will be subordinated to the complex relations of the "neighborhood" (informal groups and networks, connected by common objectives, affinities or geographical aspects, but where the elements of cultural diversity are also important); the majority of these ideas will not survive but we do witness aggregations, conversations, confrontations, and recuperations, in short, a beginning of bonding; some of them will emerge from this environment where the geography of physical space and of thought count for more than the control process.

We are therefore entering a phase of expertise, of refinement, of calculations and of plans. The idea becomes a project in search of a suitable outlet. Specialists, collaborating in the communities of practice, communicating in their own language, which leaves little room for ambiguity and is and professionally oriented, will validate each aspect of the innovative product, process or service. This is an in depth job: the project gets defined, gets concluded or is abandoned.

Then … the best ending that can come of an innovation is that it disappears and is no longer innovative but, to quote Norbert Alter, becomes a collective behavior, enters into the encyclopedia of technologies of its time, into the field of knowledge or into common use. However, its objective will be wholly accomplished only if it can engender and provoke newer innovations, either through exchanges with its public or the combination with other innovations.

Original chaos, then complexity and complication, ending in stability and a new creative chaos. This is the cycle of knowledge as described by Dave Snowden in his Cynefin (pronounce kun-ev'in) model [SNO 02].

8.4.4.2. *Knowledge – the Cynefin model*

The Cynefin model proposes four domains of knowledge, all of which coexist or succeed among individuals or in organizations, and which can be distinguished firstly by their level and type of structuring. The first domain is that of the *known*, structured, indeed bureaucratic, of rules, procedures and controls. The second is that of the *knowable*, of logical and scientific complication; this is the world of experts, where the questions posed have, in some part, an answer, and where the difficulty is to find a good specialist who will solve the problem. The third domain is that of the *complex*, where the multiple and convoluted interactions no longer allow the deconstruction of problems into much simpler sub-problems. This is the world of intuition, confidence, values and symbols, where ideas are the fruit of close collaboration between the emotional and intellectual capacities. There exists an order, a hidden organization, whose manifestations one can harness by experience, and which often reveals itself only later. The fourth domain is that of the absence of structure, of disorder, which Dave Snowden envisages as *chaos*. Creative and/or destructive, it is a synonym of rupture, of crisis, where neither experience nor expertise nor rules are of any use. This is the world of surprises, of novelty, of transience – a mirror of the first domain.

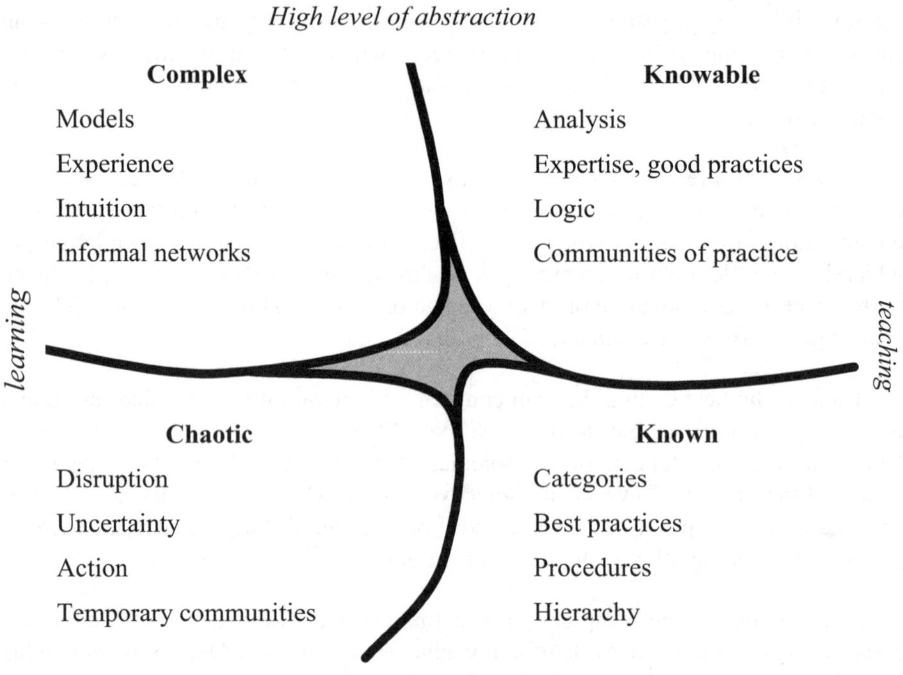

Figure 8.8. *Cynefin – the four domains of knowledge*

The space left in the middle of the figure indicates that the model is not exhaustive, but also signifies that these domains interpenetrate and can coexist in the same situation.

The domains of knowledge also distinguish themselves by the type and the intensity of their *context*. The vertical axis is that of *language* or, more generally, of communication. When two experts who know each other well, talk to each other, they exchange a lot of information in few words; their language is strongly codified, both overtly as well as covertly – overtly as it is a specialized language and covertly because they share a strong context of actions and of relations, a history which provides them with the common references and shortcuts, without themselves being aware of it. The transmission is codified, compressed, and is "at a high level of abstraction", and addressed to the restricted groups, those who share the same expertise (particularly in the knowable domain) and/or the same shared experiences (particularly in the complex domain). On the contrary, in the domains of the chaotic and the known, the language is direct and explicit, either due to a lack of common codes or because the codes are universally diffused and explained – a known, stable situation.

The horizontal axis is that of the *cultural context*: *"learning"* culture to the left, for the complex and the chaotic, *"teaching"* culture to the right, for the known and the knowable. Here we measure the nature of the interaction between the organization and the individual in the act of knowledge by their degrees of reciprocal implications. In the "teaching" culture, the transmission of knowledge is decided and can be planned: we know what is necessary to transmit knowledge to newcomers, to the next generation and to the employees to ensure their employability. We are in the domain of the known or the knowable. The explicit dimension of knowledge is preponderant. The "learning" culture begins with individuals or with spontaneous groups, who feel the necessity of learning, acquiring and consolidating knowledge. Knowledge can also take the form of a value system, of implicit rules, in fact, of belief systems. The tacit dimension is predominant.

It is important to understand the distinction between *complicated*, in the domain of the knowable, and *complex*. An aeroplane is complicated; it contains millions of parts and as many relations between the parts, but they are all catalogued. Each event that concerns the aeroplane can be analyzed, in order to retrace the sequence of its causes: a plan pre-exists in the aeroplane. However, a butterfly is complex: its components are not simple, their borders and their interactions, to a large extent, escape analysis, and the understanding of a butterfly will come more by observation, by typifying its behavior and its metabolism, and by being observed from diverse points of view. The plan, if it exists, is none other than the reflection of our knowledge of the butterfly. In other words, the butterfly system and its story will appear coherent and logical to us, but only in retrospect.

Dave Snowden illustrates this difference through the following example: when a rumor of reorganization makes its rounds in a complex human system such as a company, it begins to metamorphosize in an unpredictable manner in anticipation of the event. If you take a walk in an aeroplane with a toolbox, nothing changes.

It is by traversing the Cynefin space, passing from one domain of knowledge to the other, that we can return to innovation, and articulate, choose and combine in time, according to the contexts, the actors, methods and tools.

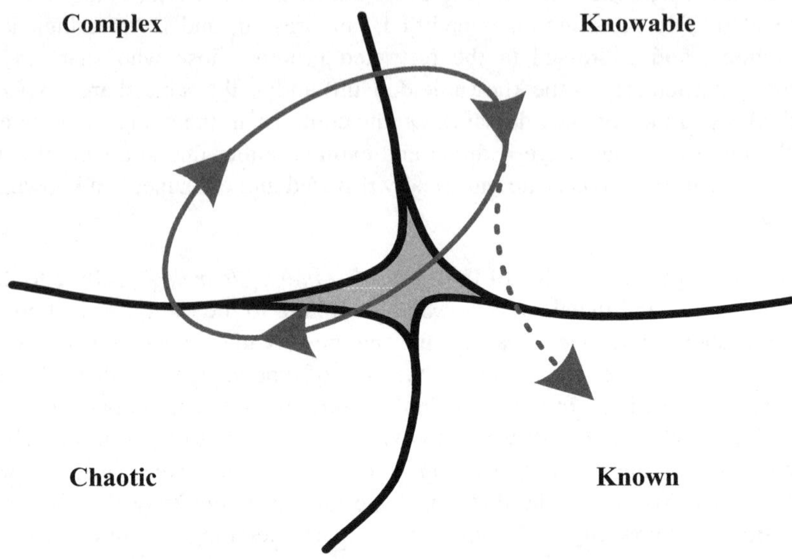

Figure 8.9. *Cynefin – cycle of knowledge*

8.4.4.3. *From the complex to the knowable*

Figure 8.9 describes a possible cycle of transformation of knowledge which is well adapted to innovation: ideas can be born, for example, in informal communities of the *complex* domain, where they are subjected to a type of self-organized natural selection (mutation, copulation, development, extinction) that is hardly measured (no formalization, no stocking). Certain knowledge will be transferred into the *knowable* domain to be formalized and will enter the field of validation. So we move on from a dimension of diverse, cross-disciplinary and horizontal flux to a vertical dimension of increasing our multidisciplinary knowledge with functional links geared towards integration. The communities of practice of the knowable domain, consisting of experts, associate with project groups; it becomes necessary to know "what we know", "who knows what", "what the organization needs to know" and likewise to develop the corresponding tools[8]. Explicit codification and validation are first pushed by group pressure, and are then organized by their managers: we move on from "learning" to "teaching".

8.4.4.4. *From the knowable to the chaotic*

The knowledge of expertise – analytical knowledge, geared towards efficiency – contains an element of self-reference and compartmentalization. The high level of

---

8 See section 8.5.

codification of language, which limits the extent of to which knowledge is shared, restricts its renewal and obstructs its crossings. Questioning certainties, the "not on anyone's side"-attitude necessary for creation, cannot be obtained either by exhortation or by injunctions ("dare to question well established truths!", "break the barriers!") having an immediate blocking effect on those whose profession is to verify these truths. The passage from the knowable to the chaotic is necessary to prevent paralysis of the organization, but it is delicate, and should remain partial. It does not generally concern the whole body of collaborators, and can be orchestrated in a cyclic manner, by well-identified events.

8.4.4.5. *From the chaotic to the complex*

The chaotic domain is one of anticipation, breaking away from lapsed models or models which are too restrictive; it is that of systems which are unbalanced, both uncomfortable and stimulating. It can be used to create novel ideas and concepts, new relations and new identities. It is these effects that will be reappropriated by the complex domain. The chaotic domain is necessary and dangerous as it is unmanageable: the ideas and the relations there are atomized. It only resolves itself by exiting: authoritarian, towards the known domain, then flexible, towards the complex domain, whose opening and diversity will reap the fruit in terms of concepts and ideas.

8.4.4.6. *Sedimentation: from the knowable to known*

In the course of being intensified and optimized, certain types of knowledge, today a matter concerning specialists, have the potential to become more widely used, at least in the organization that fosters them, and will pass on to the *known* domain, where the content, validated and put into a form accessible to all, must be disseminated and made available.

8.4.4.7. *Conclusion*

In a company, each domain favors one aspect: the *known* is the domain of the structure and its rules; the *knowable* is that of the agent and his profession; the *complex* is that of the relation and the flow that it induces; and the *chaotic* is that of pure action.

The above scenarios are surely not the only possible ones. New knowledge can be born out of the three domains (the knowable, the complex and the chaotic) according to their degree of splitting with the people and the organization that form their ecological milieu. The Cynefin model offers us a key to reading every situation and each stage of innovation. In order to know how to act, with which participants and at which moment it is important to recognize in which domain of knowledge we are situated or in which we want to be situated.

A number of failures in knowledge management explain themselves by a hitch, or by being wrong-footed: developing a culture or a method which is excellent "in itself", but which is realized at a bad moment, or in a bad area: not noticing that a border between two domains has been crossed or, on the contrary, that we have remained locked in a single domain. For instance, it is laudable to want "to avoid reinventing the wheel", but in all our earnestness to avoid reinvention, we end up inventing nothing at all. What must be done? The answer is: to play with the domains of knowledge, but not all and not all the time.

8.5. Cultures, methods and tools

8.5.1. *Where do we start?*

At the moment of undertaking a voluntary step in the direction of knowledge management, seeing that the organization thrives is of primary importance. This step itself is an organizational innovation and it is important to know which processes and which individuals will be involved, and whether this innovation is incremental or detrimental.

Ensuring that an organization thrives, seen through an optic of innovation, is to individualize the types of innovations at work in the company: proactive or reactive; organizational, procedural, of services or products; concerning the *core business* of the company, its periphery or its associated services; induced by the environment – technological, judicial, geographical – or by the market, induced by constraints or by opportunities. All these innovations coexist; they are mutually encouraging or inhibitive and together they make a company dynamic or stressed.

To examine their lifecycles, whether they are of an internal or external origin, at which level – that of expertise or of hierarchy, individual or the fruit of collective efforts - of validation, of implementation or of abandonment, is to recognize in them not only the knowledge domains covered, but also the phases of transition from one domain to the other, as well as to recognize their decision-makers, and their actors, their facilitators and their inhibitors.

Moreover, to identify their formal and informal groups is to discover their modes of functioning, their artefacts, how and by whom they were created, their members, their lifecycle, the types of knowledge that they transmit, and above all their role in the innovation cycles referred to.

We shall see the strengths and weaknesses of its organization, the unsuspected resources and the deserted domains, the dynamic forces and the resistant ones. There we shall particularly find many persons, methods and tools used counter-productively or, more plainly, inopportunely.

8.5.2. *Methods and tools for collective knowledge*

Which communities and which knowledge networks should be employed for innovation? The answers are many. First, from the point of view of knowledge, because a better adapted type of community corresponds to each domain, and, the same community can also follow the cycle and change its characteristics.

Next, we will examine the situation from the point of view of the company, its history, its positioning, and the type of innovation desired. In the end, each individual implicated in the cycle of innovation belongs to a number of communities, project groups, general formal structures, and takes on as many roles or identities. The knowledge cycle also works at the intra-personal level, in alternative states of complexity, complication, order and chaos. We are experts in one domain and novices in another; we are the chief and the subordinate, the client and the employee, all at the same time. Our aspirations and our competences will make us more voluntarily inclined to one or the other domain, but we must pass through all of them; it is not so much about doing things "at the same time", but doing them successively and in a manner well supported by memory.

A vigilant person is not the one who has a foot in all the networks – even those networks that have nothing to do with his domain – who would know to not only shoot out the flame of shock from two apparently unrelated pieces of information, but also to use his expertise to distinguish the true from the false, and to instantly engage a verification of the found information? All the domains are traversed, except perhaps that of the known, with explicit reference of its action.

8.5.2.1. *The complex domain*

The complex domain is the domain of emergence, of ideas, but also that of newer perceptions caused by the interactions between the actors; the communities here are informal, spontaneous and often ephemeral; the characteristics of the relationship between individuals are more important than the individuals themselves. Here, the sharing of real life experiences counts more than the "return of experience", and forms the basis of communities whose common codes are implicit and, moreover, are of the order of values or shared objectives, strongly bound together by the cement of confidence. The cultural or intellectual diversity of the members is desirable, but their profiles are not formalized. Voluntary services and public engagements are at the heart of these communities; the management controls neither their participation nor their content. The initiative is taken by the individual, whatever his position in the hierarchy may be. The knowledge there is especially a "process": we do not formalize that which is still ambiguous and uncertain and often tacit.

The leadership of the communities will be "natural", taken on by the individual who knows how to better manage the interactions and the new forms that these interactions create. The stake lies in recognizing, interrupting and reinforcing, actually, disseminating the emerging phenomena. This can help detect the weak signals emitted by the organization or its environment, but also accelerate the evolution and natural selection of emerging ideas. The ideas are formed by combinations of individuals and clashes of diverse points of view: no data analysis, only perceptions.

The management of the complex domain adopts firstly a cultural approach, that of reception and care; the managers must lay out the tools, the space and the necessary time for these activities.

Tools: networks enabling the mobility of users, software programs for the creation of virtual communities "in just a click", compatible with the existing systems of collaborative work and the company schedule, help tools for creativity, free access on the internet, providing private storage space for users, etc.

Space: buildings facilitating productive meetings, sophisticated locations, well equipped board rooms, business travel.

Time: to give individuals time for self-oriented spontaneous work. For example, employees of 3M can devote 15% of their time of work to develop their own ideas. This proportion is surely not possible to standardize, and not everyone uses it, but this sends a strong message from management and has brought about a number of new changes in the company.

The methods and the tools of this domain are firstly, those that accommodate the management of informal groups; simple, user-friendly software programs, the stand-along programs vis-à-vis SDI (Selective Dissemination of Information) and management; the archetype there would be the *weblogs*. Here, the monitoring is both transversal and collective.

If the communities have difficulties in creating themselves naturally, some methods of expansion, to recall an expression of Dave Snowden, can be used. The principle is that of going fishing with a lamp: it is about creating a point of attention such as a conference on a contemporary issue, a competition, a call for ideas, and waiting to see if a movement emerges, while being ready not to support it, but to facilitate it.

The acceleration of the process of formation and maturation of informal communities can also be obtained by the methods of *storytelling* (see, notably, [DEN 01]) and the software programs of network stimulation[9], which, by different approaches, tend to make common implicit codes emerge faster.

The complex domain develops more naturally in the SMEs (small and medium-sized companies) than in the big companies and *a fortiori* in institutional organizations, but the SME will find it more difficult to achieve the critical threshold of the *required variety* for their communities. The stake for a big organization is to leave some space for the informal and the complex, without formalizing or making it undergo expert examination too soon, while that of the SME will be firstly to help a good number of communities survive – which will often bring them to create inter-organizational networks – and to ensure transition at the right time (i.e. not too late) towards the domain of the knowable.

8.5.2.2. *From the complex to the knowable*

The sum of knowledge and communities in the complex domain is far more than that which is finally useful to the company; it is difficult not only to know in advance *what* will be useful to the company, but also *when* it will be useful.

A community devoted to a subject, for instance, a manufacturing procedure, can naturally move from one domain to the other due to group pressure: if the exchanges grow exponentially, the questions become redundant and the answers are stabilized, if the new roles are updated, the community can structure itself so as to give a description to such roles – for example, related to expertise – while formalizing stabilized knowledge such as creating lists of FAQs based on the exchanges in a forum.

Some mail analyzer programs can identify experts and can also help us with their references. If a person looking for expertise sends out a request, the supposed expert, informed by the program, has a choice not to make his expertise public. Only expertises made public are solicited. We note here the transition between the complex, unconstrained domain and the knowable domain where the experts are declared or named. The program here resolves the paradox of knowledge sharing: if you compel somebody to share his knowledge, he will do everything in his power to make sure he does not; if you give him a choice, he will often go ahead and do it.

8.5.2.3. *The knowable domain*

Here, the objectives are clear and the members of the community are known. The relations remain important, but the individual profile is the prime factor. The ideas

---

9 For example, Spoke and LinkedIn.

that emerge in this domain are incremental: we optimize, apply the plans of experience to improve the procedures, analyze the proprieties of new material, improve the surface state and simulate behavior with numerical models. The communities are more homogenous and are guided by common expertise; the members share a codified explicit knowledge, knowledge gained through university studies and scientific and technological works. It is more often the function or the profession of a member that determines his place in the group. These communities are created by management to ensure the sharing and coherence of knowledge within the multi-site, nay, multinational companies, or those that are issued from informal groups of the complex domain.

Leadership of the communities is generally given to experts, which often means the older members of the group. The stakes for knowledge and for ideas that have become projects is to intensify them, to validate them and to measure their proprieties and their impacts. New knowledge is formed by improvement and optimization. This is the domain of improvement and of scale factors. These communities will be as efficient and focused as the informal networks of the complex domain would have encouraged them to be.

Another characteristic of these communities is durability, which is necessary for the sedimentation of knowledge and its validation by the group.

The methods and tools are those which enable collective efficiency and cohesion. The coherence of the approaches is not left to chance but is managed in the formalized interfaces between the different professions. Time management is important, indeed, critical in the project dimension and "time to market". The communities are vertical, on the one hand, by expertise or profession and, on the other hand, by cooperation and coordination if they are dedicated to an innovative project.

The software programs are those of the communities of practice that allow the profiling of the members, facilitate content management, and functionality of advanced research, feature personalized alerts, forums of discussion and expert management of user rights, along with some functions of time and task management. Other tools allow a collaborative management of very successful projects associated to the functionalities of content management. At this moment, we witness a convergence of these products towards becoming polyvalent tools.

To form the communities, the analysis programs of relational networks or networks of affinity, according to the declared profiles, can allow the managers to identify potential members, and for the members to identify their position in the group.

8.5.2.4. *From the knowable to the chaotic*

The necessity of a periodic passage to the chaotic has been made clear. Its regular and ritualized orchestration make it possible to channel the centrifugal tendencies of the organization and to create the effects of "relaxation" (in the mechanical sense of the term[10]): we will live better in the knowable domain if we anticipate the chaotic instances and, according to its personality, better in the chaotic domain if we know that it is followed by the complex and the knowable. The transition towards the chaotic has another objective in the case of innovation, which is that of breaking the paralyzing obvious facts and of opening the breach for the very innovative ideas.

Therefore, the transition has a *feedback* effect of which reinforces the stability of the knowable and the known domains. Therefore, it must be periodic – which warrants the anticipation of its effects – and must be ritualized – which amplifies them by the realization of its change of state and the new possibilities that it creates.

A number of methods have been developed to assess this passage from the chaotic, but are not specific to knowledge management. We can cite the technique that consists of mixing two groups of experts from very different disciplines that are both concurrent, and constrained to cooperate in the resolution of a problem that is new to them.

8.5.2.5. *From the chaotic to the complex*

We have seen that the chaotic cannot be directed; only its outcome can be managed by anticipation. In creating the temporal and physical limits of the chaotic event, in ensuring that the time-space dimensions of the complex domain are opened up (for example, by avoiding making a connection between a chaotic event and the outcome of a project), we will facilitate the collection of fragments of ideas and the remnants of relations, the collective widening of the gaping breaches in certain established facts and the emergence of new shared points of view.

8.5.3. *Induced effects and combinations*

"Systems do not naturally exist, they are in the minds of people" said Claude Bernard. The same applies to models of knowledge, and it would be pointless to categorize, at any cost, the situation, the groups and the knowledge in the company, and to deduce, in an unambiguous manner, the *best practices* and the best tools.

---

12 "Reduction of the constraints by substitution of plastic deformations with elastic deformations" [Office of the French language, 1988].

The methods of elicitation and of formalization of knowledge, the ontology and methods of capitalization of experience in the "knowledge bases" have certainly shown their limitations and are hardly appropriate for innovation management. But in certain companies, their reasoned and reasonable utilization has allowed the creation of induced and unexpected effects like the sharing of real life experiences and the appearance of networks and informal groups. It is vital to seize the moment when meetings and discussions are still recent, to stimulate the communities before the deception in the tool itself and the awaited results break the relational dynamics of the community.

In the same fashion, the portals of the company, company records, the tools of KBE[11] and EDM[12], the tools of linguistic instruction and e-learning which are not *a priori* dedicated to innovation management can have positive induced effects if they are used at the right moments in the cycle and with a clear perception of the effects sought.

8.6. Key factors

There are no universal key factors. In this domain in particular, what is important for one organization ("Measure whatever you want to know!") can be hazardous for another, what is to be done today ("break the barriers!") must be avoided tomorrow, what is to be encouraged for certain pieces of information ("culture of sharing!"), should be proscribed for others.

There is a grid of guidelines for reading the reality and Cynefin is one such grid that is efficient, suggestive rather than prescriptive: it does not indicate what must be done but, more simply, what can be done, and focuses our on forms rather than categories.

Some comments can complete this point of view: dualities, yet again, that position themselves between the two poles.

8.6.1. *To share or not to share?*

Where is your competitive advantage? In your ideas or in your "time to market"? If it is in the idea, share a bit, but be careful of the risks of going stale; but if you are constantly monitoring your actions, the lack of exchange will make you blind to

---

13 KBE = Knowledge Based Engineering making it possible to exploit the rules of know-how within the scope of professional applications, especially in the domains of design and manufacturing.

14 Electronic Document Management.

changes. Widen the scope of your internal exchange, giving yourself the means for a big internal opening and an external closing; if it is in the "time to market", or better still, in the "way to market", the sole manner in which you can touch the market is by being generous with your ideas: your image will be reinforced, and you too will be more absorbent and reactive.

8.6.2. *Learning or teaching*

In *learning*, the accompaniment of the complex and the chaotic is at a premium, is not greatly affected by scale factors, and is not reusable; in *teaching*, the management of the knowable and the known can be rationalized, the marginal costs are less and the collection of everything acquired is possible.

On the other hand, *teaching* is aimed at everyone; *learning* is much less universal – customized to suit the company. The mode of *teaching* is almost permanent while, the moments dedicated to *learning* are less frequent.

The correct proportion of *learning* and *teaching*, and the correct rhythm of alternation between the two activities are the key to success, to test, to find and to diversify.

8.6.3. *Stress and confidence*

The cycle of knowledge/innovation described in section 8.4.4 is an ideal cycle, and the same succession of the domains can become a cycle of involution of knowledge, under certain conditions.

If the communities in the complex domain have, for example, been put under great stress, having previously been open and mobile, they will close up, will freeze, and will tend to safeguard their acquisitions and pass through the knowable domain like a citadel under siege; at worst, the community will split and get fragmented into isolated and competitive interests. The chaotic mode will be reinstated in the worst of conditions. A similar succession of domains can reveal two opposing realities: the difference lies in the individual and collective confidence. Confidence and stress form another duality that it would be interesting to develop; confidence is a factor of resistance to stress, while stress is the antithesis of confidence.

8.7. Conclusions and openings

In innovation, everything is a matter of rhythm. The idea you just had is undoubtedly already in the minds, or in the test tubes, of dozens of other innovators.

This is one of the effects of the inversion of the pyramid as this simultaneous emergence of the same new ideas demonstrates, and even if you are far ahead, the value of acquired or created knowledge decreases rapidly with time.

On the other hand, it is rather infrequent that your idea has the answer to the need expressed by real or potential clients. The difference between you and your competitors lies especially in what could be called perception of appearance, market expectations, the complex domain which needs some time to become efficient, but can be very reactive, in the same manner as that of intuition.

It is also possible to recognize the error and to quickly change course, which means that collaborators must be judged not by their results, but by the process[13] that they adopt and by their attitude, in the complex domain and, conversely, in the knowable domain, as far as the results are concerned.

Knowledge management for innovation is no more than management of innovation with *regard* to knowledge and its rhythms. The tools and methods evolve very rapidly and simultaneously with technologies and the increasing consciousness of the importance of network relations; the only correct method is that which we propose, a tailor-made method, in the context of its organization, with the tools that are indispensable and short-lived, and which can be reviewed in each cycle.

In her presentation of KB2[14], Anne Jubert outlines four factors of success for the community of knowledge management in Europe: passion, pulsation, recognition, and diversity. Her positioning clearly shows a rebalancing in the domains from the chaotic and the complex, moving towards those of conversation and questioning. We can also find everything that is known and knowable in knowledge management and innovation on www.knowledgeboard.com and its links.

I also strongly recommend reading the works of Jean-François Ballay [BAL 02], for his humanist and open approach, and Dave Snowden – particularly [SNO 02] for his vision of complexity and knowledge, that I presented in section 8.4.4, and which served as the main link for the presentation of the tools in section 8.5.2.

---

15 Taste for risk, initiative, spirit of criticism.
16 Knowledgeboard 2, second step of the European portal of knowledge management, launched in March 2004 (www.knowledgeboard.com).

Chapter 9

Integration of Stylistics and Uses: Trends in the Innovation Process

In order to be recognized, a technological innovation must often be known to be a perceptive innovation (sensory or cognitive). It is the case with each new generation of microcomputers or mobile phones whose appearance (form, texture, color, transparent material) and interfaces (ergonomics, wireless connections) evolve with each generation of microprocessors and optical readers. But the conception of the sensory attributes can itself generate innovation. The pearly appearance of sporting gear, the glint of the glass cabins of concept cars or the hemispherical shape of the under frames of Apple computers are some contemporary illustrations.

Stylistic innovation gathers several tools and methods, its finality being the organization of the activities of anticipation and the creation of sensory attributes of the product. Today, these applications relate mainly to four types of attributes, that is, form, color, use and texture. Gustatory, olfactory and auditory attributes currently fall under the study of other methods that we shall not discuss here.

In this chapter we shall initially introduce the theories, and define the concepts of stylistic innovation. Then we shall elaborate on the tools and methods which allow the implementation of the concepts of stylistic innovation and help obtain satisfactory results. Finally we shall illustrate our observations at the time of the conception of an automobile instrument panel.

Chapter written by Carole BOUCHARD, Hervé CHRISTOFOL and Dokshin LIM.

9.1. Theories and concepts of stylistic innovation

Before examining the tools and methods of stylistic innovation, we must define certain concepts. First of all, the concept of the universe of exchange and the influential universe that enables us to define an environment of intervention. Further, what is a tendency in product design? What is a stylistic attribute? What is a stylistic tendency? On which assumptions do their definitions rest? What are the reasons on which their constructions are based? What are the meanings of value that we will retain in this chapter?

9.1.1. *The universe of exchanges and influences*

A product comes into existence only from the moment when the market accepts it, and from then on its lifecycle continues at its own pace. Thus, to innovate and/or conceive, it is important to have an exact idea of all the contexts which encompass the product, its environment, its production, the products used simultaneously or in addition, the competing products and the substitute products. It leads us to the behavior analysis of the actors of the market, be it consumers, distributors or actors who figure in the lifecycle of the product. We will name them "users".

It is a system made up of products, of contextual and competitive environments close to its users, which we call a *universe of exchange*. In the search for the universe of exchange, there are phases that are a little peculiar, the preliminary phases of conception during which the teams contribute to the genesis of innovations. These teams seek, identify and describe the universes of exchange since when stylistic transfers, technological transfers, or transfers of use that contribute to our universe of exchange have already taken place or are already considered. These are universes of prescriptive exchange that are compared to the universe of project exchanges – that of the product and the object of the innovation – we call them *influential universes* (IU).

9.1.2. *Trends in design*

Trends analysis takes into account the transverse links between the different industrial sectors. More precisely, these links are style harmonies being found in the different fields, and showing the inter-connection through the hybridization or merging of design attributes. This transfer of data between the different fields can be observed with new technological supported devices as well. In fact, there is a massive transfer of functionalities and of interface design attributes between products like mobile phones or computers, digital cameras, personal digital assistants, car onboard computers, televisions, DVD players, compact music systems and home cinema systems.

A design trend is a simultaneous styling and technical evolution supported by a cross-sectorial movement giving its position to the product in the obsolescence cycle [BOU 97]. In fact this movement can be recognized at the same time in many industrial fields at different places. Its temporal and dynamic character allows it to be spread and to give birth to new design trends in these fields and at these places. Thus, as in the field of technological watch, the lifecycle of a product model goes through the four main following phases.

– *emergence*: in this phase we can find *avant-garde* products; these products are very innovative, and still not very widespread on the market. They have a strong identity and will have a significant influence on the market;

– *growth*: this is the phase of consumer goods; these products are widespread on the market and their dissemination speed is higher than in the previous phase;

– *maturity*: this phase is related to mass consumption products. These products have flooded the market. They are mass products which are not at all innovative. This phase corresponds to a stabilization preceding the decline phase;

– *decline*: this illustrates the end of the product's lifecycle on the market. It corresponds to up-to-date products because of their style and their technology as well.

9.1.3. *The stylistic attributes*

Even if the appreciation of a product is complete, it is normal to distinguish the structure from the mechanisms or the appearance from the use among the set of qualities of a product and the units that describe it. The product is defined as an artifact fulfilling three types of functions: functions of use, sign and the ability of being produced [LEC 92]. In design, the segmentation of work and competences resulted in distinguishing various points of view on the characteristics of the product. We can thus demarcate the sensory attributes, technical attributes and the symbolic attributes systems:

– the technical attributes include characteristics related to "productibility" (capability of being manufactured in series) such as technologies necessary to carry out the technical functions of the product;

– the semiotic attributes include values that the product implies, its "mix marketing" and its use;

– the sensory attributes include the whole of the characteristics perceptible by the five senses of the user. It is among them that one finds the stylistic attributes (visual, tactile, auditory, gustatory and olfactory) and of use (social schemes of use, scenarios of use, physical and cognitive interfaces).

In this chapter, we will use the terminology of stylistic attributes to approach the study of *forms*, their *colors, textures*, as well as the aspect of *materials*. The symbolic attributes will be primarily reduced to *values* connoted by the stylistics attributes just like the technical attributes, of which we shall evoke only those which contribute to the realization of the stylistic attributes.

In spite of these limitations, the originality of the approach lies in recognizing that perception is universal and in approaching the design of the form, color, texture, the material aspects, and the use of the product in correlation with the connoted values and technologies necessary for their concretizations. Usually, these attributes are the subject of distinct studies and act on behalf of the designers, the colorists, the engineers, the ergonomists and those in charge of the products and their marketing.

9.1.4. *Usage attributes*

As is underlined above ([LEC 92]), the function of usage is one of three functions that a product should fulfill. The usage attribute takes part in the determination of the characteristics of a product and it occupies a highly important place in recent information appliance sectors.

The usage attribute in the early phase of product design when innovations take place may include various levels: from a user interface design feature (hardware user interface development related to operations and tasks) to user behaviors such as use scenarios (user experience). Key ideas on all levels are interesting at this stage, as far as they are able to give a stimulus for the creativity of the designers (designer in a large sense; considering a multidisciplinary design team integrating product designers, engineers, ergonomists, etc. or a user participatory design team).

User interface design focuses on the design of user interfaces with which a user has direct contact and with which they interact to conduct activities. It covers product features such as data input and output methods, information architecture and its contents [BER 99]. Among these, input and output methods leave a large scope of innovation possibilities for practitioners in the early phase of product design. In the information appliances sector, a product designer's concern in the early phase lies largely on how to combine interface features to potential user behaviors.

Innovation took place in that way for the Pilot Palm ("The Art of Innovation") and proved to be a great success when it was launched in harmony with stylistic attributes. The Apple iPod had clear form factor and ease of use advantages versus its competitors by incorporating a touch wheel interface. These examples demonstrate the importance of the role of usage attributes in the innovation especially for today's consumer electronic products. They can even be a leading factor in a new product design process [LIM 03].

9.1.5. *Stylistic tendencies and use*

If artistic creation can be categorized as a school, in terms of trend and style, the conception of products does not escape the importance of stylistic tendencies. In the clothing industry, manufacturers of the textile industry structured themselves to offer coordinated products and to allow consumers to be able to harmonize various types of clothing and accessories of different origins. This trend is prevalent in other products of mass consumption. The consumers no longer look only for quality products but also for products that harmonize with other products and relate to each other. Stylistic tendencies (formal, chromatic, tactile, gustatory or olfactory) are made in such a way as to complement technological tendencies ensuring their compatibility, in fact, their connections (bluetooth technologies or WiFi, for example).

9.1.6. *Reasoning in the design professions and analogy in particular*

It is established today that it is essential to bring together several skills within an interdisciplinary design team to complete large innovative projects successfully. These teams were originally made up exclusively of engineers, but today they bring together professions which are very varied: marketing managers, salesmen, designers, ergonomists, architects, scientific researchers or artists. In order that everybody may recognized, understand and appreciate the work of the others, we believe it is necessary to communicate the procedures and logic of each profession, and then develop common methods and tools which enable the various contributions to be summarized, different points of view to be discussed and decisions to be justified.

In the upstream phases of the process of innovation, generally referred to as the preliminary design stages, we will find three major professions:

– the marketing manager defends the commercial, or even social, positioning of the product in the market and composes the various values that the product must convey. In practice, purchasing behavior changes. The consumer still worries about the performance of the product he desires, but he also worries about the working conditions of the people who produce it, the brand image of the company which sells it and the meaning that is conveyed by its formal, functional and even ethical existence;

– the Research & Development engineer is the guarantor of the technical and scientific feasibility of the product. He selects, develops, dimensions and assembles the technologies which ensure the quality and safe operation of the product;

– the designer represents the user, and using his stylistic, ergonomic, cultural and even artistic skills, he designs the interface between the product and its user. It is his "product" culture which allows him to claim this position of advocate of the user [MOL 91]. He will have to transform the values defined by the marketing manager into attributes of the product which can be made by the engineer and appreciated by the user.

At this stage of the design, where the specifications of the product are prepared and where everyone is trying to produce the profile of the future product verbally and then physically, four main forms of reasoning are employed during working meetings:

Reasoning	**Rigorous**	**Woolly**
General	INDUCTION	ANALOGY
Discriminatory	DEDUCTION	ABDUCTION

Table 9.1. *The four mains forms of reasoning [DUR 94, p.98]*

– deductive reasoning (from the abstract to the concrete) and analytical reasoning tend to support the construction of the operational specifications by the engineer;

– inductive reasoning (from the concrete to the abstract) and synthetic reasoning guide; for example, the marketing manager in the positioning of the future product in the current offer;

– analog reasoning encourages the designer to turn to other sectors of activity to transfer (hijack) technologies, shapes, colors or uses;

– abductive reasoning enables them not to envisage all the possibilities systematically, but to work only on certain options, solutions or architectures which are recognized *a priori* as interesting and promising.

These few examples are only there to illustrate the dominant reasoning employed in each profession. In order to create, each player uses several forms of reasoning, and the global process of innovation is a patchwork of these various contributions.

Analogy, which stimulates the imagination, is the form of reasoning which is most common in the creative professions, where it takes various forms:

– "isomorphism", or the "model" in mathematics or cybernetics and systemics;

– the "symbol", the "metaphor" or the "parabola" in drama, the cinema, literature and psychoanalysis;

– the "reference" or the "mix" in architecture [PRO 1992, p.59] or contemporary art;

– the "theme" or the "sampling" in music;

– the "transfer of technology" or the TRIZ algorithm in engineering;

– the "influence" or the "trend" in styling, interior decoration or industrial design, etc.

If the original contribution of the study of tendencies rests mainly on analogical reasoning, the studies of stylistic innovation evoke four forms of reasoning:

– deductive reasoning to define specifications of the study;

– inductive reasoning to identify the influential universes beginning with the current offer and to categorize tendencies;

– analogical reasoning to transfer the range of attributes in the universe of project exchange;

– abductive reasoning to create and choose solutions that are to be developed.

9.1.7. *Human values and product value*

The notion of *value* is very complex because of its simultaneously substantive and procedural character [SIM 69]. In this sense, other notions and meanings appearing under the words *design, organization* or *project* can be compared to the term *value*.

The *value* is linked to the design process, its related activities, and finally the information and decision process of the stakeholders involved in it. The French dictionary "Le petit Robert" gives the four following categories of value definitions [REY 00]:

– the first one – a *valuable* man – is synonym of the merit and human qualities recognized by the dominating current social model;

– the second one, – *use value, esteem value* or *exchange value* of an object – designates the "measurable character of an object being potentially exchanged or desired [...]. There the value is the quality of something based on its objective or subjective usefulness (*use value*), on the relation between the offer and the demand (*exchange value*), and on the necessary quantity of work for the production";

– the third one, – a system of *values* supporting decision – shows the *value* as a constituent of the decision of human systems. It is used as reference in the judgment;

– the fourth one – the measure *value* – is linked to a more specialized knowledge as metrology, mathematics, linguistics or painting and aims to designate the result of a measure.

The first and third definitions are centered on the human being in the sense that they refer to human sociological values, while the second and fourth relate to the measure of the object's qualities.

In a stylistic innovation approach (even if many measures and qualitative or quantitative values are useful in the decision process), we use more the notion of value corresponding to the first and third definitions. These are retained by the AFNOR glossary common to Value Analysis and Value Management: "the value is a judgment borne by the customer or the user on the base of his expectations and motivations. Especially its size grows when the users satisfaction grows or when the related expense is reduced" [AFN 92]. The value of a product is evaluated by the designers and the users through their own value system. In this chapter when we will speak about *values*, we will have a special interest in *human values* allowing users to judge the product.

9.2. Methods and tools of stylistic innovation

9.2.1. *The universe of exchange to the universe of influences*

In order to bring innovation in styling dimension, it is not sufficient to observe the current competitor products on the market. All these products have been designed 6 months ago, or one or two years ago! Taking the inspiration in these products amounts to designing a product that will be diffused in 6 months, or one or two years with a several years' dated view on styling trends. When the current products visible on the market have been designed, the designers who imagined the corresponding innovations took their inspiration in influence fields: the influent universes. In order to foresee styling trends, it is firstly necessary to identify them in order to then select style innovations from these universes and to transfer these innovations.

These cross-sectorial movements are of course completed by the vertical innovation which is more based on specific expertise fields. These fields of expertise are developed in the laboratories, or by the suppliers of raw materials, of components and of technical systems, in order to propose innovations in a more traditional linear innovation process.

Three techniques have been identified and implemented in order to identify influent universes:

– creativity sessions by the work team;

– visual analysis and exploration of the current product offer in the studied field;

– analysis of expert knowledge acquired by the designers considering especially their sources of inspiration.

In order to perform a firm validation of the results, it is better to use at least two different ways from the three given and to examine the results in order to keep the common core.

The first technique is easy to use and makes it possible to extract quite relevant information from the work team in a short time. It is adapted to a trends analysis performed in a reduced time.

The second technique requires more time and the participation of marketing managers who can validate the exhaustiveness of the offer being present on the market. This technique consists of a segmentation of the offer into analog product groups, that is to say in product groups having common style attributes. Then other products in other sectors having similar style attributes are gathered. These sectors enrich the list of *influent sectors*.

The third technique is longer to use because it is based on an expertise analysis by the designers, with the elaboration of a specific panel currently in charge of design and innovation generation. Qualitative interviews are done and a content analysis is realized from the verbalizations collected in order to update the information structure and contents related to the references and inspiration sources of the designers making it possible to complete the sectors of influence list.

9.2.2. *The analysis of iconic contents*

During our discussions with colorists, designers and going by our research-actions, we noted the importance of the "trends panel" tool, to represent the stylistic characteristics of an influential universe (harmonies, forms, chromatics, textures, uses and materials). The specially designed universe can be, for example, the environment of product utilization, the products of a branch of industry or a universe of exchange representing and marketing a particular position (high-tech products, products for young people, traditional products, etc.). The trends panel is the privileged tool to represent creativity (designers or colorists), *to identify, investigate and represent* these universes *in order to understand* its stylistic structuring, to *locate* the sensory attributes of the product, to reveal harmonies of color, style, use and to *communicate* their design methods.

The originator of a "trends panel" represents a universe with a set of iconic representations composed of illustrations, photographs or engravings representative of the influential universe, that has been identified for study. Once the illustrations are collected, it classifies them in harmonious groups of colors, of styles, materials, uses and values. These are the illustrations of the same group that will be assembled to carry out a trends panel. Many such boards could represent a universe made up of several harmonious groupings. To model this process of categorization of the specially designed universe, we consider the whole set of illustrations collected as an iconic corpus that we should discuss in order to understand its organization. For this, we propose to follow a theoretical course of content analysis and thus to transpose social sciences with the science of design, a method of qualitative and quantitative treatment of information. Thus, the categorization of the illustrations by analysis of iconic contents can be divided into five stages:

– preliminary reading of the illustration corpus;

– choice and definition of the units of classification (values, uses, shapes, colors, textures, materials);

– process of categorization and classification;

– quantification of the volumes of illustrations;

– description and interpretation of the categories.

To judge the quality of an iconic content analysis, we retain the same qualities as those which the categories of a rigorous content analysis must have:

– *exhaustiveness*: the whole of the categories must make it possible to classify all the illustrations describing the whole of the universe without resorting to a "diverse" category having no connection with the others;

– *homogenity*: each category should join together only those illustrations that have the stylistic attributes of use or common values. All ambiguity must be avoided;

– *obviousness*: any analyst should be able to classify the same elements in the same way and in the same categories, without any difficulty;

– *relevance*: the categories must have a direct and univocal relation with the objectives of the innovation project and the specific contents of the analyzed universes. They can be brought closer to sociological categorizations describing life styles or the purchase-behavior of the users of the product considered;

– *exclusiveness*: no overlapping must be possible between categories. Each isolated illustration must be capable of being arranged in one and only category (in practice this quality is difficult to satisfy);

– *productivity*: the categories must be elaborate in such a way that they are rich in indices of inferences, in the production of the range of principal attributes and the harmony of new stylistic attributes.

9.2.3. *Modeling of the analysis process of the tendencies of a universe of exchange*

A trends panel is generally made in order to conceive the stylistic attributes of a product. It is this finality that guides the designer all along the realization of his concept. Its construction is directed and a board of tendency cannot be used without precautions in a different context.

A trends panel is representative of a set of influential universes. The knowledge of the stylistic structure of this set must facilitate the project team to direct the stylistic design of the product; and help develop bonds between the influential universes and the product (environment of use, prescriptive universe, and marketing position) must be well explained.

For designing a telegraphic telephone harmonized with its environment of use, we have studied the interior design of private individuals and the office spaces of professionals. To design a range of shoes for teenagers, we have studied clothing, activities and accessories of teenagers (marketing position). To design an automobile instrument panel, we have studied aeronautical and domestic equipment because they are universes of prescriptive exchange for this sector of mass production activity.

By integrating the stages of categorization of the studied stylistic universe, nine stages can be distinguished to undertake a study of tendencies and to integrate its results at the time of the conception of a new product:

1.   the drafting of the specifications of the study;

2.   the research of the influential universes;

3.   the search for illustrations;

4.   the categorization (analysis of iconic content);

5.   the choice of the illustrations of the category;

6.   the assembly of the illustrations (composition);

7.   the formalization of palettes of attributes (colors, textures, shapes and uses), designation of the values, denomination and description of each trend board;

8.   the selection of attributes for the new product design;

9.   the sketching design with the selected attributes.

The first stage – the specifications – such as the last – the integration of the tendencies during draft designing – unwinds in the universe of exchange of the studied product. The second stage is where the influential universes – sources of "analogical" creativity – are sought. The third stage is one of distancing out. The fourth to seventh stages make it possible to formalize the stylistic tendencies of use that structure the influential universes. The eighth stage marks a return to the activity sector that is being studied, with structured creative proposals.

The choice of the illustrations determines the validity of the study. The iconographic sources are selected according to the universe and the relation between the universe to be investigated and the product to be conceived. The illustrations are selected to account for the diversity of the styles and the values asserted by the products and the illustrations of the universe while trying to take into account the quantitative relationship between the majority styles and the minority styles. The quality of the analysis depends fundamentally on the quality of the illustrations collected, as well as on the quality of the sources of illustrations. These illustrations must be representative of the investigated universe. If the study relates to the French, European or Japanese market, they must cover each one of these territories.

The sources of illustrations most commonly used are magazines, books and commercial documents issued by distributors and manufacturers (sometimes distributed at auto-shows, professional festivals and trade fairs). Since a few years the quality of the illustrations collected from the Internet has also allowed their integration into the study of the universes. But this quality is still modest (72 dots per inch) and the illustrations are generally scanned or extracted from the commercial CD-ROMs, for a constituted database of images of the study (the given definitions mostly range between 150 and 300 dpi).

Then the process of categorization is carried out on the whole set of illustrations collected and is completed by the description of several harmonious groups of illustrations, each one of which gives rise to a trends panel.

The illustrations of the same category are redirected to retain only the part illustrating the subject and the category then assembled and composed in order not to give the impression of a disjointed and countable unit of illustrations. Such a reconstitution of the illustrations makes it possible to create a chromatic environment, which reinforces the communication of the environment of the category in question.

Then, in order to prepare the presentation of the trends panel, the designer, the colorist, the ergonomists, the engineer and the person in charge of marketing distinguish the following attributes of the assembly among others:

– Poles of shades (colors) – representing harmonious groups defining the category. These shades are not the average color tones of the illustrations but rather the colors from a painter's palette, colors based on which the other tones of the assembly materialize. To enrich the communication of the board it is desirable that the presentation of these primary shades is representative of the color quality of the assembly as well as of their quantitative relation.

– Poles of form – represented by remarkable angles, forms that are particularly geometrical (shapes), signs, proportions, icons, etc.

– The matters and textures are described by the nature of materials, their surface treatments, their finishing, their visual aspects and also their feel and other tactile characteristics.

– Poles of uses such as that of the types of interfaces between the products and their users, the gestures and the standard scenarios of use, the activities and the contexts of use.

– The representative values of the users' lifestyles in each category. These are often emphasized by the advertising in order to make the users sensitive and to take into account the product.

As such, once characterized, the boards have nothing more than being named and described. The analysis of iconic contents and its illustration through the trends panels being a cardinal element of the communication of the step of creative design, the board names make it possible for the whole of the product designing team to conceptualize them and exchange information related to the project without having to physically refer to all the documents.

The various boards illustrating the chromatic structure of the studied universe are described by reconsidering the definitions of each category (harmonious grouping, principal attributes and values), their respective importance and the interpretation of the total organization of facts relating to the topic. For that, a description of the totality of the process is often greatly appreciated by the persons in charge of the conception of the product. Finally, the study is completed by the interpretation and the use of the results in order to direct the stylistic design of the product.

9.2.4. *The harmonies of attributes*

The trends panels are sometimes defined in terms of their denomination, the ambiance of the assembly of illustrations and by the pallets of the principal attributes which describe them. Regarding the palettes of colors, they are not just a list of colors. They are organized in chromatic harmony afforded by groups of contrast colors. Indeed, color is probably the relative medium of communication [ALB 74]. A color expresses its qualities only when contrasted with a background or a juxtaposed shade of different colorations. Therefore, the assembly of colors (or contrasts) needs a more detailed examination than colors considered individually.

By extrapolation, the harmonies of forms (proportions, compositions, rhythms, etc.) can also be very relevant variables of the description of categories. This is the objective of Chapter 20 in this book.

9.2.5. *The chain of value/function/attribute*

The adequacy of the surrounding products to the deep consumer values is a key element in the design and innovation process. Indeed these identified values correlated to the project data can be used as a support for creativity.

The synthesis of the main sociodemographic evolutions, of the lifestyles and of the consumption constitute a relevant information for the elaboration of a value-function-solutions chain based on the cognitive chain established by P. Valette-Florence [Bouchard in COC 01].

The method of cognitive chaining of means-ends [VAL 90] and of Gutman [GUT 82] makes it possible to highlight the way in which the influence of values is brought to bear on consumer behavior. This method is at a crossing point between psychology and sociology. It scrutinizes the value-attribute relationship of the product through a train of hierarchical cognitive sequences graded into ascending abstraction levels. "Product attributes, both tangible (specific evaluative and descriptive features of a product such as material, color, price, etc.) and intangible (semantic terms such as fresh, light, flowery, etc.), bring about functional and psycho-sociological consequences for the consumer helping the latter to attain their instrumental and end values" (J-M. Aurifeille and P. Valette-Florence cited by Bouchard [BOU 97]). Tangible and intangible attributes are interdependent. Consequences are considered to be functional (derived from use, from main functions) or psycho-sociological (social functions produced by the functional consequences and moulds of socio-cultural standards, e.g. a sophisticated image, high personal status). Values can be instrumental (specific behavior modes, such as courage, honesty or romantic attitudes) or end values (aims of life to be attained through instrumental values, such as self-fulfillment or hedonism).

Rokeach has defined a basis of stable values, which are limited in number. Current applications are associated with the positioning and segmentation of products: the consumer associates the attributes determining their choices with more intangible and personal advantages. Young and Feigin [YOU 75] point out that this method is of considerable interest and has a predictive aspect concerning product consumption and brand names. The semantic space can be determined by considering the frequency of apparition of individual items in the various types of chains, then by carrying out a multiple factor analysis dealing with the compatibility between individual items and types of chaining. A chaining is all the more coherent than the total number of links of which it is made up is limited. This method is fundamental for the translation of abstract values into tangible product attributes or vice versa.

In engineering design, the cognitive chain is not established by a content analysis based on questionnaires as it is traditionally, but it is built by the work team during the whole design process.

This reinterpretation of the cognitive chain method in design and engineering design turns out to be particularly interesting in order to establish a correspondence between consumers' values and stylistic or use products attributes. Indeed it allows to link coherently the conceptual space to the products [Bouchard in COC 01]. Physical attributes are connected by the work team to the semiotic they express and with the terminal and instrumental values expected by the targeted consumer.

Of course it is not the only single solution. The use of the values-functions-solutions chain is more a support in order to find out relevant relations between solutions in the product's space, values in the consumer's space, and their interpretation by the designers, according to their background and culture. This coherence between these two cultures of designers and consumers will be the basis for the creation, and transfer of value into the product.

Physical attributes	Semiotic attributes	Instrumental values	Terminal values
Savannah yellow, gold satiny, elephant, glass, etc.	Jungle, animality, ethnicity, culture, jewel, etc.	Exoticism, space, wild nature, sensuality, life in the raw, etc.	Freedom, self-affirmation, status
Knowledge of the product		Self-knowledge	

Table 9.2. *An example of cognitive chaining: the KENZO Jungle perfume bottle (after P. Valette Florence [CHR 00])*

In the previous example, we recognize the terminal values of perfume users. The marketing positioning of the brand is illustrated at the behavioral values level. Semiotic attributes describe the chosen design way in order to express this positioning under the final form of physical attributes (color, material, shape, form, etc.).

Trend boards including user's values, products and their attributes constitute a good representation of the values-functions-solutions chain. In order to do that the work team has to precise the retained semiotic products as well as the behavioral values to which they refer. The following application illustrates the elaboration of trend panels.

9.3. The step of stylistic monitoring and its application in designing the automobile trends panel

The value-function-attributes chain is an explicit formalization making it possible to link marketing and design universes. This support specifically integrates the early marketing initial brief with the brand values and more generally the dominating sociological values. In the framework of car design projects, it plays a major role in the improvement of the technological innovations. In fact it is a way of matching these innovations with real consumer needs through the elaboration of concrete physical and digital supports. For instance, in order to improve a technological innovation linked to an automotive dashboard, the technological innovation related values proposed by the marketing team were the following: family, cohesion, family unit, cocooning, peace and smoothness [COC 99]. This case study is a good illustration of the four main steps of the trends analysis. The trends analysis supported here the design of an integrated technological and styling innovation, through the following steps: (1) the elaboration of the projects brief, (2) the definition of the sectors of influence, (3) trends analysis and (4) the integration of attributes pallets in sketching.

9.3.1. *The construction of specifications and requirements*

This is the first stage of the realization of the trends panels. However, the design team previously enumerated the first specifications of the product to be designed, which encapsulates several pieces of information such as:

– the functions of the products as anticipated by its user (the needs to which it must cater is formalized during a functional analysis);

– the values that the product attributes will have to convey (this proposed case covers technicality, softness, comfort, hedonism, family, cocooning, tranquility and safety);

– a marketing description of the target users (families and group of friends in the case of our example);

– constraints of the problem arisen. These limitations of the creative freedom of the creators are often a result of technical solutions imposed (in our case, the integration of technologies of aerodynamics and thermal diffusions, pointed out by our partner and the architectures of the target vehicles: minivans, small city cars, estate cars and four-door luxury sedans);

– material, human and financial means, as well as the durations of study (a designer, an ergonomist, a colorist and an engineer for 3 months).

9.3.2. *The determination of the influential universes*

Consequently, to determine the influential universes, we use three complementary techniques:

– a creativity session, with the members of the design team, directed towards the identification of the influential universes;

– a multidimensional segmentation of the current offer according to the attributes to be conceived – form, color, use, texture. This analysis makes it possible to group together the resembling products, and those which belong to the same influential universe (Figures 9.1 and 9.2);

– a discourse analysis of the stylists, designers and experts of the concerned industry [BOU 97, CHR 95].

In order to make the study reliable, it is desirable to follow at least two of these approaches. Therefore, we implemented the first two regarding the question of availability of experts and deadlines. The study of instrument panels of vehicles of the year 1986 led us to segment them along two stylistic axes:

– a time axis with reference to the history of the automobile industry, beginning with the motor race of the 1950s (Nostalgic) up to the futuristic concept-car of the recent auto-shows (Futuristic);

– a formal axis tracking the change in form, since the horizontal instrument panels without console of the minivans (Architectural) to the concave cockpits, including roadster (Organic).

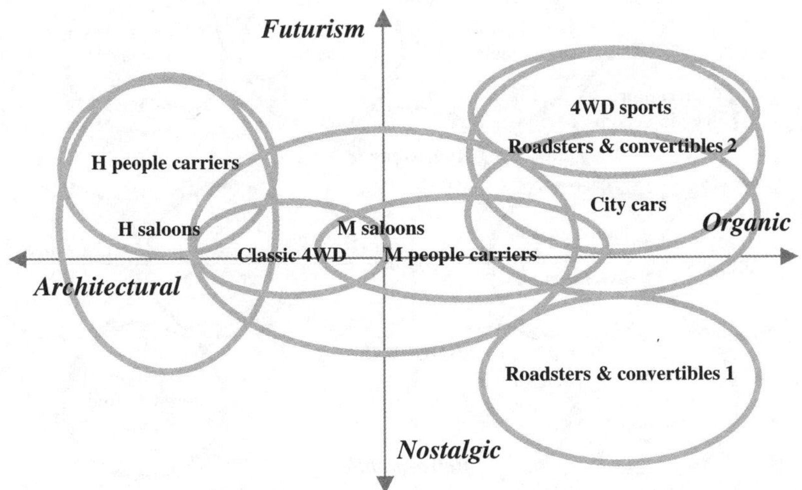

Figure 9.1. *Mapping of 86 car interiors (concept cars and vehicles sold since 1998) divided according to the shape of their dashboards*

This division highlighted groupings of similar vehicles with original attributes. The studies of these attributes made it possible to identify other products of the industry grouping together the same characteristics and thereby to help emerge a list of influential universes.

This list was supplemented and consolidated with that obtained from the creativity session carried out by the project team. Finally the whole of the influential universes was amended and then validated by the persons in charge of the study according to the objectives and the values of the project such as the brand image of the target manufacturers.

The search for illustrations

Once the influential universe is determined, it is now a question of looking for the illustrations of current products, the ambiances, or of contemporary life styles defining them. Journals, magazines, commercial documents and trade fairs constitute our principal iconographic sources. To be relevant and responsive, it is a question, on the one hand, of organizing a continuous monitoring of the key sectors of innovations such as the automobile sector, electrical household appliances sector, the sporting equipment and material industry, the furniture industry, the area of interior decoration, and the clothing industry and its accessories. Thus, the iconographies of these sectors are regularly collected analyzed and archived. On the other hand we must identify, index, select and procure the iconographic sources that describe the influential sectors specific to a project.

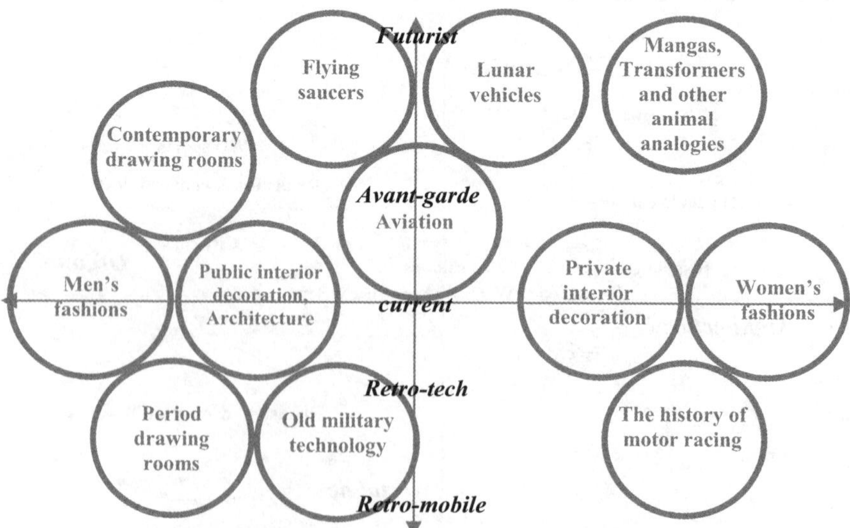

Figure 9.2. *Mapping of the universes which influence the form of the dashboards of the vehicles analyzed*

9.3.3 *The analysis of the tendencies and their descriptions*

A study of tendencies of the influential universes in the design of the instrument panels led us to determine five stylistic, chromatic, ergonomic and textural tendencies: Zen-tech, Aqua-tech, Retro-tech, Ludo-tech and Space-tech.

Figure 9.3. *"Rétrotech" trends panel with its values and its four pallets of attributes (forms, colors, uses, textures)*

The project team retained the first four; the last one, thought to be too avant-garde, was not developed. Sensory attributes representative of each tendency were identified and organized in pallets of principal attributes.

9.3.4. *The integration of tendencies in design of product*

Later the numerical drafts were carried out integrating the constraints of technology, of architecture of the targeted automobile segment (city cars A, minivans M, sedans H) and the attributes of selected tendencies (forms, colors, uses and textures (FCUT)). These drafts enabled the realization of a demonstration combining technological and stylistic innovations in coherence with the targeted segment of customers and the type of the vehicle in question.

Automobile segment concerned	Examples of sensory attributes (F, C, U, T)	Examples of functions of service	Values of tendency	Values of the project (of technology)
Minivans M and H, four-door sedan H	Textures: satin, mat, mineral materials (glass, metal) and slightly veined wood and braided fibers, etc.	To respect the obstruction and the ergonomics of the driver's and passenger's feet; to be integrated in the volume of case, to be pleasing to touch, to be esthetic, to ensure the aerialist comfort of the passengers, to avoid reflections in the windshield, to make it possible for the driver to have the equipment of the central console; to allow the driver to read the bill-posters, to allow the driver to reach the arrangement	Zen-tech: space, minimalism, sensory luxury, technicality, softness	Family, family cohesion, cell, cocooning, calm, softness, etc.
City car A and Minivan M	Colors: three harmonies, pulpits (russet-red, brown, peach, saffron); water blue (marine, electric, light cyan, turquoise); water green (pistachio, meadow, water green, light gray		Aqua-tech: fluidity, adequacy, softness, freshness, transparency	
City car A	Entertaining and amusing technologies, interfaces pleasant and full of		Ludo-tech: hedonism, dynamism, softness, safety, technicality	
Sedan H, minivan M	Simple and integrated volumes, soft and round lines, circular or linear streaks, stylistic references of the 1930 can be seen, visible screws, mechanical		Retro-tech: nostalgia, full value, modernity, esthetics, tradition	

Table 9.3. *Cognitive chaining of Value-Function-Attributes dedicated to the design of automobile instrument panel integrating an innovative diffusion technology*

9.4. Conclusion

The trends panels are tools originating from the style industry of the world of textile and fashion. Their uses as in the Value-Function-Attribute chain are developing in several industries such as the automobile, furniture and household electrical appliances industries. These tools among others constitute the methodological toolbox for stylistic innovation teams. To the best of our knowledge, there was no operational formalization, and their theoretical bases were not made explicit. We found it interesting to propose a model with pedagogical, industrial and scientific objectives: pedagogical – in order to be able to train professional designers of stylistic attributes of the product; industrial – in order to allow the conceptualizers to program and communicate better their stylistic design as a part of an interdisciplinary team; and scientific – to help build models to facilitate a better understanding of the processes of design, and to contribute to the emergence of interdisciplinary tools for collaborative work between colorists, designers, ergonomists, engineers and the marketing team during the preliminary phases of the design, i.e., at the time of the genesis of innovations. Following recent commercial successes of the iMac and evolutions of the automobile design, stylistic innovation has gained an important place within the innovation projects to contribute to the design of products that are sensitively and visibly innovative.

Chapter 10

Virtual Reality Technologies for Innovation

10.1. Introduction

Innovation never emerges *ex nihilo* from the electronic brain of a machine. It has always been, so far, the fruit of individual or team human intelligence which puts together a chain of perceptions, thoughts and ideas that spring out of information, knowledge, emotions, and many other sources. Thus, human factors are at the base of all innovation. To exacerbate this process, it is necessary to facilitate interdisciplinary teamwork by men and women of different professional and cultural backgrounds around common objectives.

Thus, the launch of a project and its leadership by an empathic, dynamic and enthusiastic project head seems to be a key factor of success. To help the project head there are methodological tools and technological means. We shall not dwell upon methods here – like Value Analysis or the TRIZ method presented in the next chapter – but we shall focus on the technologies and technical tools that facilitate and amplify innovation, especially in the area of virtual technologies.

These technologies lean on the exponential possibilities of information systems, mechatronics, and networks. The "machines" of the virtual age are based upon the representation of information synthesized in 3D, interactive images and sounds on interfaces that allow people to use their senses to interact with the new virtual media content. We no longer need to create machines using AI but rather to create Amplifiers of Intelligence to serve intelligence teams striving for innovation.

Chapter written by Simon RICHIR, Patrick CORSI and Albert "Skip" RIZZO.

This chapter deals with virtual technologies involved with the conceptualization of innovative industrial products. First, we shall present the organization of an enterprise around a digital network of information exchange that we call *the digital chain of conceptualization*, which is the overall support of an industrial product conceptualization. Next, we shall show how a team can work on a virtual project plateau around models and virtual prototypes. Finally, we shall present the virtualization of careers, and the perspectives opened by virtual technologies, before concluding our comments.

10.2. The digital chain of conceptualization in the enterprise

"Today, the enterprise must integrate the three determining factors that are the product, the manufacturing process, and the personnel working with the corporation – the 3P Concept: Product, Process, People. Communication has become a strategic means of reference, to concretize that which is abstract, and better understand why one does what he/she does."[1]

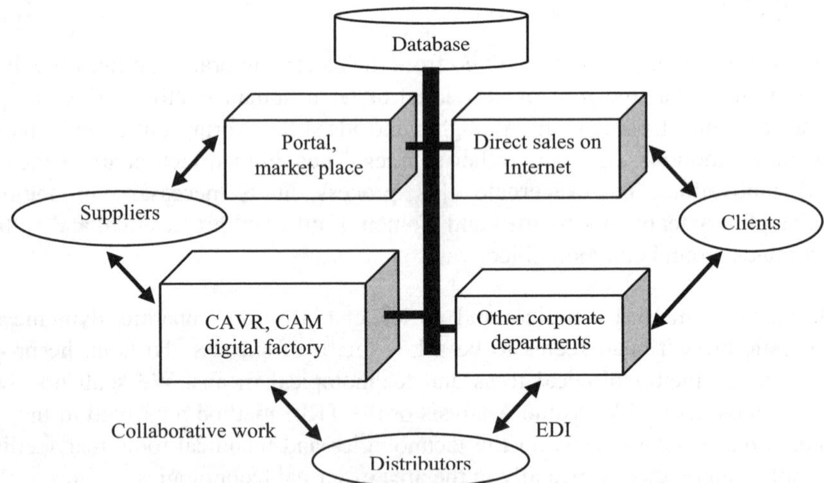

Figure 10.1. *How the enterprise is organized around an information exchange digital network*

Today's enterprise is a dynamic entity that puts the client at its heart. From idea generation, design and production, to marketing, distribution and sales, the client is what justifies and makes a company innovate. Remaining competitive is a goal attainable only if this prerequisite is in place. This is why innovation strategies –

---

1 Interview with Bernard Charles, Chief Executive Officer, Dassault System [CHA 99].

from creativity sessions to designs based on new virtual environments and open technology transfers – are enabling factors to encourage and sustain. Arguably, the brain of the company is contained in its strategy: elements of vision, culture and people, processes and systems, and the plans. In addition, the distributors and the suppliers are the arms and legs of the company.

Distributors may exchange their data with the accounting and commercial departments and thus participate in the conceptualization efforts through collaborative software that offers conceptualization assisted by virtual reality (virtual reality-aided design (VRAD)). Suppliers also participate in the conceptualization process by interacting with the enterprise through a portal or a virtual market place.

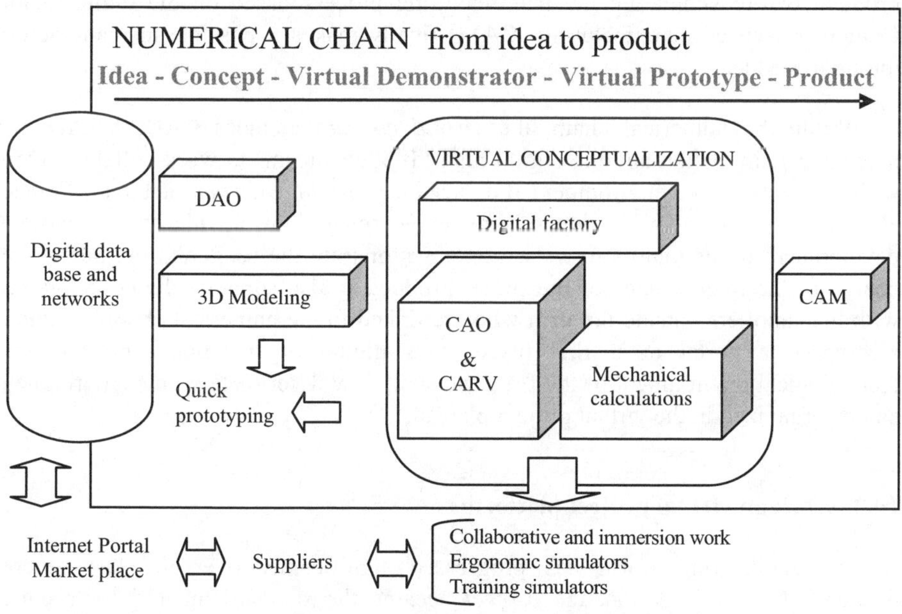

Figure 10.2. *The numerical chain of conceptualization*

Electronic data processing and networks, which is now as vital to enterprises as telephone lines or even their electricity supply, have given rise to what we call *the numerical chain of conceptualization* (see Figure 10.2). This chain represents the technological support that accompanies the conceptualization process of an industrial product, from the first idea to the sellable product. At the origin of this chain, we find the digital database that constitutes the "brain" of the system. It must be noted that in the age of the Internet and high bandwidth communications, this database may be situated on any server on the planet. Its value lies in its access, from anywhere, at any time and from any device. Often, a transactional mode is useful; sometimes access is done through mere data, information and knowledge retrieval. The interface issue is therefore important.

During the conceptualization of an industrial product, the numerical chain allows the passage from the abstract (i.e. the basic ideas of the product) to the concrete (i.e. the product that is ready for sale) [MAI 92]. We start with the 3D modeling that enables a representation of the future product in realistic synthetic images. These images could be "printed in 3D" by a fast prototyping machine.

We then go on to the computer-aided creation (CAD) and to VRAD, then to the calculations that allow the simulation of the mechanical behavior of the future product, before simulating the manufacturing process based on the digital plant. Computer-assisted manufacturing (CAM) finally makes it possible to manufacture the final product.

Within the numerical chain, the *virtual concept* includes CAD, VRAD, the mechanical simulation and the digital plant. It allows teams to work collaboratively within an immersive environment. For instance, the *Ingenierium* at Laval, France, allows the real-time projection of interactive images on a wide screen and the development of simulators to test ergonomics or train the users and the persons in charge of the maintenance of the future product. It also supports the collaboration with the suppliers outside the firm who are linked to the numerical chain through a web-based a portal or market place. This digital organization represents the technological environment of the firm; we shall now describe how this environment can be organized in the virtual project plateau.

10.3. Work on virtual project platforms

The engineering world has progressed significantly over the last several decades[2]. Over this period, the *project* became the standard method for product development [MID 98]. A project is the basic economic entity for producing in enterprises. This method-level of evolution was assorted with functions and responsibilities and was accompanied by two other major changes. Indeed, for a few years now, we have seen great progress in the conceptualization means available to industrial enterprises. The progressive introduction of electronic data processing tools has taken them from creation on the drawing board to 3D onscreen creation. This, followed by a transformation in business nomenclature has been brought about by the data management of the product under development and is often called the

---

2 We cite with permission comments by Fabrice Renaudeau from Renault who presents the evolution of design around numerical processes and of the collaborative work [REN 02].

"digital model". This representation is closer to the knowledge level and, when feasible, is underpinned by an ontology of processes and products, often shared by customers as well as distributors, in a shared quest for a leaner value chain.

Moreover, the sequential organization of the Design Office and the Process Planning Division has yielded place to a more organic organization in which all the competences vital to the creation process are to be found in the same geographical location in order to facilitate and hasten exchanges for the resolution of problems: here lies the notion of *project platform*. We have, therefore, integrated project teams within enterprises, whose tasks involve the development of attractive and economically viable products that come to the market at the right time. All this occurs in order to "dematerialize" developments and communication flows and to do away with the usual borders within the enterprise. Today, in large-sized enterprises, all developments are based on digital technologies. By this we mean that the definition files and the justification files are made up of definition or simulation data in digital data format, whose application enables a more specific and narrower data definition and a measurable prediction of the product and manufacturing process behaviors. Therefore, we can speak of a digital product, a digital process and a digital factory.

These digital "results" are the products of projects organized in platforms. The latest major development is that the conceptualization process depends on digital data produced by simultaneous project teams. Elementary procedures are then capped by synthetic and architectural processes aimed at steering the project in real time, to its pre-defined objectives. This is called concurrent, simultaneous engineering. Work clearly gets collaborative and the sharing of milestones in the development process becomes imperative. In this framework, the collaborative conceptualization transposes the functioning of the project and its production from a physical unity to geographic multiplicity. Indeed, the dematerialization of the production (product, process and digital factory) allows different project teams to work in simultaneous engineering and to keep the project coherent, although they do not necessarily share the same physical platform. In this functioning mode, the transposition and adaptation of teams is indispensable and a particular challenge is to know how to manage projects and personnel with equal efficiency on the "virtual" platform as on a physical.

The three values of the virtual project platform minimize the infrastructure necessary for the platform phase, maintain the cohesion and involvement of all teams committed in the duration of the process, and allow the different partners to collaborate within the project while piloting their own teams within their enterprises. The elements that are indispensable to the achievement of virtual platforms are of three types: computer aided engineering (CAE) infrastructure, that must be shared, allowing the pooling of structured digital data; the availability of production methods, the nature and quality of expected results that must be shared, as development processes must be, with the marking of milestones, and the management methods during the project development.

This general sharing of methods and information is facilitated by the omnipresence of digital networks that structure the digital chain of conceptualization. The Internet (in the larger sense) has become an inescapable element of industrial life. Increased computer power and data flow amplify the capacities of concept teams. These are now no longer forced to be sedentary and can swiftly cross borders virtually. Collaborative work spaces appear on professional websites. They allow work around virtual project platforms. The development of virtual meetings has seen a new rise after the recent periods or situations of troubled, difficult travels.

The coming together of clients and users in a same conceptualization process clearly shows that virtual conceptualization is a system that brings together several participants. We are no longer bound to a logical sequence of tasks; we rather become participants who function collaboratively at any times. This obviously questions the hierarchic structures and habitual modes of an enterprise's functioning that characterized pre-digital methods.

10.4. Virtualization of professions

The global enterprise is organized around a digital network of information exchange through which all participants, internal and external, collaborate and participate in its development. Off-the-self "standard" software is used by enterprises. Yet how do they go one step further? How will they retain a necessary strategic differentiator once a majority of enterprises master the *global digital organization*? How can they maintain or increase a value advantage? We believe that the strategy is to adapt software development to application designers better thanks to virtual reality technologies, and then to "virtualize" the professional methods in the enterprise.

To control future "virtual reality software packages" will not pose much problem to young employees of enterprises. This generational cohort has grown comfortable with high levels of interactivity and synthetic images thanks to their childhood experiences with game consoles that are as powerful as some professional workstations. As has often been observed since the invention of the steam engine or the car, games and toys – as indeed all fashionable products in general – by their permanent need for innovations constitute a popular vector of technology for the general public. Children natively integrate the technological innovations which belong and will belong to their universe: the microprocessor, electronic microchips, the CD-ROM and the DVD, etc. Such is the case today with virtual reality, the use of which has become somewhat more commonplace thanks to the general public simulators and video games.

Interaction with virtual worlds, thanks to ergonomically refined interfaces that are increasingly better adapted to the user, will soon be as natural as driving a car. In the context of the worldwide fusion with the digital world, which will see the convergence of television, telephone, photography and the Internet, the new designers, accustomed to "new" media, will expect more from user-friendly and interactive software which will replace the old software that made the success of professional data processing systems such as CAD, calculations, office automation, etc., possible.

It is within this framework that the virtual technologies appear commonly regrouped under the name "virtual reality". The virtual[3] reality is the transition vector between two types of data processing: on the one hand a historical data processing paradigm at first based on peripherals such as the keyboard, then the mouse, and posting data on a flat screen and on the other hand a "virtual" data processing based on behavioral interfaces using data visualized in 3D with immersive depth.

Figure 10.3. *Collaborative work in virtual reality in an immersion theater*

---

3 The virtual reality is a scientific and technical field exploiting the data processing and behavioral interfaces in order to simulate in a virtual world the behavior of 3D entities, which are in interaction in real time between them and with one or many users in pseudo-natural immersion via the intermediate of sensori-motor channels. Definition elaborated by P. Fuchs (EMP), B. Arnaldi (IRISA) and J. Tisseau (ENIB) [FUC 03].

The present challenge thus consists of preparing for the arrival of this new "virtual" data processing as it will likely displace the predominance of physical material as the primary "objects" between the user and the virtual universe of work. We will thus see an automobile research department engineer "entering" his highly realistic numerical model in order to modify it with the help of natural movements, emotional responses and/or words. The basic idea is that he will not need any training for this new software or for the effective use of the interfaces necessary to its functioning.

The human user is the core concern for this new "virtual" data processing design and gets all possible attention. The international "Cybertherapy" Congress (11th Edition in 2006) each year brings together some of the best international specialists in the use of virtual reality for the treatment of people with different psychopathological disorders. This group of researchers has acquired a profound knowledge of the interaction between man and virtual surroundings: feeling of presence, sensorial immersion, natural interactions with the virtual environments, behavioral designs, behavioral software assistance. All this knowledge is useful within the framework of the therapy and gets mobilized for the creation of virtual environments intended for the design of products through collaborative teamwork with a range of skills (engineers, designers, users, etc.). The integration of virtual reality technologies in the enterprises demands expertise from numerous scientific domains: from cognitive sciences to engineering sciences.

In the field of interaction peripherals, force-feedback interfaces constitute one of the areas where virtual reality has significantly progressed. These interface tools give virtual objects an added dimension for their perception by the user. The sensation felt when one interacts with a virtual object that is physically felt is difficult to explain without a "hands-on" trial (several displays are shown every yearly at Laval Virtual show in Laval, France). Force-feedback arms are, for example, used by modelers who can feel the sensation of sculpting on different materials … without the inconveniences. Indeed, in case of errors or a false movement, they can replace the virtual material or modify their previous action from which too much material has, for instance, been removed. Thus, optimal symmetries can be attained, and models can be modified indefinitely (see Figure 10.4).

This example shows the fundamental interest in virtual reality: one starts from an actual profession (sculptor, model maker) whose needs and possibilities of improvement are studied. One then conceives a virtual environment that reproduces the universe of this "profession" (blocks of matters, sculpting tools, paint brushes, airbrushes) and adds to this environment functions that are impossible to implement in the actual universe (symmetry, duplication, addition of matter, geographical deformations, of texture or of lighting, undoing of last actions).

For the enterprise, there exist software tools created starting from the "professions" of a production unit (such as the Delmia software of Dassault systems). Also described as "virtual plants", they make it possible to model all parts of the product to be manufactured, design the workstation which will allow an operator to assemble the product parts, to design the manufacturing workshop with all its machines and its conveyor and to simulate the production flows of different parts. Here, several professions such as production engineer and ergonomist get "virtualized" while the innovative functions they use avoid the costly expenses of the physical world.

One sometimes hears that CAD software publishers may consider virtual reality as a simple "function of interactive visualization" which they integrate in their software. However, the impact of virtual reality goes much further and often necessitates recasting the software around the natural work way for the user (his "profession"). One can even foresee a genuine revolution when virtual technologies will substitute the 2D interfaces to which we are so often attached today.

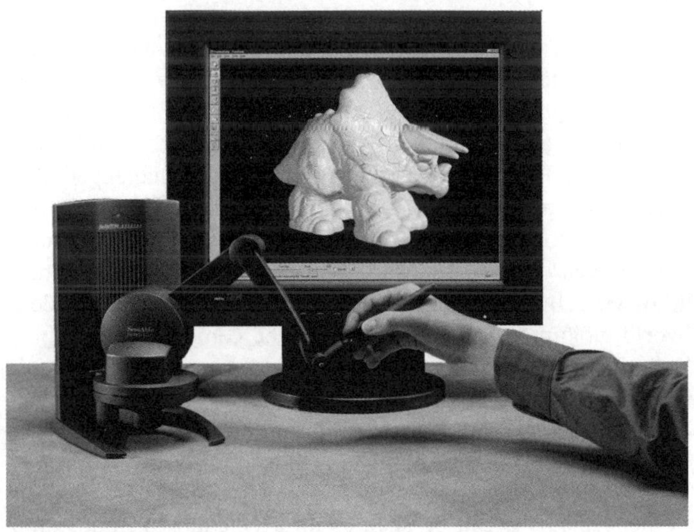

Figure 10.4. *FREEFORM system of virtual sculpture (photo: SimTeam)*

Virtual reality is also accessible for subject matter experts (SMEs). Each SME has "professions" which determine its specificity and its know-how. These professions are often transposable in virtual environments in order to optimize and improve the productivity of the participants mastering these professions (it is not a question of replacing people by machines but to provide them with amplifiers of intelligence). Moreover, the development a made-to-measure software will not necessarily be more expensive than the purchase of large general practitioner software of which one will only use a fraction of its possibilities and for which heavy training programs will be needed. The small software "professions" that one will see appearing thanks to virtual reality will be compatible between them, thanks to data-processing language and to the exchange format available from the Internet.

Whatever the name we may give to it (3D interactive, VRAD, simulation, numerical model, etc.), virtual reality will gradually invade the field of data-processing and make the computer, as we know it today, obsolete. The new generation of users is ready for this transformation. What can be more appealing to a young "gamer" (a player of video games) than to work with professional software using virtual reality methods?

10.5. What virtual environments really mean

10.5.1. *Today's challenges*

On the one hand, virtual reality (VR) poses a unique novel and complex challenge that computing science has yet to solve. VR is at the edge of perception, science (cognitive, control theory, etc.), engineering, human behavior and culture. It exacerbates usability and cognitive psychology plays a central role. It is also culturally connotated. Therefore, classical engineering is challenged. After all, what is VR? A toy, a computer, a game, a tool, a metaphor, etc.? As computing becomes pervasive, VR paradoxically strives to find its own way. Ubiquitous computing and VR have indeed been opposed by Mark Weiser: "ubiquitous computing is roughly the opposite of virtual reality. Where virtual reality puts people inside a computer-generated world, ubiquitous computing forces the computer to live out here in the world with people". So VR should not be reduced to a mere technology. It is a perception metaphor, an interaction mean; it is also underpinning *new methods and practices of interrelations* with man, nature and systems. Whereby the *"anytime, anywhere, on any device"* innovation vision tends to become reality.

On the other hand, VR can be seen as an accelerator of innovation because it makes everything come closer to the user sooner and faster:

1. *System and user mesh with intensity (sensation + perception).* Here lies the value of presence, which, for the use, may depend on the context: displacement, feeling, commitment, emotion, depth, perception of the environment, or even absence, etc.

2. *User "becomes" the system.* The user's imagination gets enhanced and may mean such things as: exploration, intention, experience building, modulation of thinking, construction and deconstruction, or time and space acceleration, etc.

3. *Focus of attention gets undivided.* For the user, this is a mind defragmentation and this can imply among other things: cognitive enlightenment, a throughput, simplification, intuitiveness, globalization, forgetfulness and focus, efficacy, etc.

Together, presence, imagination and mind bandwidth can make three constituents for building integrated value propositions in business model based on virtual reality technologies. The synthetic information and knowledge levels brought about by VR enhance a user's understanding. VR brings a new type of cognitive reference ability in application systems, that, for instance, AI methods did not succeed in tacking smoothly.

When virtual environments to tasks and outcomes are mapped, it is useful to consider the following three levels:

1. *Identify tasks and objectives.*

2. *Perform a knowledge level task analysis (perceptual, cognitive, motion).*

3. *Match clues to technologies.*

This reveals the need for usability engineering. In the face of take-up by markets, three topical challenges face the development of virtual environments today:

– Can VR make it possible to speak in the first person? In other words, can VR enable a self-identification in engagement? Not that we would ask identification and authentification of a user necessarily, but that the user must feel *empowered* by the VR system with a degree of intensity such that he can accept the transfer of consciousness through the system. Further to this, developers may want to consider building personalized locales.

– Can VR generate new how-to knowledge? Two kinds of new knowledge can be generated: knowledge about "how to field VR", called development knowledge, and knowledge about how to build good and best practices, called collective knowledge.

– Can VR enable new business models? It has become clear that VR adds intelligence to communication. In a sense, VR "modulates" communication. This seems to reveal a marketing *sweet spot* that marketers may have not tackled yet.

Yet, can we tell when VR becomes effective? A virtual environment is effective when users reach their goals, when the important tasks can be done better, more easily or faster than with another system, when users are not frustrated or uncomfortable, and when there is some measurable gain in targeted real world performance. Typical computing performance metrics such as speed of computation are really not important in themselves and, for instance, computing speed is only important insofar as it would affect a user's experience or tasks.

10.5.2. *Perspectives*

In this chapter we presented what today may be the set-up of the enterprise around the "numerical chain of conceptualization", why the enterprise must promote collaborative work through a virtual project and how it virtualizes its fundamental "professions". This will help in sustaining a competitive differentiating advantage during the virtual era. We considered the challenges that the developer and the user face. Let us now see what virtual technologies hold for the future of numerical innovation.

Since people are at the heart of the innovation processes, our efforts will have to focus on them with the aim to create *intelligence amplifiers*, which means allowing people to consciously assemble ideas, information and knowledge in order to result in virtual concepts for innovative products and services. Immersion will, for instance, amplify creative skills as it can improve the quality and quality of sensation as a result of being present in a virtual environment. People can obtain quicker loop feedbacks, react to those, and therefore accelerate design and development cycles in significant proportions. If we examine the automobile sector, where design to market cycles constitutes the major competitive battle among automobile companies, we begin to take the measure of what virtual reality can mean in terms of competitive advantages: the mastership of innovating time-to-market processes. Incredible gains can be obtained by those who will master the virtualized processes of engineering.

The immersion of people in a virtual environment can be split up into sensory immersion and psychological immersion. Today, sensory immersion is mainly based upon vision. Although we once believed that the virtual reality helmet was the panacea of the field (see Figure 10.5), enterprises nowadays prefer giant projection screens or "CAVE"[4] that do not limit users' movements in the virtual environment (see Figure 10.3). Nevertheless, new HMDs are now better engineered and offer a low cost and portable solution for some applications.

---

4 Large cubes of which the 4 to 6 faces are projection screens which allow the immersion of one or more users in a virtual environment.

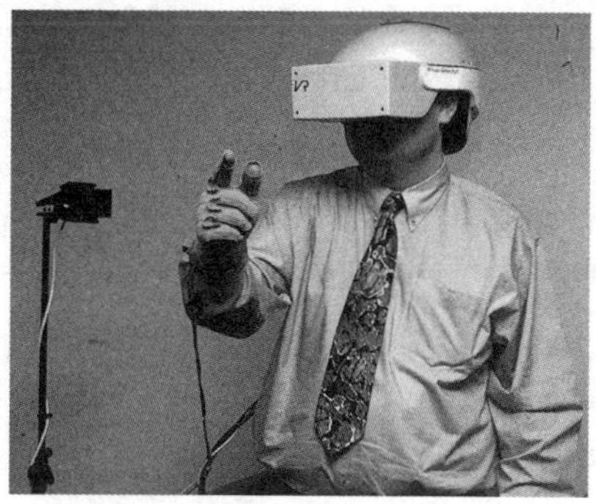

Figure 10.5. *Virtual reality immersion helmet with movement-sensitive glove*

The sense of vision is often completed by that of audition because the spatialization of sound today is very well controlled even if it requires additional processing and computing power. However, new developments in audio rendering reduce the costs and the complexity for delivering compelling 3D sounds in VEs.

Haptics and force-feedback sensation have also been the subject of much research. The feeling of contact, burden, kinetics or deformation of an object does build the notion that the user is in the presence of a virtual object. The current goal is to increase the field of maneuver for the user which usually remains very limited with a force-feedback arm (see Figure 10.4). This can be achieved, for example, through systems with nylon threads such as the Spidar of the Precision and Intelligence Laboratory (PIL) of the Tokyo Institute of Technology (see Figure 10.6). The sense of light touch is also the subject of researches but it is more difficult to apprehend, given the complexity of the human skin.

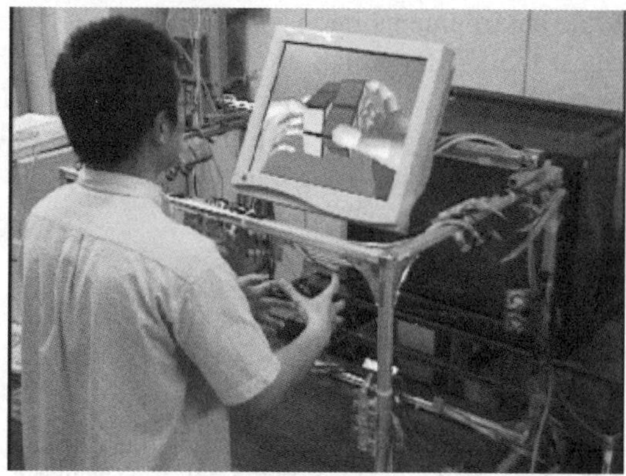

Figure 10.6. *Force-feedback mechanical interface, the Spidar 8 (photo: PIL)*

Smell is also a very powerful sense that supports immersion. We simply have to remember our reaction when, in the street, we catch the smell of a croissant shop equipped with computerized diffusers of synthetic smells. As for a perfumery, research makes it possible to constantly broaden the range of the available smells. These mostly arise in the form of oils diffused rapidly by computer driven fans. The question of spatialization of the smells like that of sound is yet unsolved.

The first interfaces allowing us to simulate the taste of a virtual product have just appeared in laboratories. They will perhaps someday allow us to test virtual dishes and to obtain complete sensory immersion for users of VR.

The psychological sense of presence is what is commonly used to describe the intimate feeling that users "exist" within the simulated situation of the virtual environment. This sense is closely linked to the cognitive schemes[5] integrated since childhood and to the users' imagination. Video games constitute a good example of psychological presence that it is possible to trigger in users. The need for psychological presence can even become pathological in certain gamers, or even, according to a recent and worrying Japanese study, modify the players' mind, called *game brain* and trigger violence in their behavior.

Fortunately, the psychological presence in virtual environments has many positive effects, which explains why the subject is intensively researched in the field of the psychology for the treatment and rehabilitation of certain pathologies. The treatment of phobias (fear of spiders, fear of the void, fear of flying, agoraphobia, etc.), memory disorders (following a head injury for example), behavioral disorders (bulimia, anorexia nervosa, etc.), rehabilitation of people with post-traumatic stress disorder (PTSD) (soldiers returning from battle) are some examples of clinical targets in which interactive virtual environments have been used. The results of such therapeutic research also benefits from other fields. They indeed make it possible to understand and validate environments intended for other uses. Which are the colors, the sounds, the images, the smells which are going to support the well-being and effectiveness of users in order to enable them to carry out their tasks under optimal conditions? In addition, which are the sounds, the images, the effects, the angles of vision which are going to give rise to strong emotions?

---

5 The notion of design is the one proposed by psychologist Jean Piaget [PIA 79]. According to him, by analyzing the birth of intelligence of the child, particularly in its sensori-motor dimension, design constitutes a means of the subject through which situations and confrontation to objects can be assimilated.

The study of feelings of sensory and psychological presence of the user of the interactive virtual environment arose from different fields of scientific research: psychological, neuropsychological, industrial, etc. Increasingly, worldwide research laboratories have studied these issues. Their efforts, in our opinion, are going to allow the development of powerful tools which will make it possible, in the field of innovation, to amplify the creative capacities of the ideating teams.

10.6. The challenge ahead

This product of which I always dreamed is there, in front of me. I catch it, I feel its weight, the touch of its material, I try it, I test it, I assess it sensorially and psychologically. If it suits me, I will buy it… as soon as it will really exist. The claim: "I dreamed about it, X made it" can be fulfilled. The interactive virtual environments, inserted in the *numerical chain of conceptualization*, allow the sublimation of the innovating capacity of conceptualizing teams. Virtual reality is not a simple technology, it is a method of "perception" and interaction, which underlies the arrival of new data-processing paradigms, liable to invade our daily life in the next 5 to 10 years, and to justify new methods and practices of relations between people and system, people and nature, people and their peers in long-distance and long-term relationships.

Will VR someday understand the user? In the IT business, usability is crucial as many potential customers are simply afraid that they will not be able to "figure it out". Who will use the system is as important to the development team as to have a realistic view of the users. Users' characteristics encompass factors such as age, education, language, motivation, domain knowledge, knowledge of computers, etc. Moreover, usability covers these three aspects of:

– efficiency: requires an appropriate level of ration performance/resources;

– effectiveness: task is completed to an appropriate quality level; and

– satisfaction: the one the user experiences in doing the task.

If users cannot complete their tasks efficiently and effectively, the system is somewhere faulty. The usability principles, as derived from Universal Design Principles, cover these 10 points:

1) Real fallback and realism – the metric that gets used all the time.

2) Navigation – *I can find my way around.*

3) Functionality – *I can do what I need to.*

4) Control – *I am in charge.*

5) Language – *I understand the terminology.*

6) Help and support – *I can get help when I need it.*

7) Feedback – *I know what the system is doing.*

8) Consistency – *I do not have to learn new tricks.*

9) Errors – *Mistakes are hard to make and easy to correct.*

10) Visual clarity – *I can recognize things and the design is clear and appealing.*

With the approach of the virtual era, which will follow the epoch of communication from 2010 onwards, and according to Malo Girod of Ain [GIR 05], these questions will probably be solved. The process of global virtualization of numerous professions has already been undertaken with commitment. This new phase will likely experience many more flourishing innovations than the previous one which, however, saw the birth of electronics and trips to space being born. Let us not doubt the capacities of virtual technologies in order to endorse this challenge.

Chapter 11

TRIZ: A New Method of Innovation

11.1. Introduction

Due to the markets' saturation and the broadening of the consumer base, companies are constantly improving their production and are launching new products and services. As such, a systematic and rigorous process for designing new products has to be implemented [MOL 70]. For this, knowledge and mastery of a large number of factors is required. Amongst these factors, the two parameters that are indispensable when the business is in a highly competitive field are the tools and methodology to be implemented. It has been noted that the tools in the "tool box" of the designer are in abundance [VAD 96] and that he has a wide choice when a new design project with the specific criteria of innovation is assigned to him.

In this chapter we would like to describe the principal components of the "tool box" that are needed to implement an innovative action. This will, on the one hand, help to determine if a new methodology is valid and, on the other hand, will prevent the designer from using only his intuition to ensure continuity in his own approach to designing.

11.1.1. *Product designing methods*

"Organizing the designing means methodically programming the process and at the same time not hindering the creativity of the designer: that is leaving one bit un-programmed, a bit of disorder in process management, a bit of chaos in the order" [CHR 95].

Chapter written by Darrell MANN and Pascal CRUBLEAU.

The combination of constant new demands of the market and technological progress forces companies to rapidly renew their products by using the best suited technologies as much as possible. For designing, engineers have explained their designing processes and have identified the progression fields [LEC 92]. The designing process satisfies three unavoidable requirements: cost, time and quality, but today, these do not suffice to have a competitive edge on an area of a given market; the new product has to be innovative [CRU 02] (see Figure 11.1).

Besides this, tools and structured processes are indispensable. However, they have to be used judiciously so as not to inhibit one of the essential components, which is creativity.

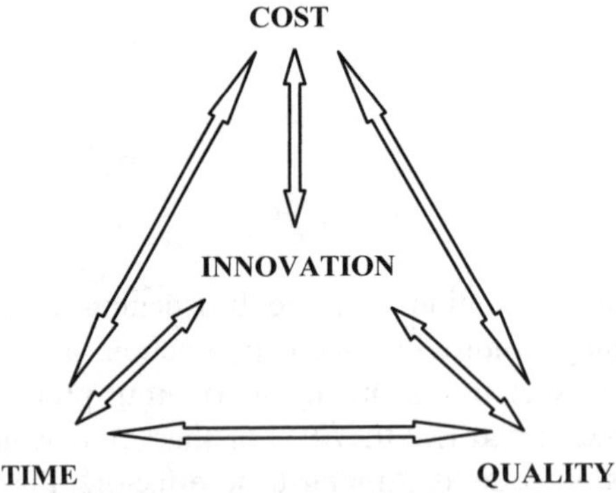

Figure 11.1. *New requirements of the market*

11.1.2. *An important stage*

The efficiency of the design tools can only be optimized through collaboration between the principal actors in the designing of products. We must not forget the thing which is the key in identifying any innovative solution: creativity.

"Creativity is a major human capacity, it consists of bringing forth new forms, [...], creativity brings out new information, sometimes surprising and disturbing" [VID 98].

For a long time inventing, or more generally creating, meant discovering. The discovery was therefore an object, or a hidden phenomenon which existed quite independently of the researcher. Inventing therefore meant finding the place or the path to follow, as one would for a treasure.

Systematic economic requirements have changed this traditional framework. Creativity is more firmly entrenched in a general designing system. The processes of creativity use various methods at different stages of the designing process. Here we differentiate systematic methods from random methods.

Systematic methods such as general combinatorial analysis, morphological analysis and discovery matrices are appropriate for simple products made up of fewer components and which can be modified in an innovative way. Using a discovery matrix for a complex product is long and meticulous, but it can, however, help explore systematically all the possible combinations.

Random methods such as brainstorming or metaphorical method reject rationality in favor of the creative powers of each individual in a design group. These methods are most productive in terms of number of identified designs if the different applicable tools are rigorously used. Just as examples, we can recommend some methods that could be defined before beginning a brainstorming session:

- no censoring of another person's statements as well as one's own;

  priority for research of verbal associations, even absurd ones;

- etc.

This method has to be developed by an expert whose competence is a key factor to its success.

11.2. A deterministic vision of future technologies

11.2.1. *General introduction*

TRIZ is an innovation technique or more precisely a theory for solving invention problems. This method has been favored by a large number of companies all over the world, whatever their sector of activity may be. It capitalizes on the knowledge of genetic analysis of the products as well as the prospective methods such as historical analysis. This step is at the same level as the creativity methods [CAV 99], the main objective, in its initial version "classic TRIZ", is to encourage emergence of ideas, provide rapid design solutions [IDE 99a] (see Figure 11.2).

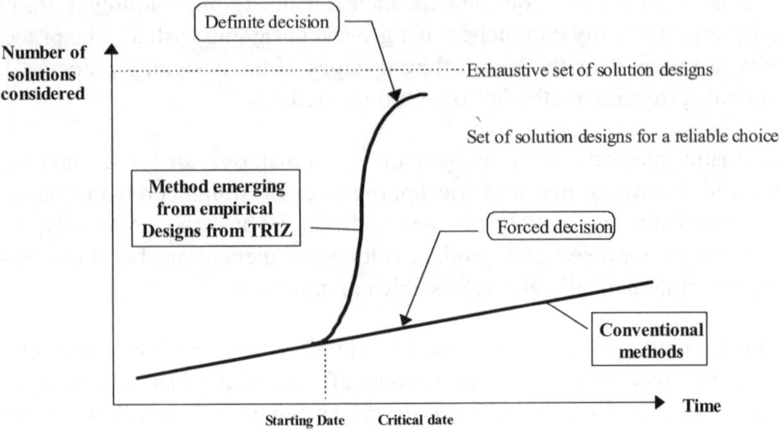

Figure 11.2. *Importance of the controled development method*

This method was created in the USSR in 1947 at the behest of Genrich Altshuller. The main factor of its success, associated to this theory by the users, is the speed of mastery of the tools. This theory is based on technological realities (see Figure 11.3) and can be applied in all the different industrial sectors.

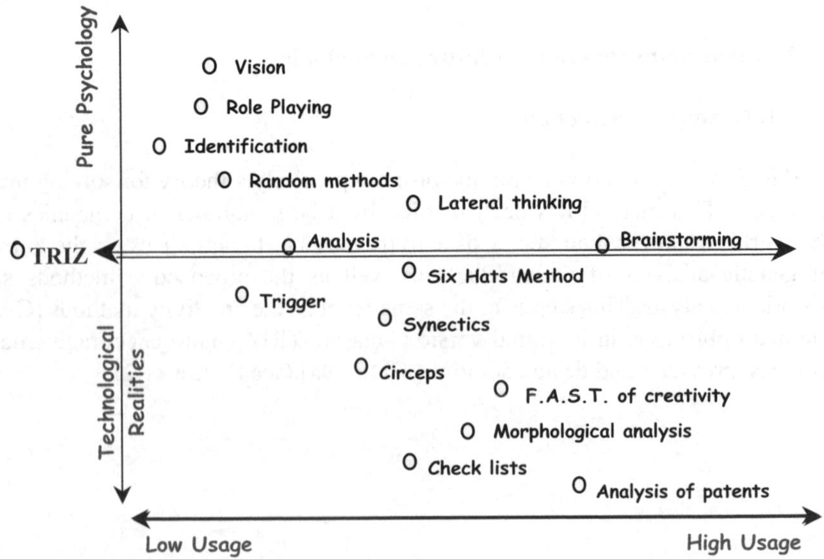

Figure 11.3. *Position of TRIZ in the tool box of the designer*

11.2.2. *Introductory ideas*

11.2.2.1. *Contradictions*

During his youth (in the 1930s), Genrich Altshuller had an education imbued with logical materialism, with the conviction that good would triumph over evil.

At the beginning of his working life, he was confronted with Stalin's arbitrary system, a total perversion of his Marxist ideal. Hence, he was consciously or unconsciously tempted to take refuge in an "intellectual bubble", a utopia. He could then apply logical materialism as a tool for resolving conflicts in a non-controversial field: the world of technologies.

In the "classic TRIZ" [ALT 88], recognized as the original version today, the solution of a problem depends on the identification and formulation of a contradiction.

There are two types of contradictions:

– the physical contradictions: in order to function in an ideal way, the system should have properties "A" and "non-A". Two modalities of the same characteristic have opposite properties:

- the same "object" should be:

Hot and Cold

High and Low

Hard and Soft

Active and Inactive

Mobile and Stationary

Big and Small

Rapid and Slow

Impervious and Porous

Thick and Thin

Long and Short

Conductor and Insulator

Strong and Weak

– the technical contradictions: an action taken on the system meant to improve characteristic "A" leads to the deterioration of another characteristic "B".

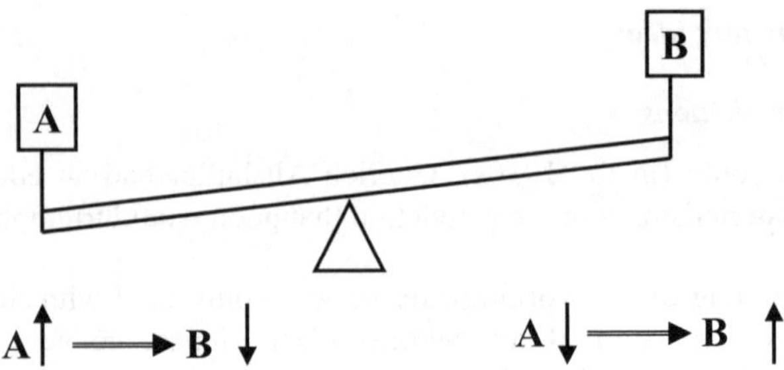

11.2.2.2. *Ideal situation*

An essential idea in TRIZ is that of the Ideal Final Result (IFR). This is the driving force of the development [SAL 99].

The majority of the systems created by man are developed to satisfy the demands and needs of clients.

In best situations, the ideal machine has no mass, no cost, undertakes the entire set of necessary useful functions and does not generate any detrimental effects.

It can be expressed in the following manner:

$$I = \frac{\sum F_U}{\sum F_N} \qquad [11.1]$$

I: degree of ideality   FU: useful function   FN: detrimental function

11.2.2.3. *Development laws*

Genrich Altshuller demonstrated that the majority of the systems created by man evolve according to predetermined laws, rather than in a random manner. These laws can be used to speed up the development of a system rather than wait for the system to evolve naturally.

11.2.3. *Postulates of TRIZ*

11.2.3.1. *Introduction*

The TRIZ method includes a set of postulates established after a deep historical study of the evolution of different systems [ZLO, ZUS 99]:

– specific technological systems and all the technologies;

– specific sciences and "the" science;

– the different social groups and companies;

– the arts;

– etc.

Hereafter we will describe the main laws and some lines of evolution such as they exist in current literature [IDE 99a] with the objective of highlighting the difficulty and complexity of using them to identify and determine the future developmental state of a given product.

In order to understand this presentation, it is necessary to link it to a set of postulates related to the technical systems so as to understand the meaning of the different developmental laws. It should be noted that some of these have a more general application, not limited to artificial systems (created by man).

11.2.3.2. *Definition of a technical system*

In the latter part of this presentation, we will propose that a "technical system" is a whole on which rests the problem to be dealt with. According to TRIZ, a technical system must contain at least four entities: a drive, a transmission, a work organ or actuator, and a control element or control organ (see Figure 11.4). It should be possible to control at least one of the entities and each of the four entities should be present in the system and should fulfill the minimum of one function [IDE 99b].

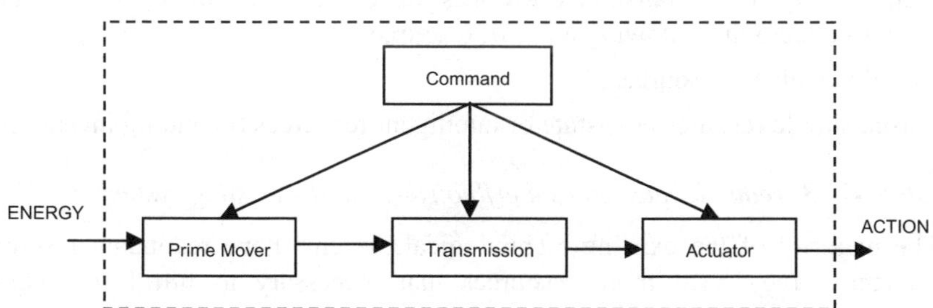

Figure 11.4. *The functional components of a technical system*

11.2.3.3. *The postulates*

Postulate no. 1: generation of change and selection

Each technical system evolves in a way that:

– different ideas are first generated leading to modifications or creation of new systems;

– a selection process retains the best system to satisfy the requirements on condition that they represent an increase in the ideality of the system. The main selection factor is the response from the market which provides the finances which are vital for the development of the system.

Two types of selections are identified – positive and negative – which influence the development of a system:

– positive selection: applicable in healthy economic situations to encourage systems capable of effectively benefiting from available resources and those which can be rapidly circulated through industry or the market;

– negative solution: effective during periods of economic recession to encourage systems capable of surviving with minimal resources and which are well protected from the negative impact of the environment.

Postulate no. 2: evolution at the cost of consumption of resources

A system is developed through the consumption of resources available in the system itself, in surrounding systems, and/or in its environment. Each stage of the development generates new resources which can be used for the future development of a given system as well as for other systems. However, the negative resources, which can generate undesirable effects, can also come from the evolution process.

In the evolution process of a system, the consumption of resources leads to a greater difficulty in the mobilization of the same. Consequently, there are a certain number of sequential transitions towards different types of resources. These transitions concern the following types of resources:

– easily available resources;

– from simple resources to astute or intelligent resources (including inventions).

Postulate no. 3: redundancies and superfluousness in the existing system

The majority of the existing technological systems have redundant resources; consequently, they have more resources than necessary to fulfill the required functions.

Postulate no. 4: co-development of different systems

Many of the technological systems are interconnected. The power of their connections increases with the evolution process.

Postulate no. 5: co-development of systems belonging to different hierarchical levels

The systems belonging to different hierarchical levels (a system and its sub-systems or a system and its super-system(s)) highly connected in their evolution and evolve in mutual coordination (co-evolution).

Postulate no. 6: long-term forecast as against short-term forecast

The short-term development (improvement) of a system depends first on its inherent resources. Long-term development, including the emergence of new generations, technological advances, etc., depends upon evolution of the entire technology and/or of the market rather than of the specifics and resources of the given system.

This postulate can explain why many forecasting techniques proposed in the period between the 1950s and the 1970s were inefficient.

Most of them believed that forecasting had to be carried out by those who were experts in the system and had to be based on the information collected about the development of this particular system.

Postulate no. 7: there are a limited number of type of solutions that will be appropriate for a particular function

A function can be realized by a limited number of distinct paths based on the usage of known resources. New types of resources will appear as a result of a discovery.

Postulate no. 8: alternatives in the development

There is more than one alternative, but they are a limited set of almost equal directions through which a given system can be developed from its current position to the next. This development is based on the deployment of different types of resources. The "winning" development is normally the one which emerges first and which attracts the highest number of human and financial resources.

Postulate no. 9: standard process for resolving problems

There are common methods, based on the laws of development, of solution of problems and of improvement of a system. These laws can be revealed through the analysis of the history of inventions and the licencing of knowledge of the innovation to be collected and transferred.

11.3. Conclusion

Right up to this day, the TRIZ method is particularly well adapted to resolve recurrent technological problems. Its principal advantage is the systematic approach that it recommends; it is based on a vast corpus of information requiring a large investment if one desires to assimilate the entire theory. However, our scientific techniques and our industrial applications have made it possible for us to assimilate this theory in three steps, which go from minimum information to fully autonomous exploitation suitably adapted to the considered sector of activity.

Different software tools co-exist with complementary functionalities and facilitate the introduction of this theory in industrial activities of all sectors. They contribute in stimulating innovation, even if they do not always facilitate the usage of all tools of TRIZ. The principal users of this theory recognize that the most promising developmental thrust is the knowledge field of the developmental laws.

Chapter 12

C4 Innovation Method:
A Method for Designing Innovations

12.1. Introduction

Innovation is the indispensable means by which companies can distinguish themselves from their competitors [BIE 94, DEL 00]. Research in productivity and subsequently in quality was the vision after the end of World War II, and the trend continued up to the 1980s and 1990s. The objective has always been to produce more in less time [ZIR 96], but the advent of new modes of communication, the improvement of transport or, better still, the political upheaval world of recent years has compelled companies to devise different modes of differentiation. As a consequence, a need has been felt to expedite production of new offers of products or services, while at the same time decreasing risks for the companies. Therefore, it appears important for the company to draw, as soon as possible, a first estimate of the success that an innovative offer could have on the market.

This gains more importance when we get to know the changing nature of the market and notably that of the clients. Like the economic context, the social context too evolves to a large extent. The client is an actor with multiple facets that one should be aware of in order to capture him. He desires to consume better, also in a 'more useful manner. The "use" notion does in fact signify something important. This can indeed be considered an action – the act of making use of something. This is about the function, the destination and the use that we would make of it. The conception of an adapted offer must be accompanied therefore by the

---

Chapter written by Olaf MAXANT, Gérald PIAT and Benoît ROUSSEL.

comprehension of its usage. In other words, the actions have to be observed and interpreted – actions that are to be put to use by the product and its environment. This is why we often connect it to the notion of usability discussed at length in ergonomics [ROU 96]. In reality, the idea of "use" covers a very large domain that is studied through different perspectives, the sociological, anthropological, semiological, etc. "Use" also embodies the individual dimension, as in a group of individuals forming a social group.

With this notion of use, it seems necessary to examine the concept of what is called acceptability: for this we shall start with Nielsen's factorization analysis (see Figure 12.1). Without going into the details of each factor of acceptability, the factorization analysis causes the emergence of two types of factors: the "technical-economic" factors (cost, reliability and compatibility) and the "Human and social sciences" factors connected to human behavior. The factors of acceptability are therefore multiple and belong to different systems.

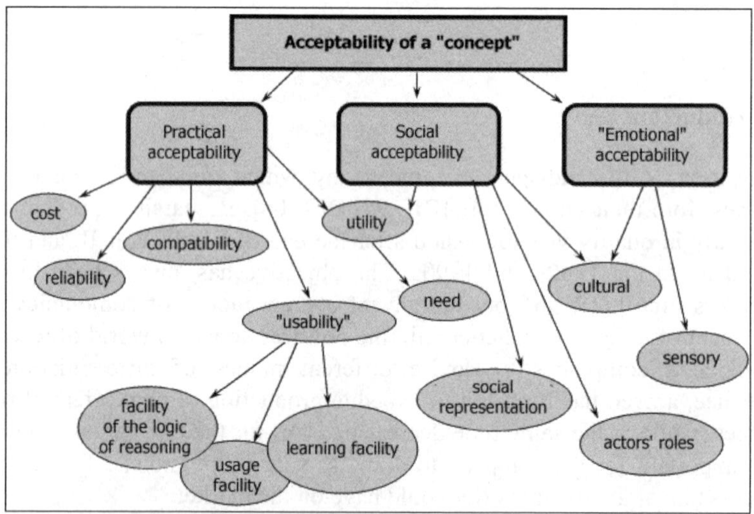

Figure 12.1. *Acceptability factors*

To understand the client and the context, and to analyze the technical state of being from a domain demand the application of multiple kinds of knowledge that cannot be filled by a single person alone. The process of innovation is a multidisciplinary process. However, this multidisciplinarity must be managed, facilitated and encouraged. Without a process of collaborative conception, the contribution of information and knowledge linked to each discipline can remain relatively limited. For Boulier [BOU 02], it is this decompartmentalization of methods and disciplines that prompts the favoring of collaboration in a multidisciplinary environment. It is, besides, in this sense that we prefer to use the concept of inter-disciplinarity that "is a form of a more integrated cooperation, worked out in view of resolving a common problem" [DEN 03].

In short, the company then needs to innovate, in accordance with the fluctuating needs of its clients. For this, it must modify its approach by favoring the inclusion of disciplines related to the study of man and his environment. In order to achieve this objective, it has an interest in ensuring collaboration between the totality of its actors. At the same time, the company must take into account the time and financial constraints related to the current socio-economic system. To that extent, it must have the wish to evaluate the users' acceptability concerning its future offers, in order to minimize its risks and costs.

In this chapter, we propose to discover the works undertaken at EDF R&D, which are aimed at responding to these statements. We shall see how virtual technologies help us to evaluate the representations of reality which does not exist as yet, and how the conception of these representations supports creativity.

In the first place, we shall establish the industrial context of these works. We shall specify the basic structure of this research, that is to say, the support device for the project, called CREATEAM®. We shall examine in detail one of the approaches used in this device (the C4 method) whose objective is to generate, formalize and evaluate the concepts of offer in an upstream phase based on the use of a process of innovation. This approach, and particularly the part concerning the contextualization of the concept, will be illustrated through its application in several projects. Finally, we shall conclude by specifying the tools that we currently put in place to implement and facilitate the application of this approach.

12.2. The approach of innovation in the commercial domain of EDF R&D

Electricité de France (the French Electricity Board) has been a public sector organization since April 1946. It works towards the objective of producing, transporting and distributing electric energy to all French people and companies. The necessity to construct such a reliable tool of production has resulted in favoring the realization of massive nuclear power production. Clearly, the company's actions are firmly grounded in technology. Its historical status of monopoly led the company to impose mass-market policies during the 1950s and 1960s, which were relatively risk-free.

Its former status does not leave the company with the option to propose to its clients offers other than the supply of electric energy: therefore, there is no "EDF-stamped product" as such. In reality, since the end of the last century, the company, anxious to develop the use of electricity in better conditions, has been working in partnership with certain industries to be in a position to offer electrical products, especially those that help emphasize the image of an invisible energy.

However, the new opening up of the energy market for the more energy-guzzling professionals began in France in 2000, and was carried forward in July 2004 with a body of professional clients, to finally end with the complete opening up of the energy market in 2007 as predicted. Therefore, in the face of this new socio-economic order, EDF is not alone anymore and must brace itself to face up to the competition.

Research & Development must take place naturally in order to play a major role in the context, even though it has always played a crucial role in the company, especially since the advancement of nuclear energy. Initially turned towards technology, it has ever since largely enriched itself. Indeed, the need to resort to disciplines that are centered more on the study of man and his environment has been felt. It is about ergonomics, sociology or semiology yet again, or actually, design. Sociology was first used in the study of opinion (about the nuclear image) then ergonomics was used in the conception of interfaces in industrial systems. It was only towards the end of the 1990s that they progressively (and only partially) turned towards the conception of mass-market products and offers. They are now participating in the development of innovative projects.

However, it appears important for us to specify that the role of the R&D remains relatively upstream in the phasing of an innovation process. Only occasionally does it intervene in the real development of an offer. In fact, this work is generally carried out by another structure of the enterprise or with external support.

With regard to its role related to innovation, the diversity of its existing competences, and also the reforms in policies that the company makes (the opening up of its market), the EDF's R&D division wanted to implement an innovation support device (CREATEAM®) in order to better arm itself to face the new challenges.

This device is aimed at facilitating the proposition and evaluation of the concepts of offers of tomorrow and the near future in support of the commercial domain of the EDF Group. Its second role is to become a field of methodological reflection on innovation, to capitalize on the works (and not only the ideas) resulting from innovation projects and, to encourage inter-project interactions. Initially it had to build itself up by proving itself. Finally, this dual system responds perfectly to the ambiguity built around the term "innovation" as described by [DEL 00], as it "covers the process and its results, both at the same time".

12.3. The C4 method

12.3.1. *Overview of the method*

The team CREATEAM® uses several approaches. The one that we present here was implemented for a period of four years. It was used more or less for its integral nature vis-à-vis different projects (about 10).

The method is composed of four steps: the phase of comprehension and reformulation of the demand (comprehension), the phase of creativity (creation), the phase of contextualization of the concepts (contextualization) and finally, the phase of evaluation of these concepts (confrontation).

Even though it is divided into four steps, the process is based throughout on a single group project, which signifies that the actors not only position themselves in terms of their own competences but also contribute to the design by putting forward new concepts. This choice of "double cap" must ensure a better collaboration between the participants of the project. It can also guarantee the integration of the different forms of occupational knowledge, carrying it throughout the project.

12.3.2. *Phase 1: comprehension of demand*

That which is called "demand" refers to, in reality, a good number of forms; it could be a general issue for innovation research on services for particular clients, or still, a question concerning domestic electrical equipment. In all cases, the scanning of the context and the reformulation of the demand are necessary.

The objectives at this stage are many. The process begins with the phase of exploration, analysis and comprehension of the demand. These phases are based on the upstream studies. They concern the technical domain (the state of the art of the existing and the possible), the domain of marketing, and gain sustenance from the studies of human and social sciences *in situ* in order to better define the real need (ergonomic, sociological, anthropological intervention, etc.). This is the translation of the problem in the universe of the designer. A meeting of sharing completes this preparatory work. Each competence presents its method of intervention (the tools used) and the conclusions of its studies (the problems and the needs met) and illustrates the body of work by means of specified and real examples. The objective of this meeting is also to equip the team with a Common Referential of Innovation (in the words of Leborgne [LEB 01]) while extracting the vocabulary of the field studies regarding use that will complete the knowledge of each one and also present the concepts of the discipline involved in the study, so that each one can appropriate them. The system of "rules and examples" that belongs to the Common Referential is therefore presented. At the same time, this exercise must prepare meetings of creativity, since each actor *"liberates"* himself from his knowledge while presenting it to others. Therefore, he will not seek to present them forcibly during the creativity session, something that could harm its good performance: this is also, what we call the phase of purge. Finally, this is the first step of the construction of inter-comprehension. From these studies and each one's own knowledge, the group builds up, through micro reiterations, the problems and needs that it will transform into entry points of creativity.

12.3.3. *Phase 2: creation*

From the definition of the real needs and expectations, a creativity session is carried out with the project group in its totality (fully respecting that the ideal number of participants for this type of exercise is 7–10 persons). The work session relies on the tools of classic creativity (brainstorming, matrix of discovery, etc.). These tools are selected in view of the objectives and the upstream reports. After having produced the maximum of ideas, a first filter is applied by the group. This filter is exclusively conceived for each project. Once the ideas are selected, it is necessary to further formalize them (at this level, they are still summarized, generally in only a few words) in order to express their common vision to the group. These new micro iterations come up to construct the concept. To facilitate the exploration of the concept, we propose to rely on a tool.

Figure 12.2. *Extract of a form of IdéeFix*

This concerns an idea sheet to which we added the relative fields at the time of use. Even though the use has to be at the heart of the discussions, the technical aspects and marketing must also be considered. This idea sheet, which we named "IdéeFix", is the first Intermediate Object of our reasoning design approach. This is a federator-element in its construction and a point of reference for the follow up of the project.

We shall qualify this idea sheet of situational concept as *concept* because the definition of the offer is not advanced enough to concretize it in material form (in any case, in a complete manner) and *situational* because a concept relates to a particular reality according to the situations. For this we propose to use an analogy with the help of the data analysis tool, "WWWWHW" notably used in quality methods. It demands merely to inquire on a situation by asking oneself the following questions: Who? What? Where? When? How? Why? Thus, each situational concept is described by:

– who? The target of the offer;

– what? The form and appearance of the offer AND its principal function;

– where? The place of procurement of the offer AND/OR the place where it is put to use;

– when? The moment of use;

– how? The code of use;

– why? The benefit for the user.

The situation thus defined must offer a single point of view to the whole project group, the evaluators and other intervening actors. The adopted language is that of the usage (we must remember that it is defined as a Common Referential of Innovation). What remains, is to specify how to define the components of the situational concept. More than "speaking about the same thing", the situational concept must facilitate a more explicit rendering.

12.3.4. *Phase 3: contextualization*

The IdéeFix form overlaps the world of the designer and the real world. It consists of elements of two worlds at the same time (for instance, technical information meant for the world of designers and information relating to use meant for that of the users). To be able to evaluate this idea, we saw that it would be ideally necessary to move into the real world.

However, the reality can actually be achieved only with help of entirely developed concepts, which is not by our choice. Therefore, we must give a vision of the reality, reproducing the context of use (earlier on in this chapter, we insisted on the necessity of having to do a specific study of it) in a realistic manner, sinking in the concept therein: the virtual means constitute the privileged tools to bring this about. One should come back to the notion of "realism": we qualify in the "realist" a representation in which the user, who is subject to evaluation, goes on to find a universe and familiar situations (particularly the Social Scheme of Use).

For this, based on the elements defined in the idea sheet, the project group will render the situational concept dynamic. The group will develop a "scenario" of the use of this concept. It will simulate its utilization by grounding itself in different modes of possible virtual representations (cartoon strips, short films, animations, photo stories, etc.).

This "dynamic situational concept" can be presented to different interlocutors subject to evaluation.

Virtual representations will be the mean to transform the static concept into a dynamic concept.

In the first place, this representation is aimed at evaluating proposed concepts through the potential users. However, this mode of representation must also allow the transfer of this concept to other universes such as that of the company's decision-makers or still, to the potential partners.

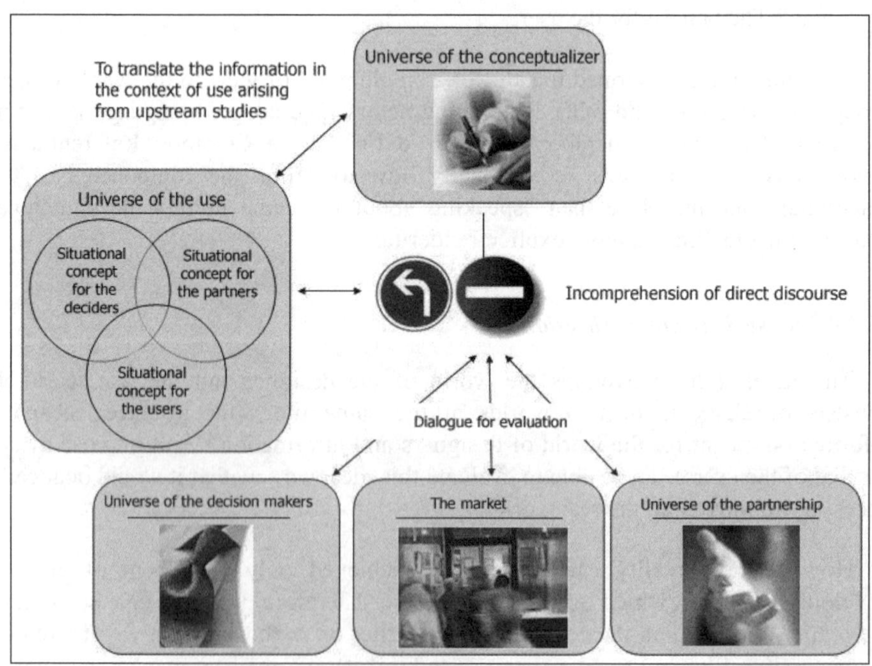

Figure 12.3.

12.3.5. *Phase 4: confrontation*

In order to recover a maximum of reactions and verbalizations regarding the acceptance of the concept, the dynamic situational concept is projected to the potential users. Group effect can play a non-negligible role; therefore, it is preferable to organize the projections and to carry out the evaluation with a group of about 10 people.

This mode of evaluation, although similar to the principle of the tools used in marketing such as "concept text" or "concept screening", brings in an additional dimension linked to the working of a realistic situation, and allows the user to identify himself better with the situation and make the best of it.

This evaluation either makes it possible to validate a concept or it does not, as we shall illustrate later. A concept can also be redefined best by taking into account the statements resulting from the verbalization of the users.

12.3.6. *Modeling of the process*

We propose to formalize this process under the formalism that evokes the TRIZ theory [CAV 97], while completely conserving the specificity of the process (see Figure 12.4). In fact, our design process seeks to be maximally based on our reference domain: the everyday reality drawing on the universe of conception as soon as possible.

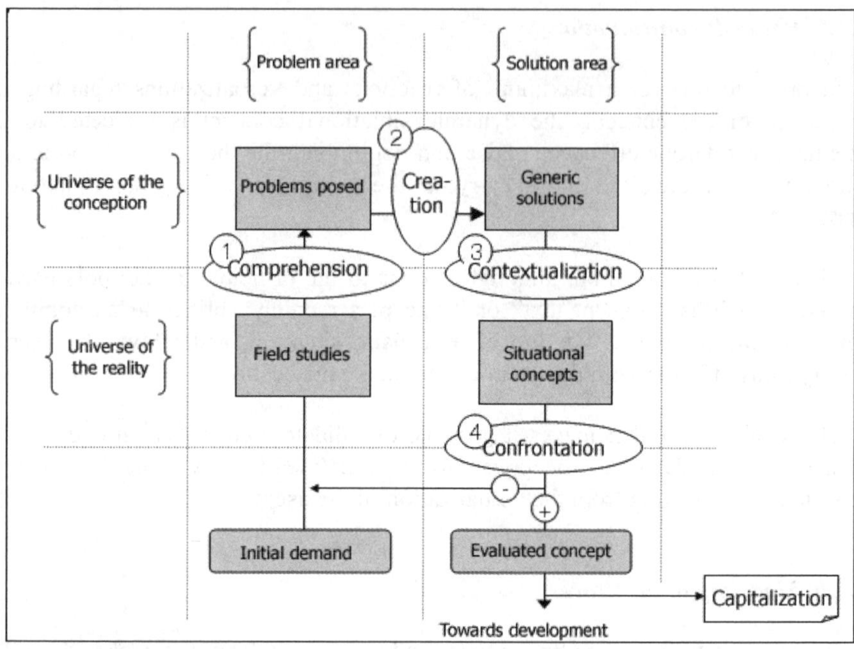

Figure 12.4. *Schematic representation of the support model of our process of validation according to [MAXANT 04]*

12.4. Diverse experimentations of the process

12.4.1. *The "New Offers" project: contribution of the dynamic concept in comparison to the static concept*

EDF's real offer in the residential sector proposes Vivrélec for the new one and a renovation for the existing one. The "New Offers" project enters into the framework of prospective reflections, brought about by marketing management, on the evolution of offers for a period of 3 to 5 years. It concerns new and existing housing and enters into a scenario of evolution, indeed a complete break away from existing offers. The new offers should have a strong component – "thermal comfort".

At the beginning of our intervention, the project had already started on a large scale. The Absys society had already carried out an important work of technical and societal economic forecasting. Our intervention had aimed at generating ideas of offers, more particularly targeting the aged and those with a reduced mobility, starting from scenarios[1] to contextualize them and finally putting them face to face with the potential users. Six concepts were presented to the aged or disabled with a specific layout: firstly, in form of a succinct descriptive card, then in the form of an image of the single product and finally, through cinematic animations of about 30 seconds each. In the end, the order of presentation of the concepts was changed, following the sessions.

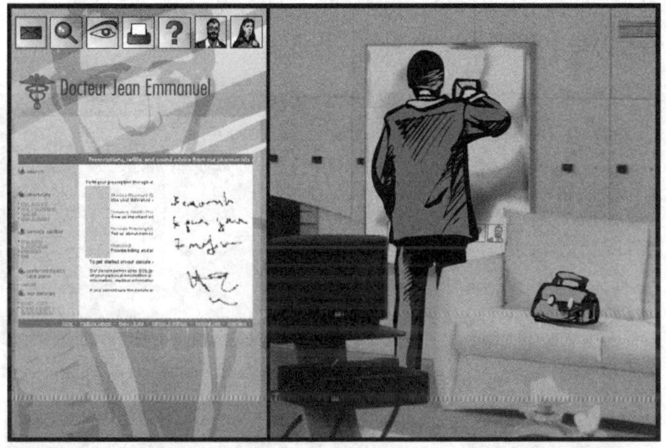

Figure 12.5. *Image extract of a dynamic situational concept (animation) on the theme of domiciliary maintenance*

Of the six concepts, four are considered as accepted or acceptable, one was rejected in the context of a situation of use, demonstrated by the projection of the animation, and another one was rejected by the assimilation of an already existing product (functional and formal assimilation).

---

1 The scenario recovers two aspects: in upstream, it is about a description of live situations (in a large sense, because it is not forcibly about the offers to the particulars) in which the offer of the product/service "runs" itself. In downstream, it is about to create a scene chosen for the presentation: we shall discuss, to raise the ambiguity, the scenario for the upstream aspect and the scenario of situational concept for the downstream aspect.

12.4.2. *Collaboration with the Studio Créatif of France Télécom: towards an evaluation of service*

This project is actually a collaboration[2] between Studio Créatif[3] of France Télécom and CREATEAM® of the EDF's R&D wing. Both organizations have common objectives, namely, proposing concepts of new offers. In relation to the whole, the factors at work are of the same spirit, as both rely on a process based on use. However, differences in the methodological approaches remain; probably due to each company's own strategic approach. The works of Studio Créatif were undoubtedly, an important inspirational source for our works.

Figure 12.6.

The objective of this project was the methodological collaboration on the innovative upstream process, but it appears also important for us to show the interest of a creative work between two companies positioned in a relatively similar context. This project is not one of a kind, at least from an industrial point of view, since there is no real demand. However, an industrial objective has been fixed: to auto-generate

---

2 Collaboration under the form of "contract of research".
3 Studio Créatif of FT R&D has been around since 1999: http://p-www.rd.francetelecom.fr/studio-creatif; we can also find a description at the following address: http://www.fing.org/index.php?num =1918,3,9,11.

the problems and to find solutions to these problems. The resulting concepts could either be included in a future project or would initiate a new project.

Moreover, in a real socio-economic context, the role of the service-oriented projects are gaining more and more importance. It is for this reason that it seemed to interest us, from the methodological point of view, to work on the means of representation concerning services (inherently immaterial although material media generally accompany the service offer). This representation besides resumes the principle of upstream evaluation of an innovative product since at this level, the product is also immaterial.

Faced with these objectives, we applied a process that was essentially unusual. A group made up of ergonomists, sociologists, semiologists and designers defined certain themes of work. A single session of creativity generated hundreds of ideas of products, offers and services. Then we chose about 15 ideas which we assorted into two large domains: "the communicating house" – a presentation of eight concepts favoring the domestic life, and "Domino" – offers of services making it possible to optimize the resources of the energy flux and communication. The group that was of a multidisciplinary character carried out a rather exhaustive work of scripting. In order to give life to the concept of "Domino" service, we chose the narrative style and included witnesses. The basic context was a television program with an anchor person who would string together the interviews of the three actors: a user, a representative of the local people and a distributor of services, all concerned with either the contribution of services or their diffusion.

The resultant video of these works was shown to a group of decision-makers of France Télécom's and EDF's R&D divisions. One of the points that we will retain from this projection is the credibility accorded to the concepts in spite of their succinct description. In the majority of cases, the reactions considered the services as if they were already existent, and the EDF/FT association, brought about by the proposition of services, appeared quasi-natural.

12.5. Some new tools to facilitate the collaboration and the contextualization; towards an instrumentation of the process: "IdéoFil" and "StoryoFil"

Creation and contextualization are two phases that can easily be supported by the collaborative tools. They should ensure a more rapid appropriation of the process and a far greater collaboration between the different actors. This instrumentation consists of two key points of the process that are collaborative constructions of static (form IdéeFix) and dynamic (animation) situational concepts. The need and the specifications of these material media have been described in terms of the use while using this process through application projects. After having formalized the functional characteristics of these tools, we explored the existing one. In the case of IdéoFil, we found that the existing products could not live up to our expectations, so we decided to develop our own application. In case of StoryoFil, several products partially do meet our expectations. Some adaptations are proposed.

12.5.1. *IdéoFil*

IdéoFil® is a collaborative information system. It essentially allows the sharing and capitalization of information (actually, knowledge) that the projects contain, to formalize and to capitalize the idea sheets of the projects integrating the process of CREATEAM®. IdéoFil claims to be an information systems support of upstream innovation. Two great changes have influenced this domain. The first one concerns the nature of exchange: we have gone from the domain of a product to that of information. The exchanges became dematerialized. The second change is only a consequence of the first; since the end of the 1990s, computers came to accelerate these exchanges. The first great software in this field, Lotus Notes, appeared 10 years earlier; we then talk about CACT/CACW (Computer Aided Collaborative Task/Work). Serge Levan [LEV 99] draws up a list of four categories of the work tool CACT:

– tools of basic communication (to help circulate information between two colleagues): email, chat, electronic white board, video conferencing and Instant Messaging;

– work sharing tools (working on the same document): the sharing of the application, divided editing, allied forums and tools;

– tools of access to knowledge or "knowledge management": digital libraries, "peer to peer" tools, portals, the mapping of skills, telephone directories, mailing lists, FAQs, etc.;

– tools of workflow (allowing the coordination of a team): tools for synchronization, task management tools, other shared tools.

What remains is that the CACT tools are not always as profitable as desired in terms of time. Most of them is very academic in nature, whose purpose is often different from that of the commercial world (the caricature of the two worlds could be thus described: a quick job for the commercial world/a good job for the academic world). The CACT tools are vital, not always for reasons of confidentiality (the users agree to share their finalized documents but do not necessarily agree to reveal their agenda). The CACT tools must remain simple and user-friendly and not slow down the progress of the project (especially when one is involved in innovation projects where creativity has a preponderant place).

IdéoFil must allow the construction, the finalization, and the collection of the IdéeFix sheets. Coordination (as in the management of the agenda, task management, etc.) is an essential means, but is something that can be accessed by other means already present in the majority of companies (Lotus Notes, Microsoft Outlook, etc.). It does not appear necessary to us to weigh down the tool with such functionalities.

The objective is, as reflected in the remainder of our work, to encourage the exchanges between the actors. However, these exchanges are not only verbalizations but also reactions that are recorded in IdéoFil. These exchanges can now pass from the synchronous mode (at the same moment and the same place) to the distributed asynchronous mode (at different moments, different places) [DEF 02], favoring the integration of experts in the construction and the evaluation of an innovative concept. The movement of the offers is besides facilitated, as this tool allows a communication (thus a transfer) of ideas in a way more interactive than that of the documents.

IdéoFil is under development and must be tested in its current version on intra- and inter-departmental projects. A second version has been envisaged: its objective would then be to become one of the collaborative tools in the commercial field and that of R&D, and could be used as a link to the Trade Branch which is in a better position to develop these client-bound offers.

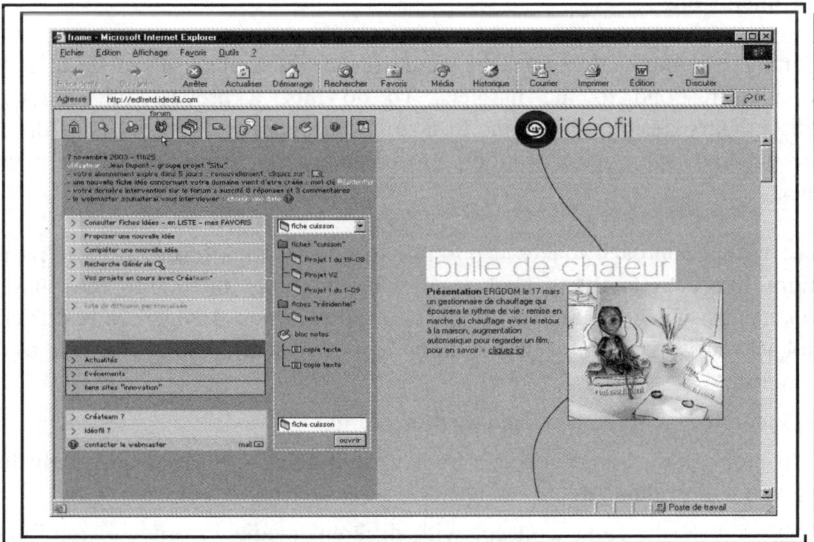

Figure 12.7. *The IdéoFil homepage*

12.5.2. *StoryoFil*

StoryoFil® presents itself as an aid to creativity and a kind of utility that is complementary to Idéofil. It facilitates formalization, in the form of scenarios of concepts resulting from creativity, while being liberated from the more restraining factor related to the realization of storyboards, to have knowledge of drawing/sketching. We must recall that the participants do not come from the creative professions, much less graphic creation. Its objective is thus to facilitate the generation of dynamic situational concepts.

For that, we take the help of a professional tool designed for the creation of *storyboard* largely used in the domain of cinema. There are several softwares of which there is one that particularly meets our expectations: QuickStoryboard®. With the help of a library of images, it makes it possible to easily place a decor (an environment of use), one (many) character(s) (users) and objects; everyone can appropriate the tool. The characters and the objects are easily adjustable (rotations on their own axes) and dimensional. To each photograph corresponds a descriptive text. Therefore, this is a powerful tool of construction of context of usage of the imagined concept.

12.6. Conclusions

Commercial innovation must be based on the idea of "use" to be in a better position to respond to the markets. To do this, a process of collaborative interdisciplinary design allows one to take into account the real needs of the users and to generate appropriate solutions, to contextualize these contexts in a virtual manner.

The virtual approach makes it possible to give various forms to these proposal solutions likely to facilitate evaluation without going much into the phases of developments, which are always expensive, and in which it is generally difficult to move backwards. Three objectives can be identified while taking recourse to these various forms:

– to propose the same concept in various forms;

– to propose the same concept from various points of view, according to the actors who will evaluate it;

– to propose the same concept with different media in order to reinforce its comprehension.

Chapter 13

Creativity World

13.1. Introduction

In this chapter we shall develop the approach to the most fundamental phenomena of creativity in relation to its environment and human relations. We shall see how creativity is an internal alchemy originating from fundamental sciences, social sciences, techniques and concepts. We shall accompany you in this alchemy on this simple process of the law of creativity connected to its environments (matter, energies, solutions etc.), and in interaction with the human being.

Through his experiments and research, man tries to artificially create the conditions of creativity particularly in the commercial milieu, and this, he does in a more or less skilful manner. He adds up the conditions, laying them one beside the other so that they are as close as possible to creativity but without creating real links between its ingredients – which can only give average results. He tries to modify the external environmental conditions (exogenous conditions), whereas the solution lies in the modification of the internal environment (that which is endogenous), which means searching within himself. In this manner we propose to examine how creative processes are generated.

13.2. Reflections on creativity

In common parlance "creativity" implies the capacity to produce artistic works such as painting, sculpture etc. In that case, creativity is the outcome of manual work by which a sculpture or a painting is created as a form, hitherto non-existent,

---

Chapter written by Michel SINTES.

instantly becoming something unique in the hands of the artist. The concept of creativity underlines the realization of something singular, novel and original which strives towards a certain beauty, indeed, excellence.

In the industrial arena, creativity is assimilated in the techniques and methods of functioning that allow one to be as innovative as possible in the creation of new products.

This creativity, where does it come from? Does it ensue from our determination? From a pursuit of an objective? As a result of chance? Or from something beyond us? Why are men always searching for this creativity? Why do they always want to innovate, to do better, to bring forth new ideas?

Creativity refers more to the functional processes of a human being than to a vague method. In fact, this process determines the manner in which the individual relates to his environment. What is it that actually makes us think that some methods, techniques or know-how will be appropriate and will favor creativity? Each one could experiment that it is important to have specific conditions that facilitate the birth of creativity, that it is not a matter of chance but, on the contrary, a product of the capacity of our brain to spontaneously create new ideas in concordance with our objectives.

We perceive that a stress-free environment, an aura of confidence, research work in the form of games and a positive temperament are the conditions that nurture the spark of genius. It is when all our intellectual, emotional and behavioral potentials are exploited to the maximum that we accomplish exceptional things.

13.3. A human concept

The notion of "concept" is related to the human model where man is in possession of various cycles of functioning that are identical to the process of creativity. For example, let us consider the functioning of our brain that comprises four determining sectors: the left-brain, the right brain, the limbic brain, and the reptilian brain. Each of these is responsible for the accomplishment of one part of the process of creativity which begins with the occurrence of an idea or an intention followed by the conceptualization of the same idea, and motivation or the emotional feeling, and finally the process of putting it into action to enable the concrete realization of the idea. If we work out each of these steps in an intelligent manner, we can succeed in getting exceptional results.

For instance, we have been responsible for having used this approach within the framework of a company project. Technical problems persisted in the working of a particular product. We formed a team of about 10 engineers and worked out an action plan that would spread over several days, which we executed following the logic of the process of creativity with exercises of animation conducted in a positive environment. As a result, after a few days, innovative ideas enabled us to overcome the problems.

This approach can also be used, for example, in team coaching, individual coaching, and health management. We present in Figure 13.1 the five steps of creativity, which are: Idea, Thought, Emotion, Behavior and Result. In the following paragraphs we shall develop each of these steps.

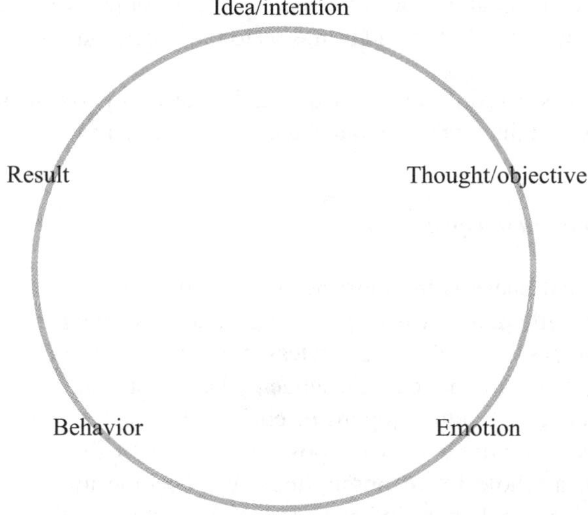

Figure 13.1. *Cycle of creativity*

13.3.1. *Idea/intention*

This logic of the human model is the representation of the functioning of our brain. The right hemisphere which is the seat of intuition and our feminine traits is the generator of ideas, intentions and also of our creative capacity. It is from this part that our new ideas and intentions emanate in relation to our life. They are transformed into the need or desire to realize.

It is our capacity to play, to give free expression to whatever comes, that favors the birth of new ideas. Knowing the need to take time off when in a regressive state vis-à-vis an objective, knowing how to clear our head and approach situations in a relaxed state puts us in a better position to open up and listen.

13.3.2. *Thought/objective*

Our left brain, responsible for our masculine aspect, then takes over the baton in the relay and starts working on the creation of the objective, a stage of conceptualization that makes it possible to go into the details of defining things more precisely. Our entire intellectual potential gets activated, taking into account our competence, our history and our memory that accumulates our good and bad experiences. It is the sum of all these factors that enables our left-brain to steer us in a particular direction rather than the other. Our real-life experiences and the knowledge acquired are those that determine our capacity to conceptualize. This reflection enables us to identify the objectives and the choices in our thoughts. The clarity of our choices allows a projected action, an objective or a need, to be the launch pad of the ensuing step and brings us to the execution stage.

The process is certainly not so sequential. The two parts of our brain function simultaneously and the total potential is used at the same time.

13.3.3. *The emotional aspect*

The emotional phase is the most neglected aspect in the technique of creativity and in our daily life, and yet is of great importance. Nothing is possible without it. This phase governs our feelings and releases emotional reactions to our own ideas. Has it ever happened to you that you suddenly have a brilliant idea? At that moment, how would you feel? Excited, joyous or euphoric! This is normally a very pleasant moment where everything appears possible as this really opens doors; we are presented with a whole lot of possibilities. We become aware of positive energy which leads to the realization of our project, as if the emotional aspect made the engine start and helped us put the body in motion [UES 00] and move in the direction of concretization.

Our emotional aspect also reacts to our thoughts. The way in which we visualize life or our everyday reality, positive or negative, generates various emotional states. To be aware of this established fact determines the taking over of the responsibility of managing our everyday life. Indeed it is our capacity to manage our different states of being that determines the quality of our results. We have the potential to do it and it is our sole responsibility.

Our behaviors and our results are equally influential. We can be happy or proud of an action or on the contrary be depressed by a failure or at any given time of our life, feel ashamed of our past behavior in particular circumstances, something that is sincerely regretted. Each thought or personal act is colored by an emotional reaction that enables us to gauge the feeling of the situation where we are pitted against our own selves at a given moment. It is a fabulous capacity that we have, which allows us to modify a situation created by ourselves, by placing us at the core of our values and our inner truth.

The emotional element reacts especially to our environment, among others, to the relations which we maintain in our professional life, our emotional life or our leisure. The happenings in the social scene and the unforeseen situations of everyday life have a significant impact on our reality and trigger emotional reactions that reveal fear, anger or joy. These reactions are the interpretation of our personal position with respect to our past, our past experiences that have been more or less successful and our projections of our future aspirations in terms of the needs or the projects to be accomplished. Actually, our emotional aspect is a magnificent indicator, which positions us vis-à-vis an event. It is a sort of a compass that indicates what is beneficial or harmful for us, and enables us to be conscious and to avoid aberrations in life.

The emotional aspect enables us to make choices, in choosing that which is favorable for us and is in our interest and also respects others' interest; it helps us take the right decisions while making professional choices or in the execution of a project which one initiates or in which one participates. Without this potential we would be lost in life, without any means to get our bearings. Without it, our important decisions would be the fruit of a single mental reflection, cold, which could leave us in conflicting situations or entail bitter failures.

13.3.4. *Behavior*

Our motivations in all areas of life, professional as much as emotional, in our leisure or our daily interactions, are conditioned by our emotional state. This state of mind controls our motivation every single moment. It releases energy which we are able to engage in a given action. This state has a direct consequence on our behavior and our attitudes, and influences the way in which we carry out our actions.

The gestures that we use to carry out an action also depend on our practical experiments, our competences and all the experimental knowledge accumulated during our lifetime. The quality of our intervention will determine the quality of the result.

This stage is crucial, because it is the hinge connecting the emotional aspect and the result; it is a stage of realization, of concrete application of an idea or a product. Man shapes matter in order to get a result; this is a transformational stage.

13.3.5. *Result*

What is a result? Product or object, it is the finality of an objective. Generally, it is concrete and measurable. It is the result of a whole process that starts with the birth of a need and ends with the satisfaction of the same. This product is the result of the total sum of the stages from the appearance of the idea up to its concretization.

All the stages of this process constitute the cycle of global creativity of the human being. Why global? Because each one of these stages also contains a cycle of creativity, single and specific to each level, called a mini-cycle and described below.

By way of respecting the unfolding of this process, in the awareness of each one of its stages and its own logic in terms of the result obtained, one can attain success or failure. The totality of this unfolding is measured in terms of quality by the indices of success that make it possible to approach creativity with the concept of excellence.

If one compares this process to a logical and uninterrupted chain, complete in itself, the solidity of this chain will correspond to the value of the weakest link. Clearly, it is the value of the weak link which gives the value of the result and thus of the whole. Therefore, it is important to consider all the links with the same interest. Independent of this logic, it is significant to notice that each one of these stages is influenced by the others. Let us consider the example of the result. It is obvious that the value of the idea, manner of conceptualizing, the emotional state which controls our motivation, or our behaviors and actions, directly influence the chances of success of our result. But this operates in all directions. A bad result will influence our thoughts and the vision which we have of ourselves, or will put us in a negative emotional state, which will in turn have a negative effect on our next result.

The unit is systemic where each element is fundamental to the other, and all the stages are of equal importance. Be it our intellectual thought or our emotional reactions, our capacity to let go the hold or the judiciousness of our behavior all complement in full solidarity to the richness of our success, as if a cobweb were connecting all the points of this schematic diagram, and that the perfection of the weave were guiding us towards the performance.

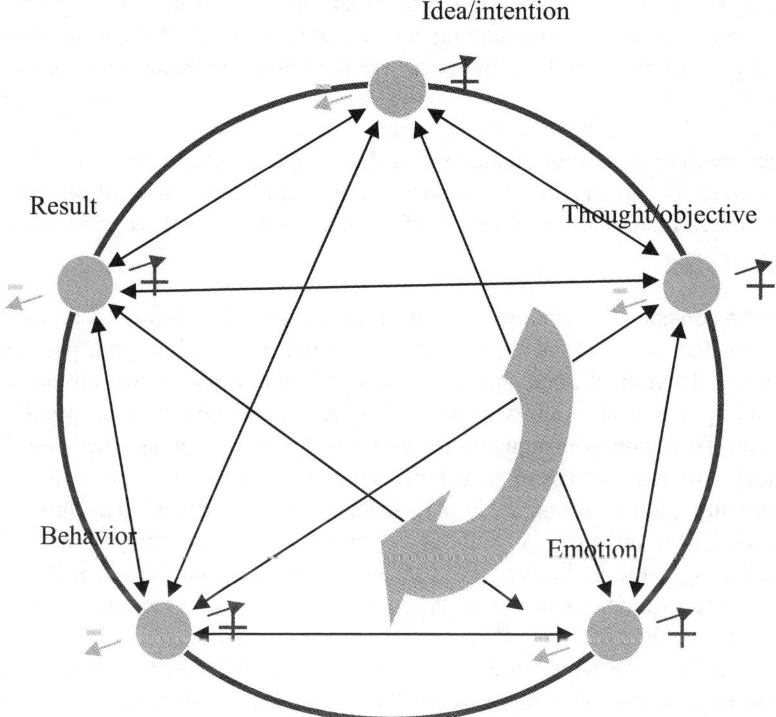

Figure 13.2. *The web of performance*

13.3.6. *Mini-cycle of creativity*

Each one of these mini-cycles represents a process of creativity in itself. That is why it is significant to understand the different stages, in order to get a grasp of the points to be improved upon and evolved through a process of transformation or change.

The cycle of creativity is a process of transformation between a conscious part comprising an intention and another part, an unconscious one, which is the primary material before having to undergo the transformation.

A sculptor has the intention or the idea of producing a masterpiece. The matter, the stone for example, which is the unconscious part will be cut with the intention of changing its state from being a simple piece of stone to that of a valuable article. The act of transforming, considered as work, will bring great additional value to the original object. The greater the additional value, indeed, the more exceptional the value, the more one will speak of a great creativity. This realization appears to be a unique and new object, not having an antecedent, being born as a result of a particular intention carried out by the work of a conscious being on an unconscious matter.

The creativity is always characterized by its unique side, which is precisely what characterizes its performance. Otherwise one would be left in a situation where there is none other than a reconsideration of form or, a more or less good copy of the reference object.

Let us consider the emotional cycle: imagine that, in relation to an unforeseen event, you feel sad. This sadness is an energy that will belittle you, perhaps make you recoil into yourself, and shut you down. It is the impact of the environment on your body [LAR 90]; it refers to an exchange, a communication between you and the event. To regain your equanimity you will have to express what you feel and thus spell out the sadness. The act of verbal expression will have the effect of making you regain your energy. It is as if the energy of sadness were thrown out of your body. Up to that point, it is about a communication between you and the event. The following stage is that you need to understand why you felt sad. Perhaps you found yourself ridiculous in this situation, perhaps you were lacking the confidence to react in the right manner. If that is the case you can conclude that working on your self confidence would prevent the problem from recurring. In fact, from the next similar situation onwards, you will be the master of your reactions and will be happy for this success. Well, this feeling of joy will end the transformation of this event and, at that point, we can say that the process of creativity has come one full circle.

It is very important to understand this process, because to feel joy is to give energy to our life, it is the engine of creativity and at the same time its success too.

13.3.7. *The scale of values*

If we consider the stages of the process of creativity, which correspond to the progression in our brain and also to our process of realization right up to its concrete implementation, we would realize that for each of them can be applied a value scale in terms of quality or the level of performance.

What makes an idea be declared brilliant? It is its singular character, the fact that it is novel and that it brings to the fore an innovative solution. This can be measured by evaluating the consequences of its application in the company: it would have obvious and positive repercussions on our evolution; it would besides be a matter of reflection or better existence for man. Perhaps this would have an exceptional dimension in terms of artistic or scientific realization. Evidently, it would fill in a void, a vacant space, because it would be much awaited and would fit in exactly into the puzzle of human society or life. It would be coherent in its place; it would give meaning to the whole of human activity and would partake in its evolution [LEN 02].

We are now going to show how the different scales of values work at each stage in the cycle of creativity.

A thought or the conceptualization of a thought can be evaluated by the complexity or the richness of the engineering that it comprises, by the quality of the knowledge applied that leads us to this concept or is itself the finality of a philosophical or scientific research. It can be extraordinary by its simplicity, and take part in the development of a tempting project. It makes it possible to define and lay down a conceptualized objective, coherent and clear in the human mind, that which can be the starting point of the act of realization. In the context of a company this corresponds to a clear and precise objective of which the majority of parameters are managed.

The emotional aspect is an index that places us in the context of feelings in relation to a situation, an environment or our relations with another person or a group of people, society for example. It can give us the complexion of a relationship, and indicate if we can measure up to what is happening. In the opposite case, we have the possibility of building up a more positive situation, by conflict management for instance. Our emotional state conditions the totality of our being (remember the drawing of the cobweb), and thus assumes the role of the management of our energy. It is the fawcet of our motivation and enthusiasm that we can engage in a project. This scale of values ranges from fear to joy while passing through the stages of anger, sadness, gaiety, etc.

The most positive emotion is joy, and it is in this state of joy that we revel in the maximum of our potential and energy. It is enough to compare a child at play with a person who takes the subway to work everyday to acknowledge the vast difference between these two situations. When children play and laugh, they never feel the fatigue. The difference between these two situations is not due to age, but mainly due to the emotional dynamics. So, to say that it should work while playing is something that we cheerfully overstep.

The right gesture at the right moment and an action adequate enough to obtain the best results is what we associate with best behavior. It is the entire set-up of our professional experience and the others which constitute our potential for action, for the right competence, and those that ensure that the actions carried out bear maximum fruit according to the priorities one gave oneself.

Our attitudes in the action are also of consequence, particularly in the exchange and communication with the partners of the action. They influence the manner of transforming situations and the handling of problems and either facilitate their resolution or not. Being a good listener, respecting different points of view and showing a positive attitude help to forge a healthy relationship. The uprightness of our behavior translates the value of our professional or artistic competence.

The results of the whole are measured using precise parameters that one sets according to the sphere of activity and desired quality, which accommodate the evaluation of performance. When the expected result is below one's expectations, it is necessary to reconsider all our steps, analyze them, identify the lapses and improve upon them. It is a question of taking responsibility for our actions and bringing about the necessary changes.

If the result is good, it is a success, which is part of the fulfillment of the need, and generates happiness; it is an achievement that makes way for the final bouquet, the fireworks and the celebration. This can be given a concrete form in the company by means of a patent or by bringing forth an exceptional product. The evaluation of the success is arbitrary; it depends on the context, the stake and the finality of the result.

When the result is strong, it will be followed by a period of recession, of retreat and even that of bereavement of the successful situation. It is a period of slackening of all that was brought to play in order to succeed. A void must set in, the situation must be accepted. It is this empty space in the biological cycle that fosters the birth of another idea or intention.

13.4. The state to being one with the environment

Man is governed by the principle of creativity, and in order for creativity to exist, there has to be an intention to create, which belongs to man, and an environment to be transformed. Without this, nothing exists and life stops.

The life of an individual is meaningful when they interact with their environment; it is the crossroads of exchanges that concerns all relationships [GIN 87], particularly that of a couple, within the home and the work place. Each time he is in a situation of communication, he responds, positions himself and interacts with the group; it is a manifestation of life. And without realizing it, in an unconscious way, he unfurls the process of creativity in a more or less happy manner according to how well disposed he is to give prominence to himself and be aware of what is occurring in terms of content and process.

Man is confronted with the environment in two ways. The first is more voluntary, i.e. when he makes his choices, decides on such and such an action consciously. The decisions he takes and the manner in which he conducts himself have a positive of negative impact on his environment. His actions follow a process that is more or less veered off from that of creativity, because each action is an act of creativity, man is thus both the performer and the administrator of his life.

It often happens that certain situations, meetings or events occur due to what is known as "chance" where apparently, the individual is neither the seeker nor the initiator. In this case, man reacts to the events, but it is the same law of creativity which is at work, at the least it would be so desired. The difference is that the interaction with the environment intervenes in its unconscious field and brings to light more hidden parameters, which unveil unanticipated actions and potentials. It is in the unconscious that man draws upon unknown material to build up his creativity and consequently his future.

In this context, the environment is our inexhaustible source of experiences and possibilities of creation. If we consider each meeting that we come across with due importance, it would turn out to be a gold mine of evolution, of new prospects. It is necessary to be firmly footed in the conscious in order not to get lost in this environment, and thus to be able to remain oneself, to maintain one's integrity and free will, to enable oneself to carry on with confidence and to surf along on the waves of life by expressing the joy that entails.

Our environment is really the unconscious raw material, which only requires to be transformed. However, in creativity, there is not only the creator who remains constant and a matter that transforms itself: both evolve, get enriched and acquire a new character. It is a case of the relationship between two people and in particular that of a couple, where, each day a reality is created which evolves with time and where each partner changes with time. When the relation becomes difficult, it is precisely the quality of conflict management that puts the process of creativity back in place and makes an all-win solution work, which is neither the single-person-oriented solution nor a compromise which gives rise to two frustrated individuals.

The characteristic of a creative solution is certainly the fact of being new, of coming out of the unknown and especially of being able to satisfy all the partners involved. For the inventor and for the beneficiary or client, it engenders immense joy and a sense of great satisfaction. The phenomenon can also work in another direction, i.e. a profound joy can give rise to a brilliant idea. This is what is happening today in the case of companies conscious of this process. These companies are sending their engineers to creative orientation programs where games occupy a significant place. But a lot of parameters are still forgotten, which makes me say that one is far from having finished creating or inventing new things; it is really one of our capacities that knows no bounds.

13.5. The age of networks

Today, in the age of the networks, man sets up a communication system on the planet, such as the Internet and all other information processing systems, with an aim of facilitate exchanges. Why? What is important to understand is that man is in possession of an exceptional richness. His potential is boundless but he is not completely aware of it. And so he projects and builds, outside himself, ideas or scientific performances that are actually lodged within him. It is for this reason that his environment calls out to the process of creativity, to help all his hidden richness emerge out of the unconscious. By creating all these networks, he creates in his environment the conditions that facilitate the convergence of more and more diversified relations and new ideas. It widens its cultural environment to avoid national saturations or fallacies, because, today, the whole planet has become our research laboratory, and more and more different opinions are necessary for us to advance further in evolution. By creating this mine of opportunities, man gives himself materials to make it possible for innovation to generate solutions that will enable us to find solutions to the major problems of the world.

More than the technological networks, it is the human networks, as we have seen in this chapter, which use these technologies that develop a collective creativity. Websites dedicated to creativity, for example that of the creativity day[1], forums, mailing lists and, more recently, wiki and weblogs show tremendous creative development through these networks.

The networks announce the coming of a new era, that of creativity, on a planetary scale, where one day, we will realize that each man has his share of creativity to put in the common cauldron, and will take part in a shared success whose result still remains to be discovered.

---

1 See www.creativityday.org.

PART 3

Innovation Management:
Which Factors Underpin Success?

Introduction

This Part introduces the human and social dimensions through the study of psychological aspects and the assessment of skills within innovation projects. We develop the financing structure and industrial protection build-up then introduce the applications founded on the theory of remarkable angles, on decision support and the development of innovative services.

Change is provoked through psychological factors that can also slow down the progress of innovation projects. We outline the seven basic indicators to assess innovation capacity at individual or collective level and how to manage and value intangible assets of organizations when enhancing enterprises' innovativeness capability through cases studies such as SKANDIA, Dow Chemical and GrandVision.

We then develop the different steps for financing innovation from seed to transfer of equity. This path is required when creating value, growth and employment, yet it contains several sources of risk. We then detail the managing of intellectual and industrial property in a global context where rule changes becomes mandatory. The discussions particularly apply to the protection of software, to networks, databases, knowledge bases and digital creation.

Three cases deal with real products, virtual reality products and services. New products, their appearance and proportions are subordinated to harmony rules that are analyzed with theoretical aspects and through examples. We reveal how to save time within innovation processes through the use of virtual reality technologies: these enable the use of alternative design solutions based on intermediary virtual representations that provide decision support facilities.

Finally we introduce recent innovations in the world of the Internet and networks under the aspects of tools, services and new practices such as weblogs.

Chapter 14

Psychology of Innovation and Change Factors

14.1. Introduction

Innovation and the development of research in the numerous institutions of a country can either develop or get stalled depending on culture, conditioning and mentalities. In the chain of innovation, these obstacles can intervene at any level of a company or a public organization: at the level of the creation of new products or services, internal or external communication, general strategy or at the level of internal organization (pyramid, matrix or network). Innovation therefore does not depend on the creation of original products or services alone but on all the various activities or departments (management, human resources, marketing, accounts, communication, etc.).

Financial resources are not the only input required to improve innovation and research; it is above all a question of human behavior, attitudes and mentalities. And as regards innovation, this psychological dimension makes all the difference. This dimension refers to our culture that we will define as the combination of our beliefs and unconscious psychological processes that condition our emotions, our thoughts and our behavior.

Thus, in order to create and develop new technologies, it is also necessary to explore and modify one's cultural habits, preconceptions, fears, illusions and ways of thinking. Now, our French cultural psychology has not really prepared us to face the obstacles that arise in the creation and the promotion of strong technological and scientific products. One must have the courage to face the limitations of one's

---

Chapter written by Laurent DUKAN.

beliefs and personal and cultural prejudices. As these obstacles are largely unconscious, it is a question of becoming aware of the same and of patiently learning to rise above them. For that, it could be useful to get the help of one or more coaches or/and consultants.

Depending on various cultures, some of the obstacles to innovation will be more marked, others practically absent. The maintenance of the homeostasis of our cultural behavior is partly realized by the organization of our society and our institutions. The French specificities like the specialized schools for higher studies called the "*grandes écoles*", the organizations of academic research (research organizations distinct from universities) or the hierarchical organization in companies are some of the structural elements reinforcing our cultural habits and our mentalities.

The analysis that we are going to present here constitutes a diagnosis grid that is based upon the key factors of innovation: our relation with risk, the unknown, the future, complexity, conformity, originality, failure and success, methods of reasoning, recognition. Many other criteria could have been selected, but we felt that the above were the most characteristic of innovation and essential in order to analyze the capacity of innovation of people in a given culture. The analysis will be carried out from both a psychological and scientific point of view. Stress will be laid on the analysis of the French context, but this analysis grid can be used for all countries and all sectors.

Whether one is an executive in a SME, a start-up, a multinational, an administration or a laboratory, we are no longer protected from direct or indirect competition. This increased competition goes through innovation. Amongst the current revolutions, the fields of science and technology and the following domains for example are crucial:
- micro-electronics and bio-electronics;
- digital technologies;
- third and fourth generation internet;
- biotechnology, and nanotechnology;
- "one to one" marketing technologies.

All these fields are at the same time the driving force and the result of an appropriate strategy of innovation being able to base itself, on one hand, on the fundamental research carried out by multidisciplinary teams and, on the other hand, on the social management of technologies based on new needs. The concept of networks takes its full meaning in this context as innovation results at the same time from information networks and human networks immersed more and more in the complexity of multi-dimensional, multi-territorial and inter-personal relations.

14.2. Innovation and research

To innovate is to design and develop new products, new services or original marketing methods. The development of a new packaging or the transition to e-selling (selling through the Internet) are both innovations. According to the systemic researchers of the Palo-Alto School, we are in this case, in the presence of a level 1 change that corresponds to an improvement in the existing system. This is what happens when new computers arrive in the market: their microprocessors are faster and their hard disks have a greater capacity but their principle of functioning remains the same. On the other hand, with the first audio CD, we were in the presence of a level 2 change [WAT 75]. It was a question of a conceptual change and not just of an improvement, as there was a transition from an analogical world to a digital world [DUK 03].

Today, it becomes more and more risky to innovate by limiting oneself uniquely to the improvement of the existing. Competition has become stronger and is now global. Consumers are not confused and have become more and more demanding as regards new inventions. They are looking for innovations incorporating level 2 changes. For that, companies are obliged to outsource bona fide research projects, both in the fields of applied research and fundamental research. For the past several years, innovation and research have become inseparable. It is now necessary to think more and more in terms of "innovation by research". This thus supposes important changes as regards partnership, budgetary distribution, recruitment, organization of services, internal and external communication, international exposure or knowledge management. Even if the organization, whether an SME or public institution, decides to outsource its innovation activities through research to university or private laboratories, these activities necessarily imply a profound change of mentality, if only to accept this need to innovate and to invent differently.

14.3. Change in mentality

The mentality of a population can be defined as the set of beliefs, mental reflexes and behavioral attitude shared by the greatest number of people. These beliefs vary from one group to another according to various cultures. The mentality of a group is often unconscious and directs its behavior strongly. It affects at the same time its survival and its identity, implying the exhibition of defense behavior in the case of threat: fear, escape, aggression, grabbing of power, manipulation, seduction, values, rigidity, certainties, etc.

To open up more strongly to innovation, but also to research, implies for certain companies a true change of mentality. This evolution must apply to all the involved actors: directors, managers or executives, employees, customers, suppliers.

The changes in mentality that are indispensable to innovation and scientific research cannot be achieved overnight. Observation shows that this process generally involves a long time period. It is implemented chiefly due to the necessity to adapt to a changing environment and under the threat of the worst. Sometimes, the renunciation of one's beliefs is impossible. Individuals could even "prefer" to disappear rather than to place their beliefs or views in perspective and modify their behavior. This shows how high the stakes are.

To innovate and to invent new products or services demands a lot of efforts and very often different ways of thinking, feeling, acting and communicating. It is not always enough to be conscious of the necessity to change mentalities, to be able to achieve that change. To know that it is necessary to change is one thing, to really change one's behavior and way of thinking is another. One does not change by decree, nor by simple decision. Fortunately for us, numerous tools that are very effective in accompanying change are at our disposal today [FIS 86] [WAT 86].

14.4. The principal cultural indicators for innovation

We have selected seven fundamental cultural indicators in order to diagnose the development of innovation in an organization. According to the action taken on each indicator, we can determine actions to be taken consequently.

14.4.1. *Fear and taking risks*

Fear is a fundamental indicator in the evaluation of the capacity of a group or of an individual to innovate and manage risks. To create something new is to place oneself directly in danger: that of losing what one has invested, without taking into account the loss of credibility, the loss of motivation and all the efforts to adapt to the requirements of innovation.

In France, our cultural habits acquired right from the time we went to school and upheld by our family and professional relationships often tend towards fear. This attitude is rather frequent. Many managers, employees, directors and executives seem to be afraid. Among the fears related to the act of innovating, we can mention: the fear of failing, the fear of loosing one's investments, the fear of not being up to the mark, the fear of being criticized, the fear of change, the fear of succeeding, the fear of the complex, the fear of being rejected for being considered as different, the fear of the unpredictable.

These fears can be true allies in numerous situations as they allow us to be careful. In other cases, they lead us towards failure. Adapting the work, at an individual level as well as at a collective level can in that case prove to be very useful. It is a question of learning to innovate with calculated risks. That means that one would be obliged in any case to take risks, particularly as risk is always subjective.

14.4.2. *Conformity and originality*

With the question of conformity, we are at the heart of innovation and what determines it. Indeed, to produce something new and original is not easy when one wishes at the same time to be like everyone else. The stakes are high, because to be like everyone else means to belong, to be accepted, and to be liked. Now to innovate is to produce something new, different and hence to show oneself as being different and to risk exclusion.

This fear of not being a part of the group, of being non-conformist, is sometimes reinforced by the fear "of being punished" because one has dared to be different. The punishment expected is the failure of the project. The words that come to mind are then: "that's what happens when one wants to do something difficult and when one wants to be better than the others!" This belief, very often shared, reinforces the idea that nobody's mind should exceed its limits and that one must not stray from the group. This burden of conformity has always been present in several countries at all times. However, one can observe important variations from one culture to another and especially from one field to another.

In France for example, conformity in the professional field is even stronger as it is in reality, masked or hidden. Whether you are a salaried employee or a professional, this will not necessarily prevent you from innovating, from proposing new concepts, new methods. But the punishment could be terrible if you failed, because you dared to leave the group and be different. Who do you think you are? You wanted to be original and you find yourself labeled "eccentric" (*original* in French). Is it an accident of the French language? Certainly not.

To innovate is going to be all the more difficult as it means producing something original and at the same time running the risk of exclusion. One thus needs courage in order to innovate [PAS 01].

14.4.3. *The unknown and the future*

The unknown and the uncertain are central and intrinsic to any innovation. Insecurity is a direct consequence of innovation. In France, we find it very difficult to deal with uncertainty. The unknown frightens us, often greatly. This need for security, at all costs, imposes upon us a lot of prudence and control. But the problem is often an excess of prudence, of control in our search for security.

"The unknown" is basically neither positive nor negative. It can frighten us as much as it fascinates us. Venturing into the unknown can be dangerous and the source of serious problems but it can also give us a lot of pleasure, money, security, success and recognition. And it is for this reason that thousands, even millions of individuals undertake innovating projects every day.

For many of the actors involved in the process of innovation, the unknown can be perceived as a source of dreams, passion, satisfaction, hope and fascinating mysteries. The unknown is what makes them live and gives them great energy. It is often researchers and inventors as well as all those who are fascinated by their works that react in this manner.

Scientific research is the door to the unknown. It is the bearer of all these dreams, but also provides concrete answers for a better life and better health. It is therefore a matter of perception, of knowing how we wish to see things and what we really want to project in an innovating project. To imagine the worst, to right away see the negative aspect could perhaps be a very limiting reflex for some. Indeed, in order to prevent the worst, these people take draconian and inappropriate measures to check and protect themselves. Such controls at every step, in order to alleviate their fears, can compromise the very existence of the project. On other occasions, one sees these same people rushing head on into very risky projects. This could seem paradoxical, but under the surface, this "excess of prudence" and "blind rashness or foolhardiness" are closely linked. When an accident takes place, the experience only reinforces our initial belief that the unknown is dangerous. Just as in the case of anything that is new, the unknown is also dangerous. We were quite right to be that prudent. And it is back to square one.

The fear of the unknown, of the uncertain and of the risk is found in our relationship with time. It can explain why some people turn more towards the past and others look towards the future. In the field of science and technology, the American culture seems to be largely turned towards the future, whereas in other fields, the social field for example, it can be very conservative and unadventurous. It is thus not a coincidence that the majority of science fiction films are produced in the USA. With the past on the other hand, we already have information. One has experienced it, one knows, one is thus reassured.

Now innovation is deeply inclined towards the future, even if one uses the knowledge of the past in order to innovate. For a long time in France, the past was considered a value in itself, representing nobility and respect. There is nothing reprehensible in that. On the contrary, the history of science and technology, that of

Newton, of Descartes, and more recently that of the discovery of the DNA or of data-processing, can be an excellent stepping stone to innovation. But the problem is that giving too much importance to the past and too much of conservatism can be detrimental to the future.

The consequences of mentalities that look towards the past are now well known. Let us talk about the difficulties that one should anticipate, the multiple transfers in decision-making, the recruitment of the same reassuring and predictable profiles.

14.4.4. *Complexity*

By definition, a system is complex when it is composed of a large number of elements in mutual interaction (at least several hundreds) implying multiple loops of feedback and integrating contradictory and paradoxical aspects [MOR 81]. Such a system thus cannot be totally understood. It invariably possesses aspects that we fail to comprehend.

Innovation that implies a level 2 change supposes that we should have worked more and more with complexity. At the level of discovery, we must understand the complexity of certain systems in order to modify their functioning. Thus, understanding the inner molecular mechanisms of the DNA is indispensable if we want to develop new techniques of gene therapy. As numerous diseases are based on the action of several genes, each one possessing several mutations, complexity is right at the heart of this effort of understanding and of action.

In the case of inventions, we have to create artificial complexities. In data processing for example, the elaboration of intelligent operating systems necessitates the creation of thousands of code lines regrouped into modules in mutual interaction. Faced with complexity, some people are fascinated. For them, it represents secrets to be discovered, mysteries to be solved. It brings marvelous things and hope. And the obstacles to be overcome could be perceived as a means to grow. For others, complexity is synonymous with the loss of control, the sentiment of helplessness, possible failure, intense efforts to provide and time that is too long. These feelings are often unconscious but they guide behavior and decisions taken with regard to innovation. An overall feeling of fear and uneasiness can simply appear. The adoption of a strong policy of innovation is then either delayed or revoked.

At the beginning of the 21st century, we find ourselves at an important crossroads because we have entered the era of information, of knowledge but also that of the complexity. This passage is crucial and difficult. Entire populations take refuge in the search for the simple, simplism, facility and appearances, but the complexity in which we find ourselves immersed, continues to grow. Viruses continue to evolve and become more and more difficult to eradicate. And there will always be new viruses and new diseases. The planet is also sick and its malfunctions are complex. The economy has become global and has always been complex. The demographic development of the Earth poses complex problems.

In fact, everything has become complex. Innovation is no longer an optional luxury, but a vital necessity in order to live in increasing complexity. We are facing a great challenge, that of intelligence. To refuse such a challenge would be suicidal [MOR 00] [GAR 85].

14.4.5. *Mechanistic, systemic and complex thought*

In order to innovate in complexity and by complexity, one needs an appropriate mode of thinking: systemic thought associated with mechanistic thought. This is what one calls "complex thought". Innovation requires the consolidated use of this complex thought [DUK 00].

Unfortunately, in France, as in many other countries, it is the excessive use of mechanistic thought that seems to prevail. This implies compartmentalization, exclusion and isolation that slow down research and innovation. The difficulty in developing truly interdisciplinary projects is significant. The research projects in cognitive sciences, in artificial intelligence, but also in genomics are fundamentally interdisciplinary. A serious study of our modes of thought is indispensable.

Mechanistic thought has existed for centuries and is at the base of scientific thought to the extent that the two are often confused. However, we will see later on that this is not at all the case. Mechanistic thought is characterized by the use of reasoning that follows the logic of exclusion of a third party, linear causality and the isolation of elements. This mode of thinking has been constructed on the basis of phenomena studied by science in the 19th century: mainly classic mechanics, thermodynamics and Boolean algebra.

The logic of the excluded third party corresponds to binary thought. Things are white or black, true or false and there cannot be intermediate truth. Any third party or third choice is thus excluded. If this mode of thinking simplifies the decision making process and makes it more rapid, it can also lead to a simplistic view of matters and tends to over simplify reality. Now innovation needs subtleties, multiple and intermediate levels. It must even come to terms with the contradictory and the paradoxical.

Linear causality implies a viewpoint according to which, cause and consequence are distinctly separate. Although it is very useful while searching for an explication or while trying to comprehend a situation, it becomes very limited when one has to describe overlapping phenomena.

To reason by isolating elements from each other has allowed great advances in sciences. But coupled with linear casualty, thought begins to seek explanations solely within elements and not at all at the level of relations or interactions between these elements. Pushed to an extreme, the environment is ignored whereas it can play an important role in what occurs. All the characteristics of mechanistic thought are neither bad nor good in themselves. As mentioned earlier, it is the excessive and unadapted use of this way of thinking that can pose a problem in the case of innovation as well as in many other aspects of everyday life.

Within the context of an innovating activity, when it is a question for example, of developing a new therapeutic molecule, it is clear that precision based upon mechanist thought is essential. Nevertheless, used exclusively and excessively, this way of thinking can quickly generate the following symptoms:

division of services, isolation of teams, weak exchange of ideas and information, the practical non-existence of interdisciplinarity, internal competition with all its associated negative aspects, reinforced jealousy, seclusion in categories.

These consequences are sometimes fatal to the development of real innovation.

Systemic thought is in a way the complement of mechanistic thought. This way of thinking has always existed even though it may not have had a name. However, it is particularly in the 1960s and in California, close to Palo-Alto that the systemic approach could be developed [WAT 75]. In France, the work carried out by the researcher Edgar Morin allowed the establishment of strong bases for the study of complexity [MOR 81]. The principal characteristics of systemic thought are as follows: the vision system and primacy of interactions, feedback and the paradoxical logic. This mode of thinking was constructed on the basis of phenomena studied by modern science that was in a period of rapid growth at the time: let us mention particularly physical quantum, the chaos theory, cybernetics and the theory of information.

To connect elements between them, to take into account the context is often most essential in order to create something new. This is how new ideas can appear. Interdisciplinarity, transfers of concepts, reasoning and methods permitted the development of biomathematics, the development of systems of recognition of speech through a network of artificial neurons, or in social sciences, the systemic approach of Palo-Alto associating psychology, cybernetics and mathematic logic.

Feedback is at the heart of the phenomenon of regulation and adaptation for numerous systems. Inventions and the discoveries cannot be carried out without taking it into account. The blood pH, electronic circuits, the management of new technologies, the assimilation of new products or services or technologies by a population; all these systems and many others possess loops of feedback that continue to function in spite of external disturbances that can modify them. To innovate is to take into account the existence of powerful and hidden feedback so as to adapt ones strategy consequently. Every change in a structure generates reactions that strive to maintain the original structure. To innovate implies strong changes in terms of organization of work, of hierarchy and of communication.

By integrating the paradox and the contradictory in its vision of the world, systemic thought is finally much closer to what actually occurs in reality. In order to stimulate innovation, it is also necessary to deal with the irrational of pre-existing paradoxes but also of those that this is going to generate. Thus, the easier it is to communicate with modern technologies, the less one communicates [WOL 99]. In the same way too much information kills information. Too much control on individuals generates exactly what one tried to avoid. Too much innovation kills innovation because innovation is then trivialized to the point of generating indifference.

It is thus a question of being vigilant. To restrict oneself solely to mechanistic thinking implies rigidity, power and control. Therefore there is no more place for innovation. On the other hand, a systemic viewpoint as the unique framework of reference can exclude structure, hierarchy and coherence, elements that are indispensable to any innovation. It is with complex thought, uniting these two modes of thought, that innovation and research will be able to develop in an optimal manner.

14.4.6. *Communication and recognition*

To communicate or to block information, to praise or to scorn, to accommodate or to exclude, to control or to hold responsible: all these choices are not specific to innovation. Nevertheless, they are crucial. Internally, they act directly on communication between departments or within the same department, and on the recruitment of innovative profiles [DUK 98]. Externally, they determine the quality of exchanges with suppliers, clients and possible partners.

Recruitment is a key stage in the development of an innovating project. This is also a moment in the course of which the fear of the unknown is likely to be activated. The recruiting organism will install numerous barriers in order to protect itself. The network is one of the most well known. On account of this network of barriers, the candidate will have to convince several intermediaries of his talent before meeting the actual recruiter. Such networks are often presented as a good way to find employment. As a matter of fact, they also limit the liberty of contact.

Experience shows that these barriers often result in eliminating innovators right at the start whereas the official statement is: "we are urgently looking for innovative profiles. But where are they?"

In other countries, like the UK for example, the rules of the game are different. An unknown candidate that possesses great skills can be recruited more easily. It is true that the labor laws are not the same, but that does not explain everything. In fact, for a lot of employers, being unknown does not necessarily make you dangerous. There is no initial distrust, but confidence.

To be able to give signs of recognition to a newly recruited collaborator, as well as to ones different interlocutors in the processes of innovation is also as important.

Depending on various cultures, the manifestation of these signs can strongly vary. In the USA for example, it is normal to compliment easily, to encourage and to say that the work has been well done when this is the case. To praise someone does not pose a problem. In France, we have learnt since our school days that one should try to do better and to surpass ourselves, mainly from all that was negative, like our errors underlined in red. But it was only the errors that were pointed out. That is not sufficient in order to progress. On the contrary!

Moreover, a belief often seems divided. Recognition could be a limited quantity subjected to the rules of functioning of the communicating vessels. To praise another person is something that can only be done at the cost of our personal reservations. The more I give, the less I possess. This is an illusion. As regards the recognition of human qualities, it is in fact the opposite that occurs. The more I praise others, the more I recognize myself. The higher my self-esteem, the easier I find it to pay compliments to someone else. To recognize is to trust the other, to give a chance to the other, to allow the other to be more responsible. And as far as motivation and creativity are concerned, that changes everything!

Finally, to know how to communicate is a fundamental factor for innovation. To share information is thus power, to diffuse knowledge, to give ideas, to say things. Our culture seems rather ambivalent as regards communication. The confusion between knowledge and power is perhaps at the origin of this. The predominance of mechanistic thought that divides and excludes reinforces it. Now in order to innovate, we must give up some of our power, or more precisely, our illusions of power.

14.4.7. *Failure and success*

It is normal and natural to be afraid when it is a question of innovating. Everyone knows that innovation involves a great deal of risk. There is certainly the fear of losing money, but there is especially the fear of failure.

This fear of failure is more or less well managed and varies depending on individuals and cultures. In France for example, failure is still considered bad and is not very well accepted. Failure often generates great shame. Until quite recently, some people would go so as far as to commit suicide if their business went bankrupt. It would seem that in our collective representations, failure is perceived to be the direct consequence of an inner personal defect. Some people say "If you have failed it is your fault!", as much for themselves as for others.

In these conditions, one can understand why some people want to be sure of succeeding even before commencing. This fear of failure is so strong and widespread in France that it affects each one of us in different degrees. One needs to recognize the problem in order to solve it. The bolder and more rash we are, the stronger its presence is.

Success is not necessarily better accepted in France. Even though those who succeed have a better experience than those who fail, success can however give rise to certain difficulties: jealousy, resentment, aggressiveness and mockery. Not hiding one's success is sometimes interpreted as a provocation, a display. Convocation ceremonies for the awarding of degrees are rare in France. One must think of those who did not succeed. There seems to be a great deal of confusion in distinguishing between "equality" with "uniformity", sometimes calling for an equalizing action at the bottom.

In Japan, the shame and the fear of failing still seem to be stronger than in France. Some observers think that it is this fear of failure that explains the strategy of development of numerous Japanese enterprises: reproducing products perfectly, bringing about improvements in the product, and not really innovating. On the other hand, success is a great honor that deserves to be recognized.

In the USA, one can observe the opposite. Failure is not really a problem. It is the sign that one has tried, it is the proof of experience that has been acquired. The innovator who has failed is thus an innovator who is going to succeed since he now knows the difficulties that he will have to deal with and the traps that should be avoided. Failure provides the ideal context in which one can succeed.

Inversely also, success is greatly admired, even glorified. Numerous festivities are held in the USA to celebrate success and these are often caricatured by the

French. We are at the heart of psychological relativism, but the consequences on innovation policies are quite evident.

Certainly, this description of the American, Japanese or French is above all stereotypical. It describes the general tendency in the culture of these countries. It is at the level of school and in education that we must search for the roots of these cultural tendencies [MOR 00].

14.5. Conclusion

For French companies, the challenge for the next few years is to develop a policy of innovation, whether it concerns a product, a service or marketing. For that, a closer interaction with the world of research is desirable and cannot be overlooked, particularly in the case of certain sectors. However, for political, cultural and historical reasons, the world of the academic research and of private enterprises communicates very little.

There is thus an urgent effort to be made on several fronts. This effort implies going beyond our cultural limitations that express themselves through prejudices, beliefs, illusions, ways of reasoning and feeling. It is a question of being alert as these cultural reflexes are very difficult to eliminate. Sometimes one can see them controling our behavior after years of silence, like a dormant computer virus.

It is through regular and continuous practice that firm believers in innovation will be able to carry out their project, by limiting risks. Why is it so important to go beyond these obstacles in order to innovate? Quite simply because:

– the survival of companies now depends on their capacity for innovation;

– the economic growth of the years to come will depend on research and technological innovations [AGH 04];

– deadly diseases such as cancer or aids continue to cause millions of deaths in the world and this is intolerable;

– new viruses and diseases will continue to appear in the years to come;

– scientific and technological progress can help us to resolve certain ecological problems that seem to become more and more serious and widespread. Finally, it provides the means for a better life for man.

Chapter 15

Intellectual Property for Networks and Software

15.1. Introduction

Intellectual property for networks and software is presently a major issue and a live topic for debate the world over. It raises the question of compatibility between the monopolies vested by intellectual property rights and the actual circulation of intellectual properties and services.

Given the magnitude of the subject and the variety of discussion contributions during the debate, the present chapter will deal mainly with the principal problems in using the intellectual properties for networks and software in the innovation services without getting into the details of electronic commerce and payments, and electronic signature which, though connected to intellectual property, would require several chapters to be dealt with.

Importance will then be given to information regarding the intellectual property tools available to the writers of innovation support software and the rules related to the intellectual property that the users of networks and software should follow. It is important to highlight that the question of intellectual property for networks and software should be dealt with straight away at an international level even if the practical application of the rights of intellectual property is handled only at the State level.

Chapter written by Sylvain ALLANO.

Finally, it should be noted that this contribution is certainly not a legal or a theoretical study on intellectual property meant for law professionals but just a simple explanation of the issues related to the intellectual property in an era of networks to be used by non-professionals.

15.2. State of the problems and the protagonists

The basis of intellectual property involves recognizing and conferring the creation and invention to their creators and inventors. Intellectual property rights are conferred on these creators and inventors on a temporary basis by the State with a view to encourage creations and inventions and to stimulate the innovation activity.

The term "creation" applies not only to intellectual work involving mainly the literary and artistic creations but also to the graphic, industrial (design), numerical and software creations whereas the term "invention" applies to the appliances, systems, products and technical procedures and applications contributing to resolving essentially technical problems.

The communication networks can be considered either as simple supports or vectors of information transmission not taking any part in the innovation, or as technical elements participating as they are in the innovation. Thus, several innovations in the field of industrial production or e-commerce are direct results of communication networks.

The main actors involved in the networks include the telecommunication operators, access and service providers on the network, the information producers, the users of the information systems and the institutional body for control. All these actors are involved at various levels in the creation process and the management of intellectual property.

The essential questions in the present problem related to the intellectual property and the networks are those concerning the generation and the acquisition of the rights, their direct or indirect use, and the litigation that could arise out of this operation.

15.3. The main "nodes" in intellectual property amidst the networks operated in the context of innovation

The communication networks consist of fabrics and nettings of variable density having multiple nodes connected to the equipments and to the communication and information system. The creations and innovations achieved in a communication network necessarily flow through these nodes which, in a certain way, form the mandatory gateways which can be used to not only locate and date some of these creations or innovations but also to observe acts of infringement.

These nodes in the network can thus form the intellectual property nodes, and identifying them can contribute to a better efficiency in the analysis of the conversion mechanisms and application of the intellectual property rights. Several paths of identification of these nodes may be foreseen.

A first path is to follow the data and the information flux originating from the information sources used for the purpose of innovation. The intellectual property nodes are then:

– the databases of the data or the information producers;

– the operating servers of these databases capable of accommodating the search engines;

– the access providers to the databases offering expert systems or even "software laboratories" to their regular subscribers;

– the informatics terminal to the information users who can themselves be equipped with configured software tools to process the information collected through the communication network.

A second path is to follow the information flux and the data connected to creation and innovation. These information and data go through the following nodes:

– the informatics terminals of the creators and the inventors, in whose storage units are found the original software creations or the new software products or services meant to be broadcast through the communication network;

– the corporate servers through which transit innovative works from the informatics terminals of the creators and the inventors;

– the corporate databases;

– the informatics posts and terminals of the customers and acquirers of innovative products and services whose utilization depends on the granting of license for utilization.

The intellectual property nodes can also be identified following various other paths through an approach based on the structure of the information systems, the communication networks then being considered as input or output paths at the periphery of these information systems.

15.4. Intellectual property rights applicable to the context of networks

Whether it involves innovations coming from the networks and intrinsically connected to the structure and to the functioning of these networks, or innovations using the networks as just simple vectors and distribution supports, different branches of intellectual property rights are involved. Thus, it is considered that:

– copyright and related laws (neighboring laws and drawings and models), which concern the totality of the intellectual works, including the software creations and the database:

- the right to brand names covering the protection for distinctive signs,

- patent rights which concern the totality of the technical (mainly Europe) or useful (USA) type of inventions;

– in the network's context, statutory regulations related to the database as well as the rights to contracts, statutory regulations related to unfair competition and also statutory regulations related to the field names on the Internet play a role [BEA 01].

It is thus a complete palette of intellectual property tools which can be put to action in an approach to protect innovations involving networks. However, as for the states in which a protection is solicited, more importance will be given to a branch of law than to anything else, as per the statutory regulations adopted by each state.

15.5. Copyright "software" against networks

15.5.1. *The main statutory copyright "software"*

Copyright "software" is an integral part of the copyright as governed by the Code of Intellectual Property while having several statutory and derogatory specifications.

According to Article L.111-1 of the French Code of Intellectual Property, the author of an intellectual work, for merely creating it, enjoys an exclusive incorporeal property right enforceable against all. The software including the preparatory design materials are considered as intellectual work (Article L.112-2, °13). The authors of translations, adaptations or arrangements of the intellectual work enjoy the protection instituted by the Code without prejudice to the copyright of the original work (Article L.112-3).

However, several derogations to the common regimen of copyright have restricted the prerogative of the software writers. Thus, according to Article L.113-9, except statutory regulations or stipulations to the contrary, proprietary rights on software and their documentation created by one or more employees while carrying out their functions or after receiving an order from their employer are vested to the employer who is the only person authorized to carry them out.

On the other hand, except in the case of a stipulation to the contrary, which is more favorable to the software writer, a modification to the software by the assignee of the rights of reproduction, translation, arrangement and launching in the market cannot be opposed by the software writer if it is not detrimental to his honor or his reputation, or to carrying out his right to withdraw it.

To pinpoint the problem of intellectual property of software in the context of networks, it is important to distinguish on the one hand the case of circulating proprietary software on the network, and on the other, the case of the intellectual property of the software involving the network.

15.5.2. *Intellectual property of the software circulating in the network*

At all times, there exists in the communication network a large number of software that have all had, at the time of their creation, at least one author who had worked on a computer connected to a communication network node. These types of software run either in a compiled form, encrypted or otherwise, or in the form of a source code with information "in clear" about their origin, their author(s), date(s) of creation, and the holder of the proprietary rights who are attached to them. When their creation is controled by distributors having rights, a license for utilization is attached to them and the acceptation of the terms of this license is a mandatory gateway so that a user who receives one such software can store a runnable version on his work station connected to a communication network node.

In a legally controled configuration, the intellectual property for networks is then generated at a point "originator" of the software creator, and recognized under the provisions of the copyright in accordance with the statutory regulations of the state in which this point "originator" is located and in other states in which the software creator or its assigns could carry out the steps, for example in the form of a deposit in the Copyright Office. It should however be noted that in several states, recognition of copyrights does not involve non-procedural steps or official deposits but the author should just justify the date of creation of his work. This dating procedure can be carried out in various ways. In France, one of them is the Agency for the Protection of Programs (l'Agence pour la Protection des Programmes) which offers deposit services or software referencing services by giving an identifier number IDDN. There are also several online referencing and dating services for numeric creations related to software as well as multimedia.

The international conventions governing copyrights allow the writer of an original software distributed through the communication network to exercise his copyright rights on his software in most countries of the world, but in reality the actual exercise of these rights is not without obstacles. It should be noted that strong disparities exist in the "receiving" states in the exercise of software copyrights.

15.5.3. *Intellectual property for software involving networks*

There now exists a new generation of software designed according to a network logic in which the network forms an integral part of the software. ASP (Application Service Provider) software, virtual reality software tools, and also software created in a participatory way, involving several writers communicating and regrouping a collaborative work through the communication networks, can be cited as examples without limitations.

In the case of ASP software solutions, the legal status of the software is apparently simpler than the software circulating in the networks because the question of physically displacing an executable code or source does not arise, although displacing the images, which can certainly consist of original graphic creations and thereby entitled to copyrights, is possible.

For virtual reality software tools which set up several distant sites participating through several software executed in real time in all these sites, the question of intellectual property can be taken up based on the eventual co-writers of the tool and the effective position of the distributed executables. In general it is recommended to reckon the copyrights of the software tool in its totality. It is to be noted that a protection through patents can also be foreseen in case a technical contribution to a technical problem can be identified. Generally this is the case for the virtual reality software tools involving for example haptic systems (force-feedback).

The software created in a participatory way by several authors located in distant sites are considered, according to the intellectual property code, as works of collaboration to the extent that it is possible to distinguish the contribution of each one of the co-authors, each having copyrights for each one of their respective contributions. In order to be able to exercise rights on the participatory work, it is recommended to establish a co-propriety rule between the different co-authors.

15.5.4. *Software copyright limitations*

Thus, for software creations, copyrights are rights of reference, but it involves a severe limitation in as much as the protection offered is only for the expression and not for the content. In other words, it is mainly the source code of a software which gets protection and not its operation.

In the present context of the software industry, where the programming phase in the development of a software constitutes only a marginal share of the total cost of development when compared to the increasing part represented by the upward phase of software engineering, it is becoming more and more frequent to develop strategies to bypass the copyrights attached to an original software. Thus, for a similar software architecture generally prepared from the software bricks, it is possible to write a multitude of distinct source codes leading to several software products with similar operations.

15.5.5. *Software copyright*

Copyright, the Anglo-Saxon cousin of the author's rights, is mainly aimed at, as its name indicates, the right to copy a work and not at the rights of the author(s) of this work. Whereas the right of the author "the French way" gives importance mainly to the authors, the copyright is directed towards the rights of the company of production and diffusion of the works. This gives rise to significant differences in exercising the rights by the authors, mainly in controling the adaptations and reproductions of the software creations, the author's prerogatives being reduced to a minimum with respect to the copyright. It is to be noted that unlike the author's rights, acquiring a copyright in the USA involves a non-procedural step with a formal deposit at the Copyright Office.

15.6. Free software

Under the GNU manifest of Richard Stallman [STA 98], the world of free software is extended and co-exists harmoniously with software protected by copyright and rights of the author. All the experts agree that the world of free software is neither a world without author's rights nor one without money, as illustrated presently by the economic issues related to the service industries linked to free software.

What distinguishes a world of free software from that of a proprietary software, other than the libertarian philosophy which underlined it in the beginning of this movement and which was mainly a product of the "Civil Rights" movements of the USA which then spread to the rest of the world, is essentially the nature of the license for utilization of these software. Presently there is a large variety of types of free software licenses and it would be tedious to enumerate all of them. In any case the GPL (General Public License) licenses can be mentioned.

A major characteristic of most of the licenses consists of banning all appropriations and reservations by a third party of amendments, complements or improvements brought about by this party to a software covered by a free software license. Thus, some of these software present a contaminating power which results in spreading the ban on appropriation, implying generally the obligation to grant a license for the developments every time a software core covered initially by this type of license finds itself integrated, adapted or drifted in a commercialized software which should also constitute a license which includes the diffusion clauses of the "free" type.

This question of contamination is sometimes difficult to tackle for software engineering companies which, while creating new software products with a view to retaining control over the intellectual propriety rights, cannot ignore certain software bricks considered "unavoidable" but are diffused under the license of free contaminating type software. A complex analysis of the intellectual propriety rights of a composite software creation incorporating a majority of bricks having divers and varied propriety rights should be carried out. This leads to a delicate but practically possible cohabitation between an exclusive proprietary approach and a "free" approach.

The fact that many software editing and engineering companies in the world develop markets and create products and services suitable for each of these two universe clearly shows that in reality they are not opponents but correspond to expectations of different actors in the software world.

15.7. Protection through patents for communication software and networks

While, for several decades, it has been the practice in the USA to deposit request for patents relating to the inventions implemented in the form of software, several other countries still have statutory regulations banning access to the patent for computer programs whenever such a protection is solicited. This situation still exists in the Convention for European Patents [C2] and in the French Intellectual Propriety Code.

While the European Agency for Patents (Office Européen des Brevets) had in practice, over the last few years, relaxed the rules to enable the numerous requests for patent pertaining to "software" to be examined and delivered, and several legal authorities have approved this evolution, there has been a real lag in this domain, mainly due to very effective campaigns addressed to the European decision-makers and legislators against the setting up of "software patents" which could infringe upon the liberty of the creators and the users of software. Our practical experience with the young software engineering companies in Europe shows that a lack of protection through patents for their software innovations could, in the medium term, be detrimental to them because the majority of their investments, like engineering, would be without rights protection.

In Europe, as soon as a software innovation could become eligible to protection through patents and could thereby access the invention status as per the patent rules, it has to satisfy the following three criteria to be patented: to be susceptible to industrial applications, to be new and to be able to involve an inventive activity.

In the USA, where the new patent barriers are not concerned anymore with the software which are included in the patentable category, but have since developed "business methods", an essential criterion to access the invention status is to be "useful". Following this, the criteria of non-obviousness and novelty are examined.

15.8. Actors in the networks and intellectual property

15.8.1. *Intellectual property of databases*

According to the Intellectual Property Code, a database is a collection of writings, of data or other independent elements, arranged in a systematic or methodical manner and individually accessible by electronic means or by any other means.

Like anthologies or collections of writings or various data, as per the choice or arrangement of subjects, the databases are intellectual creations, and can avail of the conditions of copyright. Besides, the collation of this data is in itself an original piece of work, which is not always the case for several databases.

It is mainly to offset these application limits of copyright and to block the systematic or partial plundering of the contents of the databases that, at the European level, legal conditions were adopted. These conditions are specifically for databases which were subjected to French Law in 1998 (Article L.342-1, 1st paragraph of the Intellectual Property Code). These legal conditions aim to prevent massive lifting of data from these databases.

It is to be noted that the new architecture of the databases or of the new data access mechanisms or processing of this data within a database can be subject to protection through a patent as much in the USA as in Europe and in the rest of the world. However, in the case of an application for a European patent, the eligibility for protection by a patent implies that the subject described in the report of the invention is clearly of a technical nature.

15.8.2. *Expert systems and tools of artificial intelligence*

Expert systems, which can understand sets of rules, inference engines and databases, are intrinsically software by nature and they are expressed through a computer program which in Europe is excluded from the scope of protection of a patent.

However, these expert systems can also be considered as procedures and systems meant for piloting other systems through the execution of rules, or as procedures and systems to manage data or information. It is on this basis that several applications for European patents concerning expert systems have been made and obtained over the last decade. It is no longer the same thing nowadays and it is not rare to receive objections from the Examination Division of the European Office of Patents stating that there is an absence of inventions as this is a method for executing intellectual activities.

15.8.3. *Computer generated creations*

Creations generated by a computer can be both virtual and material. Virtual would be in the form of software objects that can be seen on the screen of a computer or of a work station. Material would be in the form of objects made from production tools piloted by computer systems programmed by software objects. These virtual creations generally avail of copyright depending on their originality [VIV 89]. If they are the result of an execution procedure that can be protected by a patent especially in the USA, and sometimes in Europe, they can also avail of a protection through a patent as a product obtained by a patented procedure.

15.9. Digital Rights Management (DRM)

Communication networks are currently vectors in digital format of diffusion on a massive scale of pieces of works of all kinds covered by intellectual property rights, whether they are musical, video or more generally multimedia [MAL 00]. These literary and artistic works, covered by French or Anglo-Saxon copyright, can also stretch to educational software, games and software packages.

The creators and producers of these works broadcast on the networks currently use Digital Rights Management (DRM) tools, which combine software technologies including encapsulation and cryptology procedures and material configuration related to digital media (such as CDs) and to devices that read these media. The DRM is now an essential corollary in the exercise of intellectual property rights on the networks, but its development in the world is often opposed by partisans of free circulation and reproduction of information and works on the networks, and they oppose the basic principle itself. There is opposition as well from the association of the defense of consumer rights which presents the usage limits induced by protection techniques implemented in these DRMs.

15.10. When the networks themselves become tools for intellectual property

The development of communication networks has contributed to rationalizing and amplifying the tools used all along the intellectual property process, right from the acquisition of rights until their implementation. Thus, the search for the past history of a patent is mainly based on an online search of databases. The producers of these databases offer tools for advanced and particularly sophisticated search that couple old technology and identification of earlier rights on targeted technologies.

Numerous regional and national offices of industrial or intellectual property matters offer procedures for electronic application for patents and registration of brand names. This should become a general procedure in the next few years and should include all the exchanges between applicants, their representatives and offices and institutes of industrial property. This evolution of procedures would lead to the dematerialization of titles to intellectual property rights just as happened in the financial world.

The networks and their information equipment have also become important tools for the detection of infringements of intellectual property rights, and can track and identify violators of patented technologies offering products or services on networks.

15.11. Enforcing intellectual property rights on the network scale

If the networks obviously contribute to facilitate the diffusion and commercialization of these technological innovations all over the world, the open character of most of these networks and their speed of circulation make it difficult to implement their intellectual property rights. The first hurdle is in the chiefly territorial character of these intellectual property rights. This implies that their prerogatives can only be implemented on territories covered by these rights.

Thus, in the case where a server, offering illegal software products to consumers living in countries where these software products include technologies covered by patent rights, shifts to a part in the world where these patent rights are not applicable, the holder of these rights has no means of stopping the exploitation of this server. On the other hand, he can get cost of cause in each of the countries protecting his rights where violations of these rights have been observed.

Another hurdle in the implementation of intellectual property rights is in the mainly evanescent or furtive nature of some acts of network infringements, making it difficult to get material proof of these violations.

15.12. Conclusion: intellectual property and the networks: an advantage for innovation

Intellectual property is currently a major problem in InfoTech companies. Its role in innovation is being presently discussed and the question being raised now is to know if the intellectual property is going to be a disincentive or an engine of innovation. The communication networks obviously constitute a development factor for innovations, which are, above all, the work of creative minds.

Protection of these innovations is essential to secure remuneration to these creators and compensation for their efforts as well as to motivate companies to invest in research and development. One definitely observes excesses and dysfunctions in the implementation of intellectual property but that does not hide the essential regulation factor procured through intellectual property.

However, at a time where certain economists and sociologists doubt the role of intellectual property in the development of innovations, it seems useful to highlight this essential regulation function of competition mechanisms, and the recognition of moral and ownership innovations.

Chapter 16

Innovation Scoreboard for Core Competencies Evaluation

16.1. Introduction

Technological innovation, which is at the heart of the economic dynamics of developed countries, is a key factor that determines the performance and competitiveness of companies. It is the outcome of a complex process that comprises both the interactions within the company and external interactions between the company and its environment, particularly, in the field of knowledge management, knowledge, and scientific and technological know-how [ALT 02]. And as Nonaka mentions [NON 97], "in an economy where the only certainty is uncertainty, the one and only source of lasting competitive advantage is knowledge".

Innovation, born out of a combination of knowledge, know-how and the source of new knowledge, new products or modes of organization, is one of the elements: what economists and accountants call the "immaterial" capital. Besides, in the past 20 years, its share in the added value of the companies has had a growth rate four times greater than that of material investments [CASP 88] [MAR 88]. It is therefore a high-priority matter "to contemplate on the use of the immaterial capital of the companies; on how to identify, organize and exploit it. The problem of the management of the 'immaterial' in the company corresponds to a major stake. Work done in this field is still modest. Besides, it refers to something that is crucial to the development of the companies" [TEZ 94]. To understand the conditions that favor the emergence, development and increase in value of this capital innovation is of great

---

Chapter written by Nathalie SAMIER.

interest to managers [LOR 95] and specialists in research and development [RIL 03], in human resources [ULR 89] and in marketing [AND 94], all seeking to influence and develop this capital to increase the worth of their company.

Certain propositions of scoreboards connecting the various factors that create value respond to the questions of the identification, organization and exploitation of this immaterial capital, which is innovation. Two models are currently developed: the North American prospective track sheet or the Balanced Scorecard of Norton and Kaplan [NOR 96] [NOR 01], and the Swedish Navigator SKANDIA of Edvinsson and Malone [EDV 97]. Culturally different, these two models integrate both the indicators relating to the immaterial assets, of which innovation is one. The first, which corresponds to strategy, connects four axes: the financier, the customers, the internal processes of management and the organizational apprenticeship. For Norton and Kaplan, the principal objective of the companies that use their model is to increase the value for the shareholder, which involves customer satisfaction, the quality of processes and organizational training.

As for the second model, it puts forth more immaterial elements, and this since the publication of its first report on the immaterial capital in 1994. The founding idea is that the values of the company rest on the human capital made up of talents of the employees and the directors, their experience and their knowledge, and on the structural capital, consisting of culture, the use of Information and Communication Technologies (ICT), of databases, customer relations, etc. Five value-creating domains are projected: finance, customers, "process", research and development, human resources, through more than 800 indicators.

In this chapter, we shall show the extent to which this second model of the scoreboard identifies, takes into account and develops all the actions of the companies which support and develop innovation, as much in organizational, technological, material and economic terms as in human terms. For that, first of all we present the principal components of the result of these actions, which is the immaterial capital. We will then study the competences mobilized by the companies to innovate, and the modes of knowledge that can result from it. Next, we shall bring to light the model of Navigator SKANDIA emphasizing the important value-creating agent which is innovation.

16.2. Locations of the immaterial capital

16.2.1. *Contribution of the theories of resources*

In the face of an increasingly complex competitive environment, and the uncertainty vis-à-vis the results of the actions engaged in and harsh competition,

Barney [BAR 86] was interested since the late 1980s in studying the resources that foster the creation of a tenable competitive advantage. As a result of the studies undertaken on the competitiveness of companies, he puts forward the importance of the resources known to be "rare". From now on, the profitability of the companies rests on the accumulation, the combination and the consolidation of rare internal resources and on the concept advanced by Hamel and Prahalad [PRA 90] on key competences. The competitive advantage no longer lies in the duo (product/market) of the company, but in the resources and capacities making way for the production of goods or services.

What characterizes the justifiable competitive advantage then are resources which are:

– rare, which creates value for the customer;

– imperfectly imitable;

– not easily exchangeable and not easily substitutable.

Innovation, as an immaterial capital, can represent one of these resources, particularly in the mobilization of creative talents and expertise, and in the combination and the development of knowledge and know-how. Its location consequently seems convenient for the organizations wishing to defend their competitive advantage.

16.2.2. *The immaterial capital: intangible investment and intangible assets*

The locating of immaterial capital can be done with the help of two components: on the one hand the intangible investment and on the other hand the intangible asset. Intangible investment is factorized into five categories [SMI 89] [BOI 93] [MAR 96]:

– expenditure incurred in R&D, which finances all work undertaken to increase the scientific and technical knowledge and to make this knowledge available to the company;

– expenditure incurred in the formation and development of human resources engaged to improve competences and in motivational activities. They vary from the recruitment of specialists to the technical training organized on-site or through workshops;

– expenditure incurred in the promotion of products that contributes to increase in the volume and the quality of market products. The adoption of a new logotype or the change of the distribution network (on the Internet for example) forms a part of it;

– expenditure incurred by the organization and management, financing the conception and design of management systems and data processing. They can relate to the purchase of software or databases;

– expenditure incurred in the development of the processes of production and, logistics.

We can therefore see that innovation is represented and evaluated through these five elements. However, this expenditure-driven approach is also complemented by the approach of the intangible assets defined by Pierrat and Martory as "a unit of inheritance of the company, which is entitled to become an asset in the tangible sense of the term when certain conditions come together and, which presents simultaneously several of the following characteristics: absence of physical substance, unspecified lifespan, uniqueness or at least strong specificity, strong uncertainty of the future and not easily distinguishable from the other assets" [MAR 96].

The authors break up the intangible assets into four categories, where one can notice the profile of creative innovation of value:

– rights and quasi-rights. They are identifiable, their value can be enhanced in a market, and they are legally protected. They are patents, trade secrets, brands, models and drawings prepared, the know-how, the new processes of production or organization (which can be sold as products or services...);

– intangible assets that can be converted into tangible assets. They can be sold as soft wares and databases;

– exploitable intangible assets. They allow the flow of income by their exploitation by a third: they are the customer files or the distribution networks;

– non-transferable structures, those which are not easily identifiable. Here we find the network of relations and information systems for example;

– and finally, persistent values such as "goodwill".

We can therefore perceive innovation through these forms of evaluation carried out by the immaterial capital, like a process connecting and organizing skills that are required for innovation, necessary for the development of other skills and finally essential for the integration of external innovations. This set of innovational skills is presented hereafter by Foray, Mairesse and the research scholars of SESSI [FOR 99], [ITS 98].

16.3. Competences to innovate

An investigation in relation to the competence to innovate was carried out by the SESSI in 1997 in 5,000 French companies with more than 20 salaried employees. The companies were questioned on the basis of a list of 73 elementary skills classified under 9 complex competences to innovate. We present these 9 complex competences by their level of internal or external interactions.

16.3.1. *Competences resulting from an internal interaction*

We find in this group three competences related to the interactions that the organization engages in with its human resources as interactions that are inductive of innovation. These three skills are noted by the SESSI: "To organize and run the production of knowledge", "To manage human resources from a perspective of innovation", "To develop the innovations".

The company organizes and seeks to produce knowledge to innovate. For that, it supports and channels creativity through sessions of brainstorming, by sparing a certain degree of autonomy to each one, but equally, it evaluates the results of this collective production of knowledge and develops a human resource management system exclusively for those employees who have contributed to it. It enhances the value of originality and creativity, particularly in the individual evaluation. Upstream, it partakes in the recruitment of specialists who are creative and have a team spirit. It assesses each one's need and proposes appropriate training to them, which, in a continuous process of innovation, also contributes to the development of innovation.

Competence "to manage human resources from a perspective of innovation" is present only in 40% of the questioned French industrial companies whereas 58% of the same companies declared they were equipped with both of the other two competences.

16.3.2. *Competences resulting from an external interaction*

Six competences are related to the interactions that the organization has with the external environment. They are formulated by the SESSI as follows: "To follow, anticipate and act on the market trends", "To include innovation in the overall strategy of the company", "To finance the innovation", "To manage and defend intellectual property", "To adopt external technologies", "To sell the innovation".

The company can thus show its capacity of adaptation to and anticipation of the market trends by gathering, for example, the reactions of the customers from the distributors or from the after-sales services. It can also affirm its will to integrate innovation as a factor of competitiveness in the overall strategy, by encouraging, for instance, the internal mobility between services, departments or establishments, or by assessing the organizational changes which the company is likely to implement. It can also develop a particularly inventive capacity for the financings – public or private – of innovation and for the study of the financial viability of the latter: it weighs the costs against the products born out of the sale of these innovations (new products or the "process") and resulting from the management of the patents, drawings, models and brands. It finally identifies technologies of the future or the present by monitoring technological development followed by a study of the possibilities of sub-contracting, co-producing or of acquiring technologies.

Only competence "to follow, anticipate and act on the market trends" is present in nearly 58% of the French industrial companies. The five others are represented by less than 40% of the responses obtained.

Several sectors of the industry have been identified particularly in the list of prize-winning companies that have the competence to innovate. The companies that fall in the sectors of pharmacy-perfumery-maintenance, automobiles, naval-aeronautics-railway construction, electrical-electronic equipment, chemical-rubber-plastic, and electrical-electronic components declare that they are endowed with, on average, almost 50% of the skills to innovate. They are also more abundant in the companies with more than 2,000 salaried employees, even more so if they belong to a French or a foreign group. But within these large companies, it is especially the competences related to human resource management in terms of innovation, to the financing and the sale of the innovation as well as to the appropriation of external technologies that are more in lacking. Competences, which compete with the organization and the management of the production of knowledge, on the contrary, are well developed. It is the process of the building up of this knowledge that will interest us from now on.

16.4. The key to the creation of knowledge

To explain the fundamentals of innovation, we turn to Nonaka and Takeuchi [NON 97], who develop a new theory of the creation of organizational knowledge founded on the distinction between tacit and explicit knowledge. The key to this creation rests in the mobilization and conversion of tacit knowledge.

First of all let us reconsider the concept of knowledge. The authors believe that only individuals are capable of creating knowledge, the role of the organization being restricted to helping its creative employees or to providing conditions favorable to the creation of knowledge. The latter is understood as "a process which amplifies, in an organizational manner, the knowledge created by individuals and crystallizes them as components of a network of knowledge of the organization. This process takes place in a community of interactions which spreads over and traverses the intra and inter-organizational levels and borders" [NON 97, pp.78-79]. In this overall concept, the authors make an additional distinction between tacit, personal, contextual knowledge that does not easily lend itself to formalization, and explicit knowledge that is codified and is transmissible in a formal and systematic language.

However, these two types of knowledge are not completely unconnected; they are on the contrary complementary, they interact, change and create in their interaction a dynamic "conversion of knowledge" according to four modes.

16.4.1. *Modes of conversion of knowledge*

16.4.1.1. *Socialization: from the tacit to the tacit*

For Nonaka and Takeuchi [NON 97], socialization is a process of sharing of experiences creating tacit knowledge such as shared mental models and technical aptitudes. It produces an "assimilated knowledge". Thus, the apprentices work with their instructors by way of observation, imitation and practice. In the framework of the development of new products and of the organization and innovative processes, the brainstorming sessions, the "camps of reflection", a concept dear to Honda, or the sharing of experiences with the customer-consumers at the time of the return of the products or in the instance of customer dissatisfaction, all participate in this socialization.

16.4.1.2. *Exteriorization: from the tacit to the explicit*

Exteriorization takes recourse to the methods of reasoning such as deduction, induction, making analogies or abduction, which consists of the use of an illustrated language; this leads to the creation of metaphors, of concepts, formed around tacit knowledge. It produces a kind of a co-operative "conceptual knowledge". The individual is thus asked to express his knowledge in the form of a metaphor or an analogy and the listeners are asked to imagine something other than what they hear in order to be able to produce a knowledge that is explicit and that can be expressed in the form of models. The "concepts-car" developed by Renault with the projects Twingo and Espace refers to this exteriorization.

16.4.1.3. *Combination: from the explicit to the explicit*

The combination is a process of systematization of concepts in a system of knowledge. The employees then exchange their knowledge by means of all forms of information transmission (document, meeting, mail, databanks, etc.). The combinatory character of this body of knowledge engenders conversion of knowledge. The daily analysis of the sales of products, in combination with the use of the tools of customer relations such as the CRM to study the rate of satisfaction, contributes to the improvement of products and to innovation. The combination then gives rise to a "systemic knowledge" which is manifested in the creation of patents, models, brands, technologies, services or new products.

16.4.1.4. *Interiorization: from the explicit to the tacit one*

"Interiorization" is based on the verbalization, the reading of documents resulting from the sharing of experiences, and rests on training guided by the act of "doing it", in relation to a new product or production process [MIC 97]. The tests of prototypes or even those of new products launched in test-research areas before being put on sale on a national scale refer to this interiorization. In this way, an "operational knowledge" is built.

16.4.2. *The spiral of knowledge*

This process of creation of knowledge is not linear insofar as here, the tacit and explicit forms of knowledge interact with each other uninterruptedly within a dynamic process as represented in Table 16.1. The creation of a new product (operational knowledge) results from the combination of new or existing technologies to produce it (systemic knowledge), itself emanating from the tacit knowledge of the desires of the customers (comparable knowledge), transformed into a concept by the process of exteriorization (conceptual knowledge). The study of the satisfaction of the customers with regard to this new product then restarts the process of knowledge creation (the withdrawal, improvement or creation of new products).

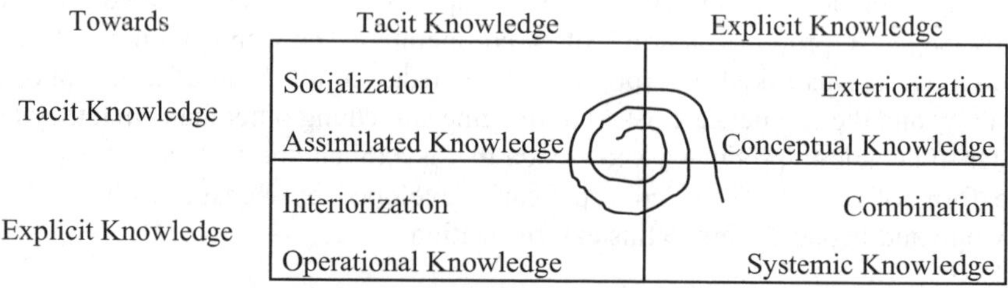

Table 16.1. *Spiral and contents of knowledge, adapted from [NON 97]*

In our previous study, we applied the concept of intangible capital to the process of innovation and analyzed innovational competences deployed by the companies and their process of knowledge creation. We must henceforth ask ourselves this question: how do organizations physically locate, develop and enhance this creative capital of value and competitive advantages? The answer given by the scoreboard of immaterial capital, modeled by company SKANDIA, seems satisfactory.

16.5. The valorization of innovation in terms of the scoreboard

The model developed within SKANDIA, the Swedish company of financial services and insurance, by Edvinsson, director of the "immaterial capital" (IC), is based on the ideal that the performance of a company lies in its capacity to create a certain lasting value by implementing a commercial vision and the strategy which that implies [EDV 97, 99]. The peculiarity of this company is its "virtual" character: it employs nearly 2,000 salaried employees responsible for more than 60,000 sale points and 1,000,000 customers. The employees therefore use Technologies of Information and Communication (TIC) allowing them to work remotely and in a network ("groupware", "workflow", GED, EDI). In the face of a difficulty in measuring the IC performance with the tools of traditional evaluation (general or analytical ledger), Edvinsson and Malone devised a model breaking up the value of the company into financial capital and immaterial capital. Their objective was to have a tool enabling them to plan, manage and adopt IC.

16.5.1. *The value of IC conceived by SKANDIA*

The immaterial capital is thus factorized into human capital and structural capital:

– human capital is the totality of individual capacities, knowledge, talents, experiences of the employees and business leaders. The question that the organization could ask itself is one of recognition of new competences and knowledge, or the emission of new ideas and their applicability;

– structural capital contributes to the effectiveness of IC. It is broken up into customer capital, represented by customer relations (analysis of the fidelity of existing customers or the acquisition of new ones and, the study of satisfaction...) and organizational capital. The latter, a true technical and logistic support for the IC, is the fruit of investments in systems and tools that make possible the diffusion and creation of knowledge. By means of group technologies such as CAPM (Computer-aided Production Management), CAMP (Computer-aided Management Production), CAMM (Computer-aided Manufacturing Management), etc., it regroups capital innovation with licenses, patents, and all the capacities or incapacities of renewal, such as the capital process which makes it possible to improve productive efficiency.

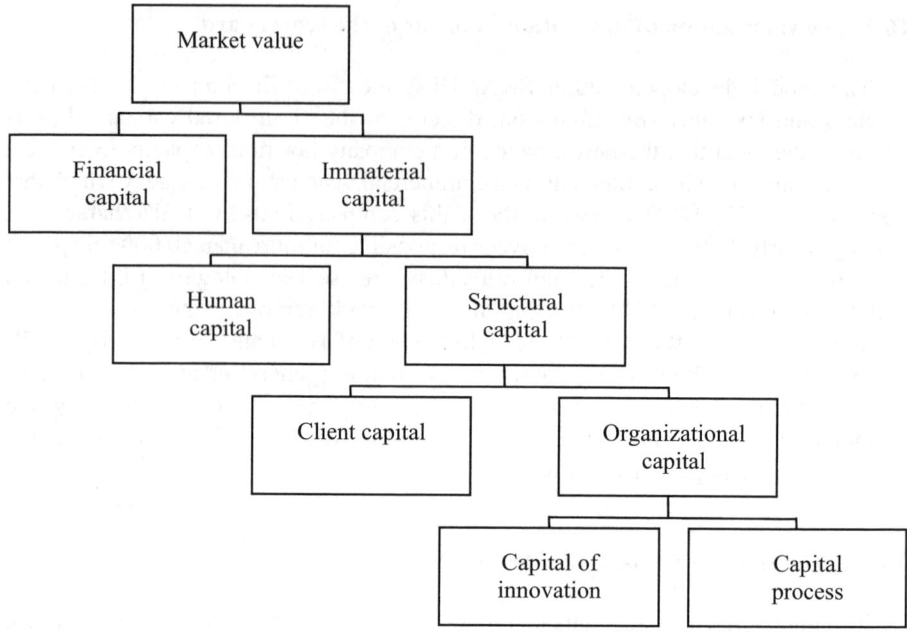

Figure 16.1. *Factorization of hidden values*

Five domains then participate in the creation of value for the company and are pointed out by nearly 800 indicators.

16.5.2. *The SKANDIA navigator*

Divided into five areas that are at the source of the value of the immaterial capital, the navigator model is represented schematically in the shape of a house (Figure 16.2). The roof keeps a track record of all past actions and financially shelters the house from the uncertainties of the environment. The walls, signifying the present, indicate the activity of the company related to its customers and its "processes". The foundations correspond to the future of the business house in the field of renewal and development. As for the hearth of the house, it represents human resources that irrigate the other fields. It is here that are born the rare resources that the company is in need of, those that it must motivate and maintain in its bosom. It is thus a combination of experiences, of the employees' own spirit of innovation and of the company policy regarding the modification or the preservation of this combination:

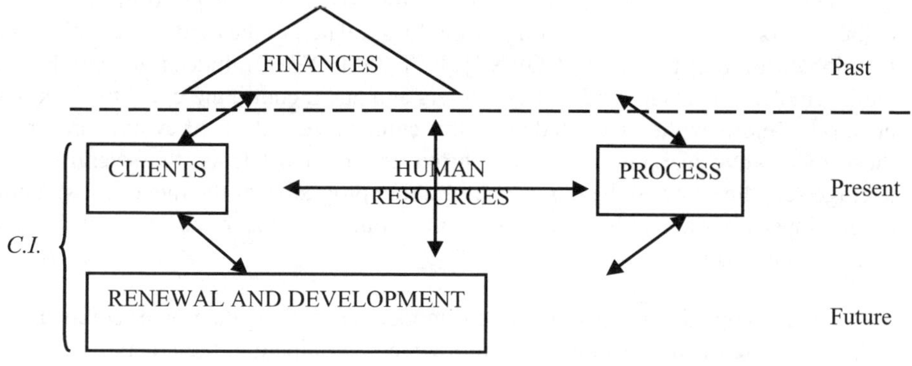

Figure 16.2. *The SKANDIA navigator*

– The financial field measures financial capitalization through, for example, the total sales turnover and new product or manpower or the strength (R&D and in total), through profit, total assets, the net return on credit and the value added by the workforce.

– The customer field evaluates the flux of relations between current customers and prospective customers by means of market shares, the number of customers, the percentage of regular or new customers, the duration of the relation, the number of customers who make complaints and the number of customers lost.

– The "process" field relates to the way in which the company uses the TIC in its internal relationship with the employees, customers, suppliers, distributors and partners. This is translated into indicators of the time taken in the processing of files, the level of knowledge in data processing, the number of computers in relation to manpower and the replacement cost of computer hardware.

– The field of renewal and development evaluates the actions that ensure adaptation in average terms, such as the investments made in the wake of competition, the number of new products and services under development and the number of those which arrive in the market, the rate of investment in fundamental research, in the design of products and in the applications, the number of patent registrations, the age of the patents, the number of licenses to manufacture, and also the number of engineering degree holders or scientists, the total manpower in R&D in relation to the total strength of the company, the number of current or future recruitments of creative talent, the ensuing training cost, and the bonus given as incentive for suggestions or at the time of the launch of new products.

– The human resources field makes it possible to evaluate the performances of the collaborators in terms of the existing difficulties, related to the dematerialization and the globalization of the result [COR 90] [ETT 92] [ETT 00]. Indeed, the employees work remotely, in group, with the customers and again contribute collectively to the creation of more explicit knowledge. In this context, the criteria of evaluation can be those of creative manpower, the new turnover, creative talents, the scientists, the average age, the cost and the duration of training programs, or the rate of satisfaction of employees vis-à-vis the policy of remuneration or the rules laid out for the conditions of work.

Several companies are inspired by this model of preparing their own scoreboard in order to update themselves on the factors which participate in the creation of value, and then to take decisions regarding general policy making which are essential: Dow Chemical of the USA, Posco of Korea or GrandVision of France. These companies defined the most appropriate indicators for their decision-making while insisting on the development of the non-financial indicators in each one of them. We propose to present two of these adaptations of SKANDIA model: Dow Chemical and GrandVision.

16.5.3. *The adaptations of SKANDIA model*

16.5.3.1. *The example of Dow Chemical*

In 1992, the Vice-President in charge of R&D of Dow Chemical, a world leader in the chemical sector based in the USA, approached Gordon Petrash, the Head of the department of expanded polystyrene insulators, after having realized that the company under-utilized its very numerous patents. The Vice-President, who had considerable experience in the correct management of intellectual property in his department, then asked Petrash "to draw more money". At that time, Dow Chemical was spending $1 billion per annum on R&D, had 4,000 researchers who deposited several hundreds of new patents every year for 2,000 products of the company. The wallet of patents reached 29,000 units, and the maintenance of their protection alone cost more than $170 million over a period of 10 years.

The first job for Petrash was to consult the patents database, the true "system of R&D information" born 30 years previously. The data was available but lacked a clear connection between them. Petrash thus undertook to constitute a team of experts to better formalize the information and build a common language. He defined along with his team, a model of six phases:

1. WALLET: states the existing intellectual credits
   - rights held; rights still in force,
   - finance departments, user departments.

2. CLASSIFICATION: to determine the use of the intellectual assets
   - the unit uses the patents; the unit will use the patents; the unit will not use them,
   - the number of licenses; number of abandonment of patents; patents to be abandoned.

3. STRATEGY: to analyze the use of the wallet and to integrate it into the strategy

4. VALORIZATION: to evaluate the intellectual credits
   - the total amount in excise duties; amount in taxes,
   - classification by order of priorities.

5. COMPETITIVE EVALUATION: to organize and compare the patents
   - number of competitor patents; importance,
   - strengths and weaknesses of competition,
   - field of application; difficulties or opportunities of application.

6. INVESTMENT: to clog the gap of discrepancies with the strategy   - internal developments, external acquisitions.

This first stage led to the revaluation of intellectual assets and to the formation, according to the actions undertaken and the competitive context. In 1996, nearly 100 people were in charge of the system with an operational budget of $1 million. The abandonment of the useless patents made it possible to save $40 million. Currently, this step is extended to other intangible assets like know-how, trade secrets and human resources knowledge.

Let us leave the sector of chemical industry and examine that relating to services to private individuals in the field of optics and photography.

16.5.3.2. *The example of GrandVision*

Greatly inspired by the approach centered on the management of immaterial capital, GrandVision conceptualized in 1995 a construct of the immaterial capital comprising four elements: the customer capital, the human capital, the memory and methods capital and the renewal and development capital. Started in 1981, GrandVision is a service company of photo and optics development. It groups together the brands Photo Service, Photo Station, Large-Optical, General Optics and Vision Express. Its sales turnover in 1999 was €0.75 billion with nearly 9,800 employees and sales distributed in 14 countries. This same year, GrandVision received the "Trophy of the decade" for its performance with regard to employment, growth and profitability. The scoreboard developed by GrandVision is that adapted from the findings of Bonfour [BON 00], and applied to the field of innovation in each brand and to the entire group.

HUMAN CAPITAL	Indicators per brand and for the group
Number of collaborators	Manpower as of 31/12/99
Number of collaborators in R&D	Manpower – R&D as of 31/12/99
Number of recruitments of young talents	Total hiring in 1999
Education	Number of hours
Age and average seniority (R&D, marketing, outlets)	Age and number of months of seniority per service
DOCUMENTS and METHODS CAPITAL	
Stages, micro-training, tools and modules of training	Number of hours of training organized by the Institute of Group Competence
Quality services	Number of collaborators
Speed of service	Work deadlines and % of the objective
Suggestions of collaborators	Clipboards, reports of astonishment for new ideas
Modules of explanation of back office procedures	Number in 1999 and the disparity *vis-à-vis* that of 1998
Formalization of the culture	Charter of values and principles of management
CLIENT CAPITAL	
Served clients	Number in thousands in 1999
Rate of issue of discount cards (to loyal customers)	Number in thousands in 1999
Satisfaction (welcoming, counseling, choice)	Rate of satisfaction or dissatisfaction
New clients	Number in 1999 (and in 1998)
CAPITAL DEVELOPMENT	
Year to year increase in the number of shops	Number in 1999 (and in 1998)
Number of renovated shops	Number in 1999
Number of new products	Number in 1999
Rate of success of new products	Share of sales of brands
Profit-sharing by employees	Total amount churned out at the end of 1999

16.6. Conclusion

The approach by the scoreboards of the immaterial capital has enabled us to understand innovation as a rare creative resource of tenable and lasting competitive advantage. However, innovation is a dynamic and continuous process, created from the combination of individual, collective and organizational knowledge and know-how. To identify it, to evaluate it and develop it, we called upon an overall visualization of factors which support and encourage innovation. The scoreboards of the SKANDIA navigator variety do not just put forth the financial indicators (results of the transfer of patents, of licenses for example). On the contrary, they put together a cluster of indicators that could take the shape of a chart of competences to innovate, multi-criteria that can be superimposed in average terms [LOR 95], on which are found:

– criteria of appraisable results like the attainment of individual and/or collective quantifiable objectives (reduction of deadlines, increase in the sales turnover related to the new products, etc.);

– criteria of observable behavior (ability to work in a team or to make suggestions for improvement., etc.);

– criteria of dysfunctions of production (rejections, complaints) and of behavior (turnover, loss of customers, etc.);

– criteria of knowledge production: the number of improvements in the production process, simplification of the prescribed rules and the number of new products put up on the market, etc.

The scoreboards steering forth the innovation must however be accompanied by a more individual, annual evaluation process or by planning, to estimate competences of the creative talents and their capacity to be developed in the organization. Face-to-face talks and one-to-one interactions then become a matter of regulation of activity (definition of the objectives, evaluation of the results, definition of the need for training and orientation, etc.) but are also a tool that serves the intermediate restructuring, enabling it to make an assessment of "immaterial" skills [PEY 04].

And if one finds Jean Bodin's expression – "there is no greater richness than men" – alluring, it is nevertheless necessary to maintain a more general, human and structural capital to support innovation.

Chapter 17

Financing Innovation

Introduction

Time, risk and cost are the fundamental components of the financing of innovation. They can be fatal to the innovator as he is bound to make heavy investments, both material and intangible, without being sure of future returns and profits. It is often "patient money" brought in by partners, "business angels" or capital risk takers that make it possible to reach the most uncertain or risky stages of the innovation. In this chapter we shall deal with the characteristics of the financing of innovation and show how a system of financing, associating private and public money was gradually set up in France, so as to accompany the innovator from the beginning of his project right up to the development stage.

17.1. Needs for financing associated with innovation

Innovation and its financing fall under the logic of investment: they require funds in advance, the recovery of which will be deferred, spread over a period of time, and uncertain. Whereas certain investments are routine and less risky (for example, renewal of investment in the scope of the continuation of the project), innovation presents in several cases a breaking situation that gives rise to five principal difficulties:

– very high costs;

– risks that arise due to an element of uncertainty as regards to the technical and commercial success of the innovation;

Chapter written by Pascale BRENET.

– a relatively long period between the beginning of the project and the return on the investment, which stretches from the stage of implementation of R&D to that of the industrialization and then the commercialization of the project;

– difficulty in making reliable projections because of the uncertainty associated with the innovation and the low visibility of the future performance of the innovating company;

– difficulty in appraising or evaluating the innovating project as this requires technical as well as commercial and marketing expertise, not forgetting the appreciation of the human factor that often plays a key role in the success of the project.

17.1.1. *Time, risk and cost of innovation*

The costs that arise due to innovation change throughout the process of innovation. One may assume in general that they increase exponentially as the project advances: the Chabbal Report [CHA 95] shows that the finance requirements of high-tech companies increase in a 1/10/100 ratio between the stage of research, that of development and that of industrial and commercial implementation. Figure 17.1 shows that the increase in the cost of innovation is accompanied by a decrease in the risks involved: indeed, at the beginning of a project, the element of uncertainty is maximum, whereas the visibility gradually improves until the commercial launching of the product in the market, for it is only at this moment that all the technical difficulties have been solved and that the innovation is validated by the market.

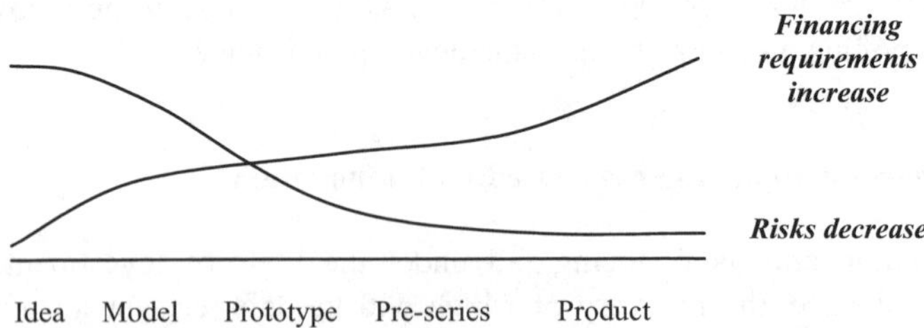

Figure 17.1. *Time, risk and cost of the innovation*

Moreover, the time span could be longer or shorter depending on the technology and the markets: for example, in the field of biotechnology, the structural constraints relating to the development of a new drug involve a 10 year time period between the discovery of a molecule and its approval by the concerned medical authorities before its release for sale in the market. In the case of the automobile industry, there is a 3 to 4 year period between the designing of the vehicle and its actual development and launch in the market, the flexibility of design being a key factor in the competition between Japanese and European manufacturers since the beginning of the 1990s.

Figure 17.1 illustrates the various problems associated with the financing of innovation: how does one resolve the difficulties specific to the beginning stages of the project which are the riskiest in the whole process of innovation? What are the respective roles that public and private funding should play in financing? How can one optimize the relay between the various categories of resource providers? How can the company divide its business plan into successive stages that would correspond to as many "pools"?

Regarding finance, innovation leads to three categories of needs: material and intangible investment, increase in the need for working capital that arises with the progress in development activity, and finally expenses related to the project (for example, remuneration of the personnel involved in the research and development effort) that could incur losses and even a negative cash flow. These three needs, which are directly related to the implementation or development of the project, appear in the finance plan. The company will have to mobilize resources in order to finance them.

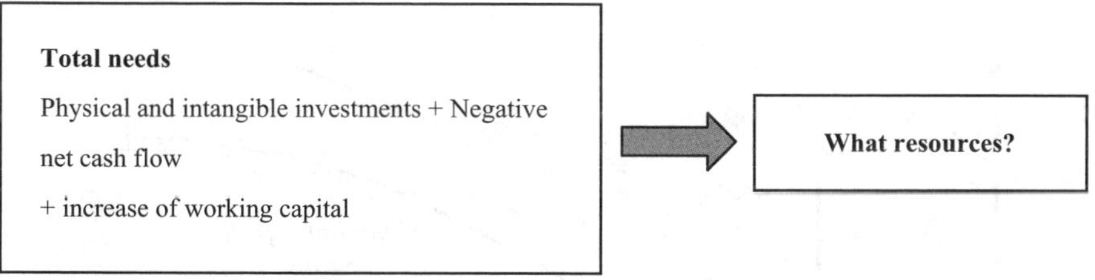

Figure 17.2. *Operating needs in the financing plan*

Investment is defined in accounting by the acquisition of goods, property and securities that are not consumed at the first use and that are meant to be used sustainably in the activity of the company[1]. Material investments represent equipment and material necessary for innovation whereas intangible investments cover expenses associated with R&D and intellectual property, marketing or

---

1 Let it be noted that in conformity with the principle of investment, companies may "activate" certain expenses incurred during a financial year if this will enable returns in future financial years so that they do not weigh too heavily on the indicators of performance in the year in which they are incurred. For example, the activation of R&D expenses is possible if the corresponding projects are clearly identified and if they have a good commercial earning capacity in conformity with the rule of conservatism that governs private accountancy.

software or training. Within the framework of a strategy of external growth or partnership, the company can also carry out investments in the form of financial participation. The investment of companies is characterized today by the predominance of intangible and financial investments, which raise the delicate problem of their evaluation.

17.1.2. *The financial lifecycle of innovation*

By assuming certain linearity in the implementation of the innovation project, it is possible to define four phases within the concept of a financial lifecycle: design, launching, development and maturity. For each one of these phases, expenditure, the level of sales and results change in the case of specific financial difficulties.

The phase of design is characterized by intense R&D activity, and by an effort to define the marketing strategy of the product. During this initial phase in the life of a project, the investments made are primarily intangible; the sales turnover of the new product is still zero. The losses incurred during this period are characteristic of this first phase and involve a strongly negative cash flow whereas the risk is maximum, as technical and commercial uncertainty weighs heavily on the innovation at this point in the process.

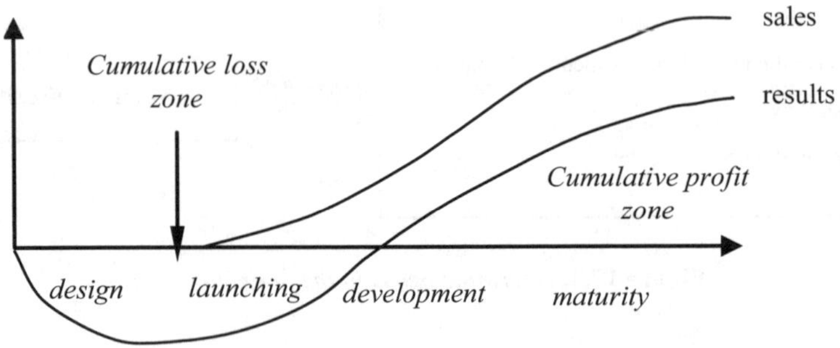

Figure 17.3. *Financial lifecycle of an innovation project*

In the phase of industrial and commercial launching, the risk is still very high but the visibility will gradually improve. The costs increase and include in particular physical investments (industrial tools) and expenditure associated with marketing (deployment of a sales team, development of distribution channels). This expenditure will be gradually compensated by the sale of the new product.

It is only during the phase of growth that the project can become profitable and release positive cash flows. At this stage, the risk becomes an economic risk inherent in any activity, whereas the innovation has proved successful in the market. If the growth is particularly strong, the company will in general continue its investments and will have to incur expenditure and meet an increase in working capital to be able to increase the production capacity, develop its marketing policy, ensure its international presence and, if necessary, ensure the technical readjustment or the improvement of its product. The search for investors able to accompany this growth is then a key element in the success of the project.

Several parameters influence the financial lifecycle of an innovation project and consequently the terms and conditions of financing of companies:

– duration of the cycle, due to technical constraints (development time) and marketing (timeframe of the commercial launching of the product, deadline for market returns);

– the economic model of the company and its capacity to generate growth and profit;

– criteria of validation of the innovation by the market and the potential growth of the market;

– importance of the expenditure generated by the innovation, particularly that related to R&D and marketing which are in general the heaviest.

The biotechnology sector is a perfect illustration of the difficulties of financing innovation: it shoulders the increasing cost of R&D; the obligation to carry out pre-clinical and clinical tests means that the duration of the projects is around 8 years; that of obtaining the requisite approvals and authorizations of the concerned medical authorities before the product can be launched in the market is about 12 to 24 months; the level of risk, higher as the process has not yet reached the development stage, forces companies to multiply the number of molecules being developed which gives rise to corresponding expenditure. The "pipeline" in Figure 17.4 represents these various constraints.

Research	Development				Approval
	Pre-clinical trials	Clinical trials			
		Stage 1	Stage 2	Stage 3	
		Failure rate 80%		Failure rate 12%	

Figure 17.4. *The "pipeline"*

Thus, the principal criteria of evaluation of biotechnology companies are the size of their R&D budget, the number of molecules under development and the equilibrium of the portfolio of molecules making it possible to ensure the regular release of new products. To survive, biotechnology companies, which by their nature are "devourers of capital" and which are generally in deficit, have to greatly resort to external financing which is provided to them by public and private investors or by industrial partners. The crossing of the critical stages of the "pipeline" determines the life of these companies and conditions the progressive arrival of capital that they need.

Data 2002 – in millions of dollars – source [ERN 03]		
	USA	**Europe**
Turnover	33.6	5,368
R&D expenditure	20.6	3,164
Net losses	1,500	1,189
Number of companies		1,351

Table 17.1. *Biotechnologies*

This situation, which is very specific to the biotechnology sector, is illustrated below with the example of the first 10 years of the financial lifecycle of a small to medium-sized company. In spite of a constant increase in its sales turnover, the company reached its break-even point only in year 10. The cumulated losses of this long period were financed by the capital brought in by financial and industrial investors.

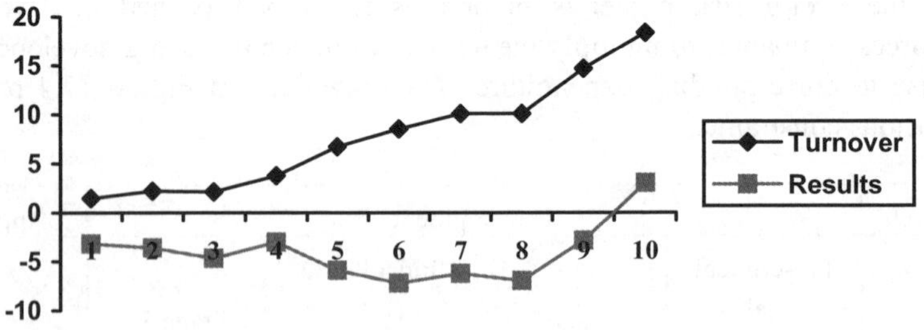

Figure 17.5. *Financial lifecycle of a small to medium size biotechnology company in thousands of Euros*

17.1.3. *The financial fragility of innovating small companies*

The financial difficulties evoked above become more or less critical for different categories of companies, depending on their size, their age and the diversity of their activity. Whereas a large company has the possibility of financing the expenditure associated with innovation because of the profitability of its other activities within the framework of portfolio management and an allocation of total or combined resources, a young company that is not yet into diversification must bear the full risk of the innovating project and cannot depend upon self-financing, which means that it has to look for external sources of finance and to convince investors.

The financial structure of small and medium-sized companies is generally characterized by meager capital stocks or funds, a strong dependence on inter-firm credit and a debt level that is often high. The INSEE statistics show that the financing of these companies is in itself a risk, since the failure rate is higher the smaller the size. These companies distinguish themselves by their low life expectancy; only half of them survive for more than 5 years after their creation (source: INSEE).

Innovation adds to the intrinsic risk associated with these companies, either by imposing a breaking situation on the existing company or by weighing upon a risk that is already high due to the creation of the company. The handicap to innovation is particularly obvious in a financial context that is hardly favorable for it, as banks offer essentially short-term finance and the majority of investors displaying certain timidity when it is a question of assuming high levels of risk over long timeframes.

17.2. Adaptation of resources to innovation: "patient" and "loseable" money

Taking into account the specificity of financing needs associated with innovation, the resources that are adapted must have three characteristics:

– the first relates to "patient" money, i.e., for which the investors are ready to wait several years before recovering their initial investment and hoping for a return on investment. In the field of venture capital, for example, the timeframe of such participation is on average 5 to 8 years;

– the second relates to "losable" money, i.e. accepting the risk inherent in any innovation. From the investor's point of view, the risk should imply a greater return on investment which is called a risk premium. Capital investors measure the performance of their investments by calculating the rate of disaster (proportion of projects in which they lose their investment amount) and by the TRI, i.e. the internal rate of return (see on this subject [ERN 02] and [AFI 03]);

– finally, investing in innovating projects requires strong expertise of the investors to appraise projects and the teams which implement them. This requires time and involves relatively high costs which can be redeemed only on future investments.

17.2.1. *Arbitration between debt and capital*

All companies dispose of resources that are provided by two categories of actors: equity shareholders and lenders. Shareholders bring resources at the time of the constitution of a company or new issues of capital which mark its development. They also enable self-financing by the company, when they forfeit their dividend which constitutes their part of the profits of the company. The financing policy of the innovating company is based on the fundamental arbitration between its financial debt and capital. Taking into account the needs that were analyzed above, it appears that the innovator, particularly for the phases preceding innovation, will have to privilege "losable" and "patient" resources. After having explained the reasons of this arbitration, we will show how the innovator can harness the possible components of a pool of varied resources.

In the debt contract, the borrower makes an initial commitment as regards the terms of reimbursement and payment of interest. Consequently, the debt involves fixed costs, independent of the economic performance of the company. This is at the origin of the financial risk associated with the impact of the debt on the returns of the capital of the company. The debt also involves a risk of liquidation associated with the capacity of the company to meet its commitments on their respective due dates.

Financial theory and analysis of investment options [VER 03] teach us that resorting to debt is convenient if the cost of the debt is higher than the operational rate of profitability: the income that the company draws from its assets must enable it to remunerate the resources which enabled it to acquire these assets. The company can then bear a higher level of debt, as its profitability increases and becomes more stable in time, and it has a better visibility of its future performances. It benefits in this case from a positive leverage effect. In a young and innovating company, which has, in principle, a low visibility on its future economic performances and that could incur significant losses at the beginning of its creation, the cost of the debt, even if it is low, is then hard to bear.

The selection criteria retained by banks and their principle of care eliminates dossiers or projects that are too risky because of the lack of visibility on the results and financial flows. In addition, when the debt must be used to finance intangible investments, the difficulty of the guarantee and the appraisal of the value that could be created by such investments arise.

Financial lever: net financial debts/equity capital	Considered a risk above 100 to 120%
Reimbursement capacity: net financial debts/capacity of self-financing	Considered risky if it is above 3
Ratio of interest cover: operating results/financial expenses	Considered comfortable beyond 3 or 4

Figure 17.6. *Banker's ratios*

Capital is a resource that is fundamentally different from debt for two reasons. The shareholder is a "residual creditor" of the company [BAT 99] and takes part in the risks of the company: in case of liquidation, he will be considered after all the other creditors and will receive the residual value of the assets after retirement of all the debts. In the worst case, the shareholder could lose the entire amount he invested. In addition, the return on capital, contrary to the remuneration of debt, depends on the level of performance of the company and the decisions relating to the allocation of profits. Capital returns are adjusting data.

A double uncertainty thus weighs on the committed capital: the risk of profitability, which relates to the level of return on capital, and the risk of bankruptcy, which corresponds to the risk that the capital is devalued or wasted (hence the expression "losable money"). The concept of duration is associated with that of investment. This requires the investor to define at the very start, the advisability of his commitment, by resorting for example to an actuarial calculation making it possible to define the current value of future flows that could be generated by the investment.

Capital bears the risks of the company. As their remuneration is more uncertain than that of a debt, the shareholders expect to be better remunerated than creditors: they expect a risk premium that implies that the projects thus financed have a strong potential for high returns. Therein lies the difference for the innovator: whereas the banker defines a fixed interest rate by contract, the shareholder evaluates the appropriateness of his participation by measuring the potential of profitability founded on forecasts of activity and performance.

Consequently, capital is an expensive ingredient that obliges the innovator to convince investors of the potential of his project, particularly with the help of a "business plan", a genuine tool for dialogue between the innovator and the investor.

When the innovator chooses to go through the stock exchange, he must give great importance to the quality of his financial communication. On his part, the investor must have true expertise meeting multiple criteria, which could be a limiting factor as the appraisal of projects is long and expensive, and often requires the ad hoc appointment of teams.

Another characteristic of capital – and not the least important – relates to the right of control that is associated with it. The shareholder owns a part or a fraction of the company, and thus has controling rights on the management of its assets. The opening or issue of capital can be blocked by the cultural reserves of entrepreneurs/shareholders that have a particular vision for their company. As regards investors, the existence of a right of control leads to behavior that can be analyzed within the framework of the theory of the agency [KOE 99]. Thus, investors in venture capital, for example, constitute what the theory of the agency calls "blocks of control" and can provide counseling and accompaniment that could often be useful for the innovator.

17.2.2. *A pool of resources*

Taking into account its need for financing, the innovator must generally create a pool of multiple resources, with a good linking of its components according to their characteristics and their adaptability to the constraints of the various phases of the process of innovation. Figure 17.8 indicates the principal categories of these resources and their possible distribution throughout process of innovation (for a statistical outline of the financing of technological innovation in industry, refer to [LHO 01]).

At the start of the process, and particularly in the case of young companies, innovation relies on resources that can bear a high risk over a long period of time. The first of these resources is often brought by the innovator, his collaborators and people close to them[2], or by *businesses angels*. Certain national, regional or local public aids add to this private capital.

Self-financing constitutes a useful resource for companies that can afford it: this is not generally the case with young companies in the process of starting their business activities, which often have to bear losses associated with the setting up of a business and have to turn to external investors for finance.

Among these external investors are investors in venture capital, which take minority shares in young and high tech companies or projects with strong potential,

---

2 French legislature has recently created a new vehicle: the Fund for Investment of Proximity, which should facilitate regional industrial investments by providing fiscal benefits.

with the intention of making a profit (due to appreciation in value) through the sale of their share over a medium term period of 5–8 years. One can classify the investment pattern of these investors into four categories: seed capital that makes it possible to finance an innovation project in the early stages of the company, risk capital that invests in young companies (usually less than three years old), the development capital that invests in companies that are at least three years old and that show tremendous growth; and finally, certain investments in capital stocks which are meant only to finance transmissions of companies and LBOs (leveraged buy-out) and tend to distance themselves from the problems associated with innovation.

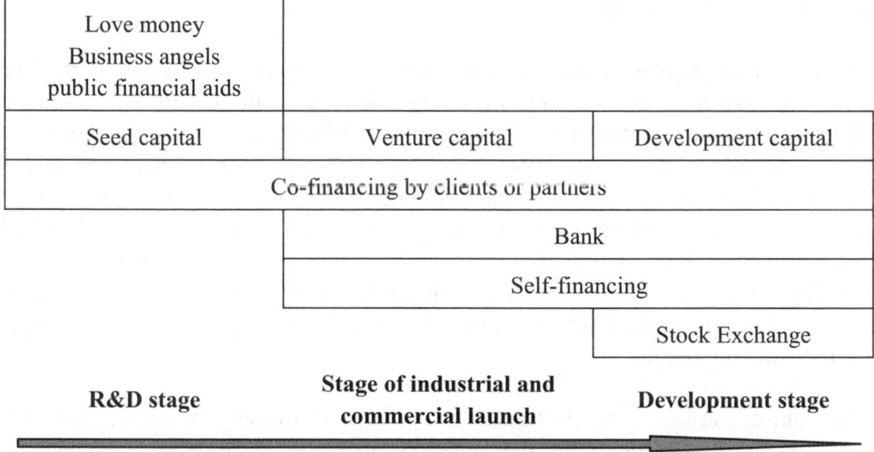

Figure 17.7. *A pool of resources for innovation*

It is only later, when the visibility of the project improves, that debts become a viable resource. It should be noted that bank credits today have a relatively low after tax cost (about 3%), which make them an invaluable resource for companies that have achieved regular growth and are able to provide guarantees to the lenders.

Lastly, the Stock Exchange can be a source of financing viable for companies that are hungry for capital. The creation of the NASDAQ at the beginning of the 1970s in the USA, and the appearance of similar stock exchange compartments during the 1990s in Europe – in France for example, the New Market was created in 1996 – allowed significant lifting of capital in favor of young companies with a strong growth potential.

Apart from the financial sphere, it is necessary to underline the sometimes essential role that customers and partners can play in the financing of innovating companies, through R&D partnerships that include clauses of co-financing or even through minority acquisitions of a holding from a strategic and not strictly financial point of view. In the case of biotechnology for example, the SMEs of the sector systematically develop research partnerships[3] with pharmaceutical groups, whose existence brings to these SME, a scientific and marketing guarantee and often determines their access to other investors.

17.3. The financial system of innovation

17.3.1. *Capital-investment*

Even though capital investment has been much talked about since the past few years, this method of financing has very old roots: in his time, Christopher Columbus obtained financing from Ferdinand and Isabella of Spain to undertake a voyage towards India by the West: then one spoke about a "loan for a great adventure". During the past 30 years, venture capital has been illustrated by many success stories: Intel, Motorola, Genentech, Biogen, Microsoft, Apple, Dell and still more recently Yahoo!, to mention only a few, that are the jewels of American technology that benefited from the contributions of venture capital. Generally speaking, three-quarters of the companies financed by the capital investment employ less than 100 employees.

In France, the legal and tax framework of capital investment was defined at the beginning of the 1970s, with the creation of the FCI (Finance Company for Innovation) statute. Since then, other legal structures have been created: the CCR (Company of Capital and Risk, created in 1985), the CFIR (Common Funds of Investment in Risk, created in 1993) and finally the CFII (Common Funds of Investment in Innovation, created in 1997). The general principle underlying these various statutes is to encourage investment in young and non-quoted companies by reducing taxation on returns or profits made on these investments.

Capital investment seems a resource particularly adapted to the financing of the innovation, for two essential reasons: in the first place, its contribution in capital that strengthens the top of the balance sheet of the innovating company can bear the risk of the innovation; in addition, this resource necessarily involves an expert appraisal of the project, which often leads the investor to help the innovator to put together a

---

3 According to the terms of these partnership contracts, the groups co-finance the research done by the SME and in return could obtain a contract for exclusive intellectual property rights. The right to exclusivity involves the payment of royalties to the SME when the new product is commercialized.

business plan or to refine or modify his project, as well as his management and marketing policy. Such advice can be useful for the innovator, as it requires him to take into account all the aspects essential to the success of his project and helps him to refine his vision of the project. This expertise and advice implies a significant cost for venture capital companies, a cost that could go up to €50,000 in the first year which explains the concept of "entry tickets" that are generally higher than €150,000 for this type of investment.

17.3.1.1. *The recent surge in French capital investment*

Until 1997, French venture capital was characterized by the weakness of the raised and invested capital and by the scarcity of teams trained in this difficult profession. The end of the 1990s was an exceptional period for investment of private capital. In 2000, the bursting of the Internet bubble slowed down the desire to invest and was the beginning of a period where investors wanted to clean up their investment portfolios. With the recent signs of resumption in investment, one can reasonably hope for the development of this mode of financing.

Several categories of actors have considerably increased their participation in capital investment since 1998. Among these are banks, insurance companies and pension funds, which represented half of the raised funds in 2002. Funds from funds became the second source of financing of the profession, accounting for 16% of the total capital raised in 2002. Next are investments by individuals (12% of the total), followed by public organizations (6% of the total). One has nevertheless observed a decrease in the absolute and relative value of the investments by industrial firms.

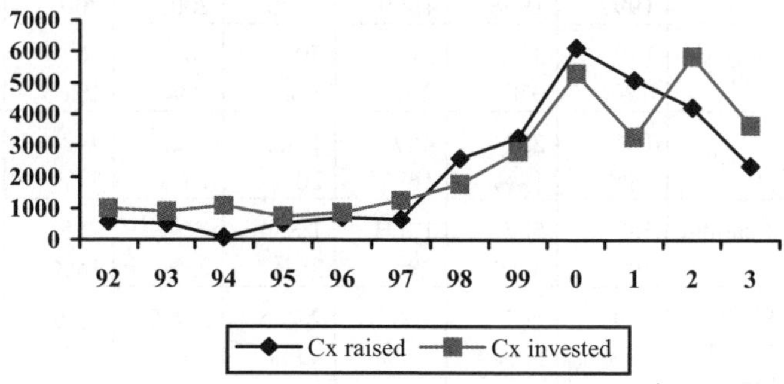

Figure 17.8. *Capital investment: amounts raised and invested: [AFI 04]*

Several factors can explain this recent development. First, there are fiscal incentives that are offered to capital investment along with the creation of new vehicles of investment over the last few years. At the same time, the state has considerably increased its financial participation in capital investment, particularly in seed capital, within the scope of the law for innovation of 1999. The recent development of Stock Exchange markets that welcome growth stocks contributed to a favorable environment for innovation and, more precisely, facilitated the outflow of capital: the multiplication of opportunities for such outflow is indeed vital for capital investment which is, by nature, temporary. Investments are also "drawn" by the multiplication of projects and the exposure that they receive – a scenario that is relatively recent in the French context – and the social and economic stakes associated with innovating SMEs. One must note the increasing internationalization of capital investment: in 2002 foreign capitals accounted for 59% of the total of the amounts raised by French capital investment. Their small growth during the last three years partly compensated for the decrease in French investments since 2001.

In the field of the capital investment, venture capital takes shares in the capital of innovating companies still in the process of creation, just about starting their business activity or companies that are in the first years of their activity. This type of investment is the riskiest and, for this reason, is directed towards innovating projects with a very strong potential for success This potential has to compensate for the risks. Compared to other types of capital investment, venture capital was the most affected by the reversal of economic situation in the year 2000 and showed radical and contrasting decrease, with extreme levels -19% in 2000 and +86% in 1998 (or +47% if one excludes two operations having generated very high returns). From 1996 to 2000, the amount of the average unit investment in venture capital was 716 KE, a net increase compared to the average of 320 KE between 1994 and 1998.

	1997	1998	1999	2000	2001	2002	2003
Seed capital	1 0%	3 0%	52 2%	70 1%	30 1%	50 1%	25 0.7%
Venture capital	166 13%	257 14%	467 16%	1,085 20%	531 16%	443 8%	307 8.4%
Development capital	382 30%	587 33%	1,071 38%	1,884 36%	720 22%	755 13%	785 21.5%
Transmission and repurchase minority ownership	709 57%	942 53%	1,227 44%	2,265 43%	2,005 60%	4,603 78%	2,015 55.3%
Total	1,258 100%	1,789 100%	2,817 100%	5,304 100%	3,287 100%	6,511 100%	511 14.1%

Table 17.2. *Amounts invested at each stage of development in ME and in percentage – source: [AFI 04]*

The amounts invested in start-ups are lower than 1% of the total amount devoted to capital investment. The starting of business activity is problematic because of the extremely high level of risk associated with it and the high costs of expert appraisals depending on the size of the project. Recently, funds exclusively dedicated for start-ups were created from public funds and supported with research centers, so as to encourage young firms. These funds currently invest unitary tickets of about 150 to 500 KE on very selective blue-chip technologies: information and telecommunications technology, electronics, and life sciences.

Development capital relates to investment in companies having reached their break-even point, and aims at financing their growth. By definition, this type of investment has a lower level of risk. It is more common in the banking sector. It involves higher levels of investment than those of venture capital, with an average unit investment of 1,240 KE between 1999 and 2002, as against 700 KE between 1994 and 1998.

	1998	1999	2000	2001	2002	2003
Entry on to the Stock Exchange	347	361	291	201	114	90
Industrial transfers	259	448	456	547	240	129
Others	630	593	487	428	417	567
Losses	138	167	164	136	155	53
Total	1,374	1,569	1,398	1,312	926	897

Table 17.3. *Transfers in numbers of transfers – source: [AFI 04]*

Transmission capital relates to the financing of repurchase and the transmission of companies with set-ups such as the LBO. These operations profit from the strategic movements of restructuring of certain large companies that are then brought to agree to the transfer of entire sections of their activities. During the last few years, some operations of exceptional size represented a very high share of the amounts devoted to capital investment, as can be seen in Table 17.4.

	1993	1994	1995	1996	1997	1998	1999	2000	2001
Risk capital	49	43	28	37	13	86	-7	-19	-6
Development capital	20	25	20	36	28	38	-1	-5	0
Transmission capital	13	33	18	41	36	20	8	11	4

Table 17.4. *TRI per year and per type of investment, in % – source: [ERN 02]*

Since 2000, capital investment has witnessed the reversal of the economic situation that weighed heavily on technological firms, the climate being unfavorable for outflows in the stock markets or industrial transfers. In 2002, the transfers came essentially from transfers made to other investors (secondary or tertiary buy-out operations) and the losses weighed more heavily than before on the total of the transfers (Table 17.1). Table 17.5 shows the strongest volatility of the TRI of venture capital compared to development capital and transmission capital for the period 1993-2001.

17.3.1.2. *"Business angels"*

"Businesses angels" are individuals who take shares in young technology companies in the first stage of their development. Like the venture capital investors, they invest in projects in order to make a profit at the time of the sale of their share that would generally be to a venture capital company. Financial interest is not the only motive of their action: for the majority, who are themselves entrepreneurs, it is a way of participating in creation and innovation. Businesses angels prefer projects that belong to their professional field so that they are able to evaluate their interest more accurately; and they also bring competencies and contacts. Their average investment per project is about 100 KE. There are no official statistics on the activity of these businesses angels that developed in France at the end of the 1990s because of attractive tax benefits and an overall favorable financial environment.

17.3.2. *Markets of growing stocks*

The New Market (NM) was created in France in February 1996. The objective was to accommodate young rapidly growing companies in the high-technology sector. Figure 17.9 indicates the conditions that companies wishing to enter this market must meet, which on the quantitative level are less constraining than those on the first and second markets. The chief specificity of the NM lies in the information requested from candidate companies: no previous accounting records are required; the company only has to present one development project along with estimated financial projections for the next three years.

It should be noted that entry to a Stock Exchange, even if it meets the needs of capital hungry companies, imposes numerous constraints on the financial communication of the company and obliges it to reach the high levels of profitability expected by their shareholders or at least to meet the targets laid down by them in their projections.

After its creation, the NM[4], in order to resemble other similar markets in Europe, quickly reached a critical size essential to its proper functioning and the liquidity of its securities. This market was, however, extremely volatile and witnessed a very marked reversal in the middle of 2000, as was the case in other growth stock markets, including the NASDAQ. Currently, the NM comprises of a little more than 160 listed or quoted companies and almost no listings took place between 2001 and 2003. On the NASDAQ, 60 listings took place in 2003, against 480 in 1999 and 408 in 2000.

Equity capital stocks > 1.5 ME
Diffusion of at least 100,000 shares for a minimum amount of 5 ME
Floating: 20% minimum (share of capital in the public domain/sector)
Cost of listing: 7% of the raised capital for financial intermediaries 1% fees New Market company Financial communication fees

Figure 17.9. *Conditions of entry on to the New Market*

17.3.3. *Public financing of innovation*

The state contributes towards the creation of an institutional and tax environment favorable to innovation. Numerous measures have been taken over the past few years:

– the law on innovation of 1999 whose objective is to encourage the creation of technological companies thanks to research, brings about a modification in the statute of researchers that favors their mobility, provides for the creation of 31 national incubators supported by the state or public research and for the creation of funds for start-ups;

– the creation of several medium of investment in young unquoted innovating companies that offer generous tax benefits on returns on investments made in these companies;

– such tax incentives also benefit individuals that invest directly or indirectly in young innovating companies;

– the state also contributes to the financing of private industrial research through the credit research tax (see www.impots.gouv.fr for more information on these fiscal incentives).

---

4 Visit the websites www.bourse-de-paris.fr and www.euronext.com.

The state also directly participates in the financing of start-ups and in venture capital investment through a budget managed by the Treasury of Deposits and Consignments (Caisse des Dépôts et Consignations), which is, with the ANVAR, the principal public operating agency as regards financing of innovation. The Treasury of Deposits and Consignments finances the incubators, the start-up investment capital, certain venture capital and guarantee capital. In all the above cases, the principle adopted is that of the systematic addition to public investment by private investment.

The ANVAR, a major protagonist in innovation in France, disposes of a budget in the range of $230 million every year that enables it to help more that 3,000 SMEs (the annual reports of the activities of the ANVAR can be found on its website www.anvar.fr). A large part of these funds are reimbursable/repayable in the case of the success of the project and this allows for the financing of innovating projects that are in the research or development phase. The financial principle that underlies this financial aid is adapted to innovation: repayment is necessary only if the project is successful (in this case, the aid offers functions like a loan without interest with deferred retirement of the loan) and the aid becomes a subsidy in case the project fails which amounts to the financing of the risk. On the other hand, the aid offered by the ANVAR cannot cover more than half of the expenditure incurred by the innovating company, which underlines the necessity of private investment in any business project.

17.4. Conclusion

Today, France has a coherent system for the financing of innovation, whose foundation has been gradually built and reinforced during last 10 years, with the development of "businesses angels", the tremendous surge in venture capital and, in general, in capital investment, the creation of markets of high tech growing firms which made it possible to improve the liquidity of venture capital and to clearly identify the specific features of the innovating company, the emergence of funds for start-ups, associated with public research centers, and finally all the measures taken by the state to accompany such innovating companies. If the reversal of the economic situation, observed from the middle of 2000, has badly affected the players that were the most exposed to the risks of the innovation, one hopes that the system of financing of innovation will emerge strengthened by this experience [SES 02].

The existence of a true system for the financing of innovation is fundamental, as stated by J. Lachmann in 1996 [LAC 96] who used the example of the flow of air in a chimney: the financing of innovation is fluid if the chimney functions well at the entry and the exit points: the entry point presents the problem of seed capital that must bear a high level of risk over a long period; the exit point presents the problem of liquidity and transfer opportunities that investors must find in order to ensure the movement or rotation of their investment portfolios.

What characterizes most of these resources, apart from the sums of money that they represent, is the expertise, the advice and the accompaniment of projects that they provide. This expertise is very useful for innovating companies. Recent facts underline the real difficulty in the exercise of this expertise: the fundamental question that is asked relates to that of possible methodologies to evaluate the intangible nature of innovation: the quality of the innovator, the potential of the technologies, the strength of patents, the value of brands, the potential of markets (one can read as an introduction to the problem of securities [HOA 01], [HIR 98]).

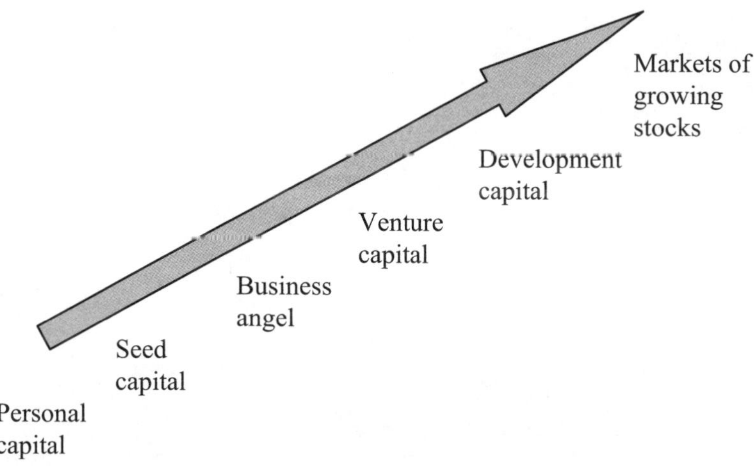

Figure 17.10. *The system of financing of innovation*

The bursting of the Internet bubble and the temporary disgrace of technological securities reminded us of the risks inherent in innovation. For the first time, the sums invested in certain projects were very high, and the projects largely mediatized in the case of quoted companies: this over-exposure should not make us forget the basic character of the risks taken. High technologies continue to make us dream – and rightly so: in recent performances of capital investment studied by Ernst & Young [ERN 02], the highest rates of return were obtained in the information technology and telecommunications sector (TRI equal to 32%) and in life sciences (TRI equal to 24%) – but innovation can also be the feat of so-called traditional industries.

Chapter 18

Innovation on the Web

18.1. Introduction

Along with email, the Web is the most widespread service used by Internet users. This phenomenon can be summarized in one formula: there is a before and after the Web. Nothing is the same today as it was before the Web.

After about 10 years of use, the Web has reached a certain maturity and today it offers services that facilitate communication and allow information not only to be shared easily, but to get it published too. However, the Web is far from being the only tool of the Internet; a number of tools are available to users (news, chat, instant messaging, Webcam, etc.). Now these services, despite their real qualities, seem especially adapt for non professional uses. That is why we will limit the scope of our study to those services on the Web that appear to offer the best from the point of view of innovations.

Two approaches can be proposed for innovation: using the Web services for innovation is a reality, as what could be more natural than using the possibilities of publishing and exchanging data offered by the Web and using them for innovation? However, there is another approach, a sort of endemic virtuous circle that has become over the years a storehouse of information: Web services are not just used for innovation, they are the cultural center for innovative initiatives that cannot be realized without the Web. This movement is at the root of the new paradigm which is in the limelight today: Open Source and its counterpart, the software patent.

––––––––––

Chapter written by François DRUEL.

This auto-maintained double movement is at the base of a certain number of innovations. Some of the most important of these are being dealt with in this chapter. As it is difficult (and dry) to be only theoretical when practical subjects are discussed, we will give concrete examples to illustrate the heavier aspects of this sector.

Sometimes two rather opposite approaches come together: certain Open Source projects are in fact supported by commercial companies. This approach creates quite a virtuous circle but sometimes also a Machiavellian one. A company can greatly benefit from Open Source software: this gives them the opportunity to obtain the services of a large number of free developers.

This happens in the case of Open Source Mozilla[1]. As everyone knows, Mozilla is Netscape's "clone" distributed under a free license. Netscape declared another version of Netscape, Mozilla, and closely followed the techniques of this version that contains a large number of innovations. Naturally, the development of Netscape incorporates most of the techniques integrated in Mozilla.

It is a win-win situation: the free developers who can participate in an interesting project and who (all things being equal) reserve the right of coding the functions of their dreams, and Netscape which reserves the right of integrating these functions in its commercial software. The famous "anti-pop-up" function was first developed for Mozilla before being integrated in into Netscape's Navigator.

From the user's point of view, the consequences are equally beneficial as two possibilities are offered to benefit from the flexibility of Mozilla or from the services of a commercial software.

In France, the company Soft Innov[2] is at the root of a similar initiative as it supports the technique R# (pronounced R sharp), a clone under free license of the language Rebol. The method is similar: SoftInnov has been using Rebol for a long time and has been facing the limitations of this language. Therefore, why not "clone" Rebol in order to improve it? Why not give this benefit to everyone? To these questions, which earlier would have been unthinkable, SoftInnov brings a positive double response: yes, we can "clone" the language (with of course the consent of its creator), and yes, R# must be made available to the community of computer users. SoftInnov begins at this point and subscribes to the principle that a language is, by definition, open and for "free circulation".

---

1 www.mozilla.org.
2 /www.softinnov.com.

On the other hand, the Web has created around itself its own paradigm, its own system – technically of course, but also economically. Marketing, electronic commerce, sharing and publishing of information… entire sections of the old economic systems have been modified by the arrival of their online equivalent. The life of the innovation too has its before-Web and its after-Web life. We would like to describe some innovations at the service of the innovation given by the Web.

These changes brought about by the Web should be compared or even contrasted with the earlier methods. They should also be studied for themselves in their own context. This is what this chapter deals with.

18.2. Distribution model: Open Source and software patents

One thing that is clear in matters of innovation is that the Web is developing in detail the new distribution model. Once again, there is a tussle between freedom of publication and protection of inventions.

18.2.1. *The clash of the titans*

Today a simple idea cannot be patented (ideas are for "free circulation"); only the processes that can be used industrially are capable of being patented. This is a well established rule.

Crucial point: a patent makes it possible to take legal action against anyone who infringes it during its lifetime, which is 20 years (for a non-modified patent, but its life can easily be prolonged: the only thing required is to "encapsulate" the earlier patent into the next). During this period the inventor can freely use his invention and its related rights (moral and pecuniary). Often inventors concede usage licenses of their patents to companies which exploit them. Royalties (generally a percentage of the sales turnover) are paid by the companies to the inventor, which frees them from financial worries in case the patent is used to manufacture mass products. At the end of this 20-year period and if the life of the patent has not been prolonged, the invention falls into public domain and can be freely used and adapted.

The defenders of the software patent wish to stretch these principles to software packages and they stand to benefit from it: more international, more practical, more modern, more flexible to use and encash than the good old copyright.

And does everything happen for the best in this world?

Well no, not really, as there is a "small detail" which makes everyone sit up: a patent does not come for free: it has to be purchased, and sometimes very is very expensive for its holder.

In fact, though copyright is very flexible and not expensive (for example, the only thing required is to deposit a manuscript of a musical score to Sacem (cost less than €10) in order to be protected under copyright laws) applying for a patent is very onerous and rather costly:

– onerous because it is formal: applying for a patent means mountains of paperwork, tons of procedures and hours of work, to such an extent that there are firms that specialize in applications for patents;

– expensive because when an application for a patent is made, the firms charge various fees and taxes and, what is more, application for the patent has to be made separately in different countries and for limited periods which can be increased by paying taxes each time (so as to arrive at the fateful 20 years and sometimes more, by modifying the earlier patent).

A simple individual cannot correctly apply for a patent without spending huge amounts (if you consider that applying for a patent for a method to protect an invention in France alone and for a period of only one year costs about €10,000).

Thus copyright protects all creations of all authors, whereas the patent is clearly adapted to company products.

Especially if the software patent is based on the principle of licenses (particularly the famous free licenses) it will be heavily questioned and certain software "majors" could use freely and with impunity source-codes under free license (precisely because this code has not been patented).

In addition, the same "majors" can patent tools which we use every day (the scroll bar, cursor, and even the window).

What about the individual developers? Well, there will no longer be free software packages, shareware, freeware and all their variants. One can almost bet on it that one would have to pay usage duties to those selling platforms (I have personally developed a small accounting software package that works under Windows and so I have to pay royalties to Microsoft as my software uses "bits and pieces" from Windows that is protected by a patent).

18.2.2. *Publication vs. patents: innovation vs. industry?*

In agreement with the current protesters, we note that if the software patent is brought about, it will tremendously change the small world of software.

Evidently, the software patent is the answer given by the industrial shepherd to his charge Open Source. This is the clash of the Titans: creators against editors, freedom against money, discovery against vested interests. Even more serious is if the software patent is applied in the software field (and so in the intellectual domain), it could hamper other connected fields especially music. When one knows how much the industrialists from the musical distribution sector fear the Internet, one understands their concerns in this debate.

Open Source is basically made up of an innovation, as its approach radically changes not only the method of designing and developing software projects but also and especially one's mindset.

This principle is not new; it has been present for a long time in the world of standardization (standards are public) and instances of standardization of the Internet (one of which is Web Consortium[3] for fundamental techniques related to the Web), also work according to the principle of openness. And one does not have to restate the success of other technologies, often the result of research and all Open Source (PHP language, the CGI, MySQL, without counting the unavoidable Linux, which of course is an Open Source implementation of the Unix system which is also an open system…).

This state of mind is somewhat altruistic as acceptance of the "liberties" of the Open Source model implies the abandoning of the most commonly accepted form of intellectual property[4]. Placing an object under the Open Source license comes back to working for collective development and abandoning individual rights in favor of community progress.

It is this approach that is problematic, precisely as it goes against the usual rules of the liberal economic model and its induced cupidity. However, in terms of innovation, Open Source is a benefit, as what is more adapted in fact to the development of new ideas and advances than the sharing of information?

One thing, however, is certain: there will be resultant changes in the small world of software, or at least the sector will not come out untouched. Open Source can be seen as a sort of pendant in the information company of anti-globalization. Richard M. Stallman, its religious head and symbolic guru of some kind of "ecolo-freak"!

---

3 www.w3c.org.

4 By an English pun, it is sometimes called "copyleft" as opposed to copyright.

In fact, beyond its pleasant or even comic aspects, Open Source asks the eternal question of the place of research in society. Since the beginning of the industrial era, research has always grown along with industry. The world of publication (especially scientific) with that of the patents. The research personnel with the industrialists. Fifty years after their birth, computers also face this thorny and painful problem.

Once again, humanity faces a dilemma opposing liberty (of creation and expression) to economy in an enormous conflict of interests. I would be tempted to philosophize with Gide: "the world shall be saved if it can be only by deserters"…

It is too early today to conclude on the thorny and important question that we have just raised: what is the future of Open Source? Will the software patent be imposed upon it? Here is a subject on which one has to be very vigilant: undoubtedly this will affect innovation in the software field.

18.3. An enormous base of information

If there has been a domain that has been strongly affected by the Web, it is that of publishing. The Web offers, in fact, a capacity for publication that is easy, approachable, immediate and pleasant.

Easy because the techniques used are friendly: the HTML language is a model of simplicity for an author at least (despite the well-known limits and encumbrances known to software professionals). Even without any particular knowledge, the rudiments of HTML can be learnt in a day. Several softwares have reduced the creation of (web-) pages to simple administrative tasks which are known by all. In addition, online publishing systems have greatly simplified the work[5].

Approachable because the necessary resources for publishing (disk-space, server softwares, databases, programming languages) are often freely accessible, Internet access suppliers (like Wanadoo) include these services in their access packages (a consequence of the commercial war that they wage in order to provide the greatest benefit to their clients).

Immediate and pleasant, there is no need to return on these fundamental characteristics of the Web. With a simple click you can create a site with a respectable look and feel.

It must be noted that publishing is all the more friendly as Internet access is simple and quick. In this sense, the great development of high output networks

---

5 For example, the Open Source Spip system (www.spip.net).

(cable, ADSL) is an important facilitating factor. Efforts of operators and access providers must be commended for the development of these technologies.

With all these qualities, the Internet has become an impressive source of information on all kinds of domains. Of course, the Web edited by companies having difficulty in providing adequate content to their clients make up an important mass of data.

However, while publishing was until now reserved for an elite wishing to publish on a limited number of supports, the limit seems today to be the creation of a bunch of specialists (more or less self-proclaimed) who have difficulty in communicating.

This is naturally true of university publications, students using their personal sites to "publish" their PhD researches; to advance their research, research personnel publishing on the net articles given to magazines, but also amateurs fired up by a subject and wishing to share their passion.

Amateurism (in all senses of the word, especially the good ones) is king. Nevertheless, data exists and is available in an unimaginable number of domains all as varied as each other...thinking of it: a billiards course, a treatise on crystallography, anarchist calendar of events, cooking lessons, marketing analysis, history of the automobile and photography lessons are only some examples in certain domains taken randomly out of curiosity.

Although traditional research for information is complicated (setting up a bibliography, collecting books and periodicals, etc.), the information research exercise on the Internet is amusing and surprising: take a theme, type the keywords in the search engines and within seconds you will have found a certain number of answers to your questions or at least indications for deeper research. In this sense, building up a dossier can be done in a few hours as everything is within the reach of a click on the Web. The consultants from specialized firms and pupils of primary schools both know this!

This is most important for the latter that their relation to information search changed forever, as one knows that habits inculcated in the youth are enduring; but we will not debate this here.

Evidently, one cannot go further without asking the important and thorny question of the validity of the sources. What credit should be given to an information apparently pertinent, but found on a personal page of a totally unknown "author"? The time and flexibility gained in searching is sometimes counterbalanced by verification and cross-checking. In certain fields, this question can even be crucial and searching on the Web should be done intelligently. The power of the

Web should be well understood as it allows the easy diffusion of various data. It is well known that the text is not the only type of data that can be published: multimedia data especially can be easily made accessible online. Technology and techniques progress at high speeds, but this is not the case in all fields, for example: law and, as far as we are concerned, intellectual property. It is evident and the media are constantly echoing it: it is not because it is possible to make such and such type of data available that we have so many laws. One of the knots in these problems seems to be exactly in this conflict between law and practice. As usual innovations and rules cross and collide against each other.

18.4. Marketing and innovation on the Web

18.4.1. *A leverage*

If there is one field where the Web encourages, favors and entertains innovation, it is the field of marketing. One cannot imagine a more harmonious marriage: marketing thrives on novelty and innovation. As for the Web, it is unknown territory (the first tests of this new service are barely a decade old at the time of writing) on which small investments made on its power of communication made it no longer necessary for the marketing professionals to earn their daily bread on this new media that was being showed off as the new Eldorado hardly two years ago.

> When one thinks of it, brand names like Yahoo, eBay, Amazon, Alapage and Wanadoo became famous all over the world within just a few years… let us recall that respectable brand names of "the old economy"[6] (Coca Cola, Michelin, Gillette, Bic etc.) required decades to reach the same status.

On the other hand, the Web also profited from marketing: the famous e-marketing is a creation of the "web generation". Evidently, derived from traditional marketing, ideas such as viral marketing, community approach, without mentioning e-business, are today so much a part of life that the Web is an integral part of the marketing mix and has become unavoidable in any communication policy worth its name.

In less than a decade, e-marketing has reached such a level of maturity that today sites on which can be found in all domains. E-marketing is no longer reserved only for technological products: cheese, bread, perfumes, cars, Tour de France, to cite a few examples, all these popular products of mass consumption are present on the

---

6 The idea of old economy which has been formed at the peak of what is called the Internet bubble can seem degrading. However, it is not so. It simply recalls that there was life before the Web…

Web with sites whose high-quality content is combined with amazing creativity in communication.

> President Butter, Sitram casseroles and many such consumer products now have Websites of a very good quality. They combine creativity and techniques to be closer to their consumers and gain their confidence even more...
>
> The Web can even give a serious rejuvenating push to brand names which have a rather dusty image: the most glaring example, shoes from Mephisto, whose site is an example of a well-made catalogue. The same for the site for online selling of the brand!

Evidently, the end of the Internet bubble has put the clocks back to the present and people are no longer betting on futuristic scenarios. Today who would still dare to bet on the end of certain sectors of the economy in favor of the arrival of certain others? To recall, we go back to sectors like that of the car industry (one car in two was sold on the Web in 2003 according to studies done by internationally renowned firms), mail order and telephones, or real estate are some fields. Illusions of the ultra-liberal model have also fallen and Silicon Valley has become the Venice of this new Renaissance. Only industries dealing in non-material goods (in particular the information and the audio-visual businesses) have been completely transformed in this maelstrom. Nobody expected such a massive diffusion of technologies from peer-to-peer and the sad consequences that are known to all.

Listening to the media, peer-to-peer would be like a new wound from Egypt, a danger that would imperil a part of humanity. Worse than plague and cholera, peer-to-peer is an enemy to be eradicated although it only concerns exploitation of possibilities offered by the technique and is not yet integrated by the professionals of a sector, which it must be recognized, is the least remunerative and has generated a well-oiled industry.

It is not because it is technically possible that it is legally permitted (even morally acceptable, which is another affair). Here it is, summarized in a short formula, the essence of the debate of which we will not engage with here because if it is connected, it would be out of context.

18.4.2. *A deep impression*

The Internet in fact has not changed everything, and although it has changed the mentality of the generation which has never known a world without email, without the Web or without online forums (a part of the working population born after 1970), it still has not penetrated all strata of the population. The different cultures have neither melted nor unified to form a single model. In the same way, despite impressive progress, computers and the Internet have not yet entered all homes. In France, at the end of 2003, the number of Internet subscribers in the residential segment is slightly lower than 10 million, i.e. less than 45% of homes[7]. Even if these figures show a good spread of the Internet in France, there are a number of inequalities: the Internet is mostly in cities and is present in the middle and higher classes of the population. Despite the futuristic optimism of the gurus who, during the magical hours of the bubble, spoke of the "6th media", the Internet is still far from achieving the reach of the press, radio, and of course, in terms of impact, television. In fact, as is known, a media is judged or seen on its capacity to be able to penetrate in depth the target populations and to provide prominent results.

The eruption of the Web has, however, changed the mentalities of the marketers and has improved the progress of the technological culture of our contemporaries at least in the developed countries of the North. And, if certain other parts of the world have been submerged by the Internet wave (Asia, notably), there still is a vast unknown territory on the planisphere of the Internet.

The Web has been and is still an important source of innovation for marketing to the point of having spawned a new branch still in an embryonic state: the e-marketing which has its referrals, its methods, and which still has to see long-term results. One can compare the eruption of the Web in the life of marketers to that of the arrival of the CAM a little less than 20 years ago: it gives more flexibility and favors creativity[8].

18.4.3. *New reflexes*

The Internet has created new reflexes for its users. Often young or even very young, this new race of users is to be considered seriously because they are the consumers of tomorrow. And it must be recalled that the reflexes that are durable are those which have been formed in youth!

This new race of consumers has integrated the Internet in its entire consumer cycle: before buying (forums, price evaluators, information site and even "online" press) during purchase (e-commerce) but also (and perhaps especially) after purchase (consumers' opinions, forums again, personal Websites, etc.). Besides, this

---

7 Source: www.afa-france.com, the trade union of French Internet providers.

8 Which is seen in a magazine (type newsmagazine) before 1983 so as to compare it with a magazine of the same type edited today and one can easily understand the progress made, for a greater part thanks to the flexibility offered by the electronic tools of creation. Industrialists of the graphic chain have had to adapt and learn these tools of a new age in very little time. A graphist trained now would be surprised to abandon the graphics palette and creation software in favor of paper and colored pencils!

generation is to say the least dubitative, even obstinate when faced with advertisements which are constantly challenged. Evidently the phenomenon is most visible in computers but it is gaining ground and is a part of a more general movement (traceability of products, transparency, respect for environment, etc.) of "intelligent" consumerism of some sort in which industrialists must justify their actions to their consumers.

> Who in France does not know the Website Rue Montgallet[9] which lists the prices of the shopkeepers of the area who specialize in domestic computers? This site, result of a consumerist initiative, is both a reflection of the market prices and a most efficient price evaluator.

These new reflexes are the consequence of a fact of which developed countries are immensely proud: the level of education of the population has gone up greatly over the last decades. So, inclined to find good data (even over information), dubitative vis-à-vis commerce, the new consumer comes to the shop as or even more informed than the sales person whose role has evolved with the arrival of the Web and is still evolving…

18.5. A fantastic tool for sharing

18.5.1. *If you don't know, ask, and if you know, share!*

The slogan of the forum of the site for PC INpact[10] could be the motto of the Internet which, we recall, has been precisely created to allow information to be shared. Today the tools for sharing are legion. Apart from newsgroups[11], the Web has become the privileged media of forums for all types.

The function of the forum is very old and existed even before the eruption of the Web which submerged the systems that preceded the Internet. One thinks of newsgroups but also of Bulletin Board Systems (BBS)[12] which have more or less disappeared today. Besides, this function was one of the first to be programmed on the Web. The importance of a forum is evident: sharing knowledge enables the whole community of members of the forum to progress. Here we find a principle fairly close to that of Open Source. Wealth of ideas is the rule: one would be very

---

9 www.rue-montgallet.com.

10 www.pcinpact.com.

11 By convention the term "newsgroups" will mean data shared with the help of the NNTP protocol. This will allow us to concentrate better on the tools of sharing on the Web, the subject of this chapter.

12 One must remember CompuServe, which was the archetype of BBS or systems of sharing data such as FirstClass, as well as Lotus Notes (whose positioning is more particularly that of groupware).

surprised by the number and diversity of forums on the Internet. All subjects in all languages, of course.

This wonderful world must however be considered with circumspection: the signal/noise ratio is often very weak and finding important information in the middle of a forum loaded with unimportant messages can be quite a challenge.

Amongst the tools that help implement the forum, one can cite PHPBB[13], Open Source solution based on the PHP language.

It is sometimes necessary to reduce the difficulty of implementation of such tools, which is why there are companies that specialize in these activities. Earlier on, these computer-based Web agencies have been taken over by the communication sector and it is not rare to find Internet services in large "traditional" agencies.

Finally it is to be noted that, as in publishing matters, for the sharing of information emerges the question of the validity of data and identity of the issuer. "On the net no one knows that you are a dog"… the joke is as old as the Web and is valid as long as the problem is not taken up[14]. Certification, electronic signature and proof are far from being resolved.

18.5.2. *Business-to-business: Eldorado or damp squib?*

The large international research firms specializing in forecasts made certain promises in those times: Internet for the mainstream is the visible face. The real market is in the world of BtoB.

If, in fact, the companies are hugely connected to Internet, the usages are quite serrated. The Intranet, seen earlier as a power tool controled jointly by its telecom, computer and communication services, are today first and foremost tools for internal communication that control the diffusion of validated data. However, one function cannot be removed: the directory, which far from being the most widespread service, is the most practical one for fairly large companies. The fact that the companies were deeply interested in Intranet was certain and this revealed a need for communication and modernization in the information systems.

After a first phase of discovery, the companies have, however, had the experience of a long-term management of solutions which have definite advantages, and are in spite of everything based on software which has to be well managed:

---

13 www.phpbb.com.
14 Although certain key actors seem to be launching some initiatives in this domain. This is the case of the Certmail project of France Télécom, www.certimail.fr.

applications to be used, access to data, the rights of access to which must be controled, etc. The greatest winner of this movement is the email system which has become a must in all business tools.

Setting apart some attempts more or less media-based, Extranets which were known as "the hidden face of the Web" have remained in an embryonic state and are often forgotten as quickly as they appear. In this category must be included electronics marketplaces, which can be classed as e-commerce activities. A lot was expected from these market places and finally nothing (or little) happened.

However, making an international tender available to the grocer at Landerneau, centralizing purchases of small and medium scale industries so as to offer them tools of large companies... these were dreams of marketplaces! However, nothing happened as in the brilliance of this new Eldorado, it was forgotten that good business is rarely transparent[15].

Companies have taken to the Internet for the advantages it brought: a "facility" that reduces certain costs. And nothing much else, this is not surprising: a company invests to increase its profitability it does not spend for the pleasure of doing so...

Nevertheless, by providing their employees computers connected to the Internet, companies are participating in the diffusion of this phenomenon in France, and it is not impossible that some homes are equipped with personal computers connected to the Internet in order to have the same comfort at home. In the same way, a new phenomenon is developing: the attraction of senior citizens for the Internet and the PC, which could come from the fact that those people who have used it in their professional lives and who now have "all the time" to give to it[16]. If the forecasting companies are to be believed, this is a promising market if ever there is one!

18.6. E-commerce: a soufflé fallen flat?

E-commerce continues on its own path noiselessly. After a slow and difficult start-up and arid early years that have brought about a sort of economic "Darwinism", it has today received a second breath of life which is more reassuring.

---

15 A paradox... at this time when all the actors were still being rocked with illusions by easy Internet, no one had really seen the peer-to-peer which is a type of marketplace... non-monetary, to the great disappointment of some industrialists!

16 Regarding this, it is interesting to note to the surprise of the ISP, a large part of the ADSL clients are the first time accessors... reassured by the package tariff of this mode of connection, certainly full of resources!

People were attracted by e-commerce just as the Internet users had been for a long time avid consumers: a PC at home and an Internet connection are luxuries that only the more well-to-do sections of the population can afford. And using these sections as commercial targets was not a very difficult step to take!

18.6.1. *Between the hare and the tortoise*

E-commerce started off like the proverbial hare: initiatives regarding on line sales were no longer counted. Everybody wanted to sell anything and everything on the Internet.

It must be said that e-commerce had all the qualities of the philosopher's stone: simple and not expensive to set up (compared to the prodigious logistical requirements of "traditional" commerce), it facilitated the act of buying and selling by the immediacy of the link between buying and accessibility. E-commerce should allow the combination of two aspects considered contradictory until now of the important trends in commerce.

On the one hand, the importance of attacking the markets of mass consumption with large economies of scale, and on the other the importance of diversifying to the maximum possible supply, of providing a personalized service to the client and in doing so establish his loyalty. The development of large trade took place in parallel with the diversification of behavior and the public and it is the Internet which allows one to solve what appears to be a major contradiction that direct marketing has taken into account but has not been able to really reduce.

Another apparent contradiction amongst the traditional trends of commerce can also be resolved: on the one hand it is important to target the consumer and to reply quickly and precisely to his demand, on the other hand one should also be able to surprise him and broaden his fields of interest. These are the two aspects of traditional trade.

On the Web, one of the risks has been to excessively target the consumer, which could hinder the commercial field and finally tire the consumer. However, the force of "traditional" commerce, which allows the generation of complementarities amongst supply offers in a commercial space has another role of presenting openings and surprises and finally profits to all the businessmen present – this is generally the principle of the commercial street, malls or departmental stores and this is the systematic behavior of the supermarkets. Electronic commerce should make it possible to satisfy the immediate demand of the consumer well and at the same time propose to him diverse offers.

Finally, no commercial development should mask the question of price. Outside particular cases and opposing markets, everything else being equal, the lowering of prices remains fundamental. By taking into consideration material goods, in trade in general as in electronic commerce, one important part of the price is linked to logistics and the transportation to sales outlets or to consumers. On a long term basis, e-commerce should allow a new reduction of prices and the question of logistics will often prove to be critical. The consumers are well aware of the force they represent and the notion is well entrenched in their minds that reduction of prices (which reduces costs) of goods sold is equally crucial for the businessmen not yet used to new forms of competition created by the Internet.

There are a number of advantages of electronic commerce: capacity to ensure a greater supply and demand balance, and a personalization of commerce which is the specific contribution of e-commerce. However, e-commerce has not been able to avoid the rocks in its path and has fallen into what could be an error in terms of commerce: with such a strong adequation and personalization, the consumer never has to face offers that do not correspond to his profile and his own demands (which is in contrast to strongly validated strategies of the supermarkets which still represent the archetype of modern commerce). The technical quality, putting together of numerous offers and knowledge of the characteristics of a large audience should make it possible to manage the contradictory aspects of targeted purchase and impulsive buying.

Evidently, imbibing all this culture of commerce and realizing finally that e-commerce is first and foremost commerce more than an activity on the Internet does not happen in a day and a large number of parameters have to be incorporated to reach the quintessence of e-commerce.

Therefore, time as well as means were needed, as the setting up of e-commerce sites that are worth a name took more time and was more expensive than what business plans made hastily in the euphoria of the Internet bubble had proposed... we therefore came back to the fundamentals of analysis and management that would conform more to the realities of commerce.

As this movement took place in a context which turned out to be finally unfavorable, the actors (not too many) who managed to get out of it came out stronger. A surprising paradox should also be noted: the big actors of traditional commerce are often absent from the list of major actors of e-commerce as opposed to new arrivals.

After some dark years, the end of the tunnel is in sight for the stronger, more professional actors of e-commerce.

18.6.2. *Incorrect good ideas for reel disadvantages*

It would take too long to list the incorrect good ideas of e-commerce. Weird price evaluations, consumer opinions knowledgeably organized, ridiculous loyalty programs, unnatural cross-selling… the list as long as that of all those who died on this heroic battlefield of damp squibs!

However, it is the e-commerce which has today generalized bad habits which have been well inculcated and very annoying. Of course one thinks of undesirable commercial canvassing… a soft French euphemism for spam which invades email inboxes!

Despite some initiatives which are noteworthy, only a few real measures have been taken to fight against this phenomenon. No real list, none or few "charts" or labels (although the emergence of e-commerce has witnessed a burst of attempts!) and few really efficient tools. It is even more regrettable that easy availability puts the Internet at the disposition of people who are not always knowledgeable in this field.

Questions related to payment have for a long time made up an important nucleus of the main problem of e-commerce. The list was long: who should make the invoices, to whom, on whose account, from and to which amount, at which cost. Naturally two actors came face to face: the access providers (certain of whom are still under some illusion) and the banks, without counting specialized operators especially in micro-payments (amounts less than €10) whose specific aspect requires the setting up of dedicated structures. One thinks especially of PayPal[17] or the more French W-HA.[18]

Then, as it happens often and notwithstanding real barriers, it is finally an intermediary solution which is applied: the number of this good old credit card whose number is transmitted through frankly rudimentary security solutions[19]: a tunnel secured in SSL, and the game is played. Despite these justified anxieties of the actors of the card, the number of transactions do not stop growing and finally, even if the actors are on the lookout for fraud, solutions for adaptation are numerous and often efficient but the politico-economic problems dominate.

---

17 www.paypal.com.
18 A system based on the technology iPin, W-HA is a branch of France Télécom.
19 After some trials at very low levels of security, the SSL generalized today at 128 bits offers, it must be reiterated, a very high level protection.

18.7. Conclusion

In 2004, we celebrated the 10th anniversary of the Web. Who would have believed it? Today in fact the Web is fully integrated in the technological scenario. Tool and engine of innovation, it is difficult to see how we would go back and do without it.

The era of maturity has now begun and one must be attentive to the ability of this technology to find its second life. Given the virgin territories that remain to be invested (television, home service, mobility, etc.), we can be confident of the future. One of the problems will be to "deinformatize" the Web, that is, to take it out of the sphere linked to purely software tools (micro-computers). The mobile phone, the wireless terminals, garage doors, cars, problems regarding directories … so many terrains to conquer which will rise (and which have already risen) from several problems. Therefore, further innovations will be required, aided by tools considered innovative today. The wheels turn, the world advances!

Chapter 19

Virtual Decision Support System for Innovation

19.1. Introduction

Innovation aims at acquiring a competitive advantage by carrying out important strategic activities, at a cheaper rate, or better than its competitors. In order to complete the logics of margin, the logics of differentiations must be examined. Concept or design becomes strategic in the innovation process as decisions taken in this phase represent 50 to 80% of the expenses [GIA 1993], [TIC 1995], [National Research Council, 1999]. However, few tools are available in the early stages of conception to help choose a product design or another one. Even fewer tools are available if the actors of design are distant.

Our problem lies in this paradox of designing: to decide the pertinence/relevance of the proposed product designs as soon as possible, while at the same time, having very few elements to be able to do so. In fact, at the stage of design creation, the product designs are not a physical reality (model, prototype). Moreover, it is necessary to integrate participants who are more and more distant geographically, who are from different cultures and disciplines, and finally even integrating the client. We face the need to take decisions based on multiple criteria by participants from many different disciplines (general management, finance, commercial, marketing, technical, etc.). Innovation project management is the objective of this chapter. How does the project manager synthesize all the evaluations of distant actors of design? Together, they have to regularly estimate different concepts of product proposed in the creation phase.

Chapter written by Emmanuel CHÉNÉ.

After having approached the management for design innovation, we characterize intermediate virtual representations in the industrial context transmissible through the Internet. We then propose a help tool for the decision through joint analysis and we experiment on SME for the creation of packaging. Finally, we formulate certain analysis and perspectives before drawing a conclusion.

19.2. From the management of innovation to the management of design

New reference points are made when the innovation inevitably plays a major role in the triptych: cost, quality, time. Innovation becomes the pivotal point of the creation of value [FOR 99]. Innovation is multifaceted according to the approach of Schumpeter[1]. However, the central process of innovation is not science, but design creation [PER 01a]. "The aim of design creation and elaboration of products and systems, is to satisfy user needs while guaranteeing adhering to respect for the environment, legislation and profitability needs of the company" [BOC 98]. As its etymology indicates, design concept has to do with the manipulation of design concepts, that is, the manipulation of a general and abstract representation of an object or a collection of objects. We see the importance of the environment in the process of designing here. In order to apprehend it better, we follow three models of designing, chronologically. Complementing the Interconnected chain Model [KLI 86], the Swirl model [CAL 95] integrates iteration loops, recaptured by the Recursive model [GEN 98] which confirms the necessity of a close collaboration between the system of creation and the environment. What Kline and Rosenberg called "research", Genelot widens to the notion of "environment". These approaches correspond to the evolution of the complexity of products, which require more and more external intervention (experts, laboratories, suppliers, clients, etc.). We propose a global process which couples the approach of Suh, of a creation of design between functional space and physical space [SUH 98], with systematic analysis [Le MOI 99]. This approach distinguishes the operating system, the information system and the decision system (see Figure 9.1).

---

1 Schumpeter (1912, 1939) distinguishes five cases of innovation: manufacturing of a new product, new production method, new organization, opening of new outlets, new source for raw materials.2 Also indicated in the bibliography by the term *Simultaneous Engineering*, *Integrated Development*, to describe the term *Concurrent Engineering*.

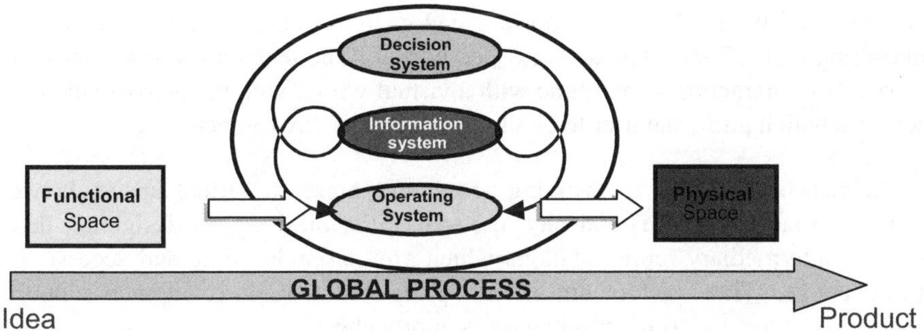

Figure 19.1. *OID Model of Le Moigne in functional space/physical space according to Suh*

The operating system evolves, on the one hand, towards a delocalization of participants of the design outside the company and on the other hand towards a modification process of designing. The different actors in the design chain can no longer hold internally between them all the resources and knowledge necessary for the realization of a product. Teams that work from a distance [SAU 00] or distributed teams [ROW 02] reflect the notion of *extended company* which brings together teams in distant locations to collaborate. The use of information technology and communication makes this exchange fluid and allows the integration of the client in the decision process. Concomitantly, teams work not in a sequential manner (or PPP: *phased project planning*), but in a concurrent manner. Where sequential organization allowed fragmented problems to be resolved in a rational way [NON 90], *concurrent engineering*[2] tends to bring them together (integrated engineering) and at the same time (simultaneous engineering) [MID 93, MID 95; CHE 97; PAW 97; AFNOR 50 415]. It is characterized by a "systematic approach which integrates the simultaneous development of products and associated processes, including manufacture and logistic support. This approach takes into consideration, from the very start, the lifecycle of the product from its creation up to its exploitation, including quality, costs, planning and user needs" [WIN 88]. From this approach we retain the notion of user needs which should be integrated in the operating system in extended concurrent engineering.

The information system must allow these exchanges to be fluid. The intermediate representations correspond to "interaction tools between the set of actors in the life of a product, in order to allow a more efficient mutual exchange and comprehension" [TIC 97]. Models and physical prototypes remain necessary, but their number is reduced greatly through virtual intermediate representations (VIR). "The notion of design also being an intermediary object in the representation of that which will be the final product solution" [YAN 01]. Thus, the buyers of the A380 airplane from Airbus based their decisions on synthetic images. The technology of virtual reality

(VR), coupled with CAD (computer assisted design) define VRAD (virtual reality aided design) [FUC 01]. The authors specify that "The techniques of virtual reality are based on interaction in real time with a virtual world, with the help of behavioral interfaces which allow the user to be immersed in this environment".

The decision system is evolving more and more in virtual space, between functional space (the idea) and physical space (the product). In designing, this is based on intermediary representations which are comprehensible and accessible to all the actors. These actors are taken up by the CSCW (computer supported collaborative work) according to two cross approaches:

– the temporal aspect of exchange (synchronous or asynchronous);

– the space aspect for the actors (co-localized or remote).

We know the co-localized space aspect, where project teams are grouped on the same project plateau[3] or in a "reality center". It is often coupled with the synchronous temporal approach. In order to obtain interactivity and immersion in virtual reality, images calculated in real time[4] are projected on wide-angle screens. The Herley J. Earl[5] room has thus made it possible for Renault to economize about 13% in six months on machining costs of models in other words, a gain of €22,000.

On the other hand, the remote spatial approach is little used to help the decision in the conception. However, it is noted in actual reality of distributed concurrent engineering, with remote/distant actors, spread over different continents. It also integrates the final client. The remote spatial approach is the main subject of the present chapter.

So as to avoid arbitrary choices which may initially preempt a promising product design, we propose virtual reality and joint analysis to be used together in designing, by using Internet as the communication media. In terms of organization, we clarify the scope of work to remote and nomadic teams in an organization of extended concurrent engineering. In terms of information, intermediary virtual representations are proposed of designs conceived very early in the designing process thanks to a model for the diffusion of these via the Internet. With respect to the decision, joint analysis is integrated so as to offer a multidisciplinary decision system, based on multiple criteria. We then experiment in SME.

---

3 At the Renault Technocentre, 7,500 persons are grouped according to the projects.
4 Between 15 and 20 images per second to obtain good fluidity.
5 Of an investment amount of €1 million.

19.3. Intermediary virtual representations in the industrial context and transmissible via the Internet

Virtual reality technologies (VRT) make it possible to propose virtual intermediate representations, without expecting a real prototype (rapid prototyping or finished product). VIR aims to create the impression of reality, for a product which is still virtual. We will talk about virtual prototyping from now on, which is a form of intermediate representation through virtual reality, dedicated to industry.

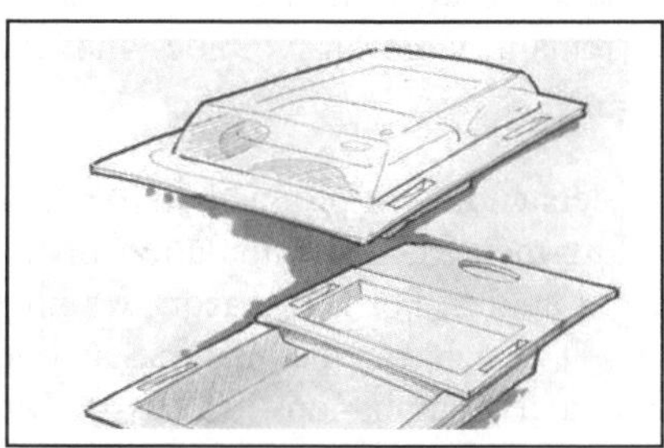

VIR 1: drawing before project

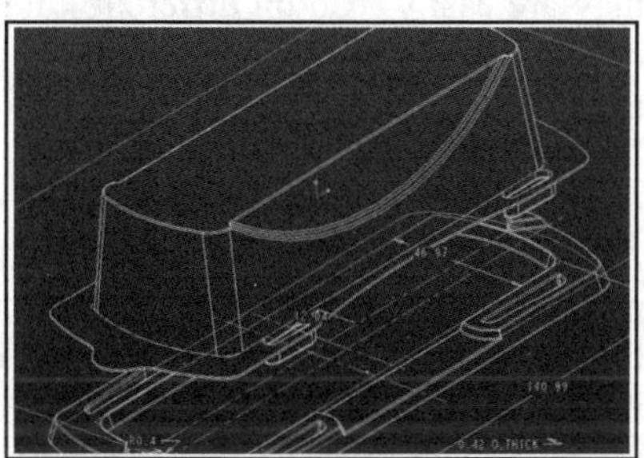

VIR 2: fixed image

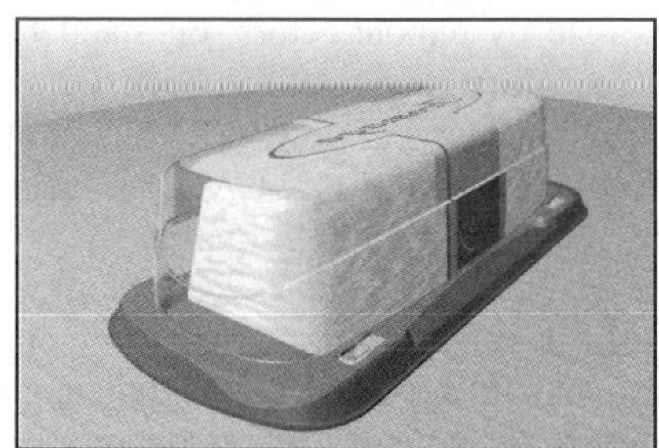

VIR 3: 3D sequential animation

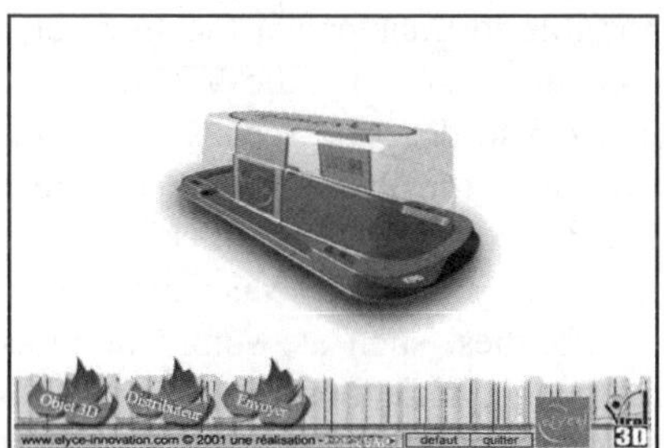

VIR 4: 3D interactive animation

Figure 19.2. *Virtual representation for the decision of packaging design*

In Figure 19.2, four levels of intermediary representations are proposed: drawings before project (VIR 1), fixed images (VIR 2), sequential animations (VIR 3) and interactive animations (VIR 4). How can these intermediary representations be diffused to teams distributed via the Internet?

19.3.1. *From VIR in fixed 2D to VIR in interactive 3D via the Internet*

We are looking to evaluate a greater number of possibilities for products through the use of VIR in the context of distance work. Communication media such as

Internet makes it possible to imagine real integration of distributed teams including the final client in the designing of products. On the other hand, this imposes a certain number of limitations.

Concerning VIR 1 and VIR 2, the exchange through Internet is easy as these VIR are of 2D type, of smaller "size" and widespread use over the Internet of JPEG and TIFF format. To gain in comprehension, sequential animations (VIR 3) have been used. They have resorted to video formats (MPEG, AVI). Also used very widely, they are easily read on different kinds of computers. On the other hand, despite the developed compression algorithms, the files remain large in volume and the exchange is almost impossible at low speed.

These 2D images do not allow an optimal understanding of products in 3D. The formats in video type are heavy and do not allow the virtual manipulation of the product for appreciation of its form from different angles. For this reason, we note the importance of development of formats in 3D which can be manipulated to improve comprehension. We call them "3D interactive animations" (VIR 4). But these formats must be integrated into a digital chain of design i.e. particularly in design system computer aided design (CAD). On the other hand, they must be possible to transfer via the Internet, as well as read by remote actors, on standard computers. For this, the VRML (virtual reality modeling language) has been set up in 1995 to describe scenes in 3 dimensions over the Internet. Now, VRML is not sufficient for the interface of the digital chain of design from CAO, it lacks interactivity in terms of kinematics and the realistic rendering quality is not satisfactory. VRML scenes are most of the time devoided of the latter technological possibilities, such as: sound or video influx, or "bump mapping[6]", "environment mapping", reflection, shadows, NURBS (curbs) etc. It appeared in the language market making it possible to go further in terms of realism, effectively giving excellent results but requiring proper plug-ins for the display of 3D. We have listed 14 appropriate technologies[7] for the display of interactive 3D on the Internet.

---

6 Possibility of generating micro-reliefs on a surface, using an image in gray which serves as a "bump". Darker parts forming creases, the clearer parts remain on the basic surface. The "real" surfaces are not modified, only their surface state and incidence of lighting are modified. Hence, for example drops of water on Elyce cans and on the thermocol packaging are carried out through "bumpmapping".

7 Technologies analyzed in Chéné *et al.* [2003a]: 3D Anywhere from 3Di, 3space of TSG (Amapi), B3D of B3d, Cult 3D of Cycore, Kaon of Interactive inc, Director of Macromedia, Mendel 3D of Duran Dubois, Pulse 3D, Qedsoft, Shout 3D, Eon, Viewpoint, Virtue 3D and Blaxxun 3D of Blaxxun interactive.

19.3.2. *Characterization of virtual intermediary representations in the industrial context and its transmission via Internet*

In order to characterize virtual intermediary representations with Information and Communication Technology, we propose the 3i model, based on three fundamental ideas: infrastructure, information, interoperability.

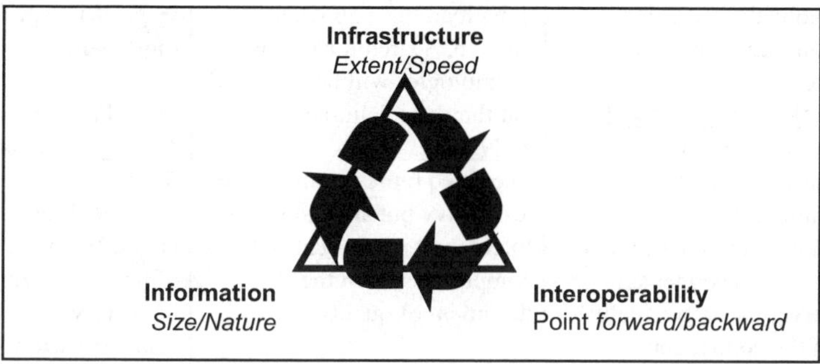

Figure 19.3. *The 3i model*

The 3i model proposed in Figure 19.3 is based on three fundamental notions which are infrastructure, information and interoperability. In Figure 19.4, the characteristics and limitations of these three basic ideas are enumerated.

Characteristics	Exchange network characterized by its extent and speed of exchange. ▶ **Extent**: infrastructure materials: LS, cable, ADSL, Wi-Fi, satellite - **Intranet**: internal LAN network (*Local Area Network*); - **Extranet**: external private network WAN (*Wide Area Network*); - **Internet**: external public network ▶ **Speed of exchange:** - **bandwidth**: absolute maximum capacity of transport of information	Size and nature of the exchange ▶ **Size**: (size of file), is the function of technology and compression: - bitmapped/vectorized - compression/degradation ▶ **Nature**: representations: - **Fixed Image:** less heavy, good graphic quality, limited comprehension - **Sequential Animation** in 3D: pre-calculated visualization of an animation, manipulation limited to a trajectory - **Interactive Animation** in 3D: increased functional comprehension, manipulation	Comprehension between client and server, linked: ▶ **Backward: format of file/Client** Protocol TCP/IP¨(portability) ▶ **Forward: format of file/CAD** - **General:** readable and et comprehensible through many applications. In 2D: JPG, GIF, TIF, BMP, etc. In 3D: DXF, IGES, STEP, etc. - **Specification:** to one application. In 2D: PSD (Photoshop)

	- **flow**: relative quantity of information transmitted in a unit of time	and zoom possible	In 3D: PRT, ASM, DRW (Pro-Engineer) 3D distant: (fixed, sequential or interactive)
Limitations	▶ **Extent**: the limitations are from international, national/state or local community exchange, etc. ▶ **Speed of exchange** Bandwidth 56 Kb/s (Internet) to 1Gb/s (intranet). Flow: function of the no. of clients connected, server capacity, quality of the connections.	▶ **Size**: strongly correlated to quality. 1 Mo constitutes the limit of download via a 56 Kb/s modem. Search for vectorial technologies will be preferred but their use is limited. ▶ **Nature**: The fixed representations are less heavy but are insufficient. Interactive images carry comprehension at the detriment of quality.	Material Interoperability Software Interoperability: ▶ **Backward: format of file/Client** Protocols TCP/IP(portability) ▶ **Forward: format of file/CAO** - **General:** open but not very efficient. Transfers of bits of information but no history. - **Specification:** efficient but Plug-in necessary for reading and/or writing. Difficult to manage for clients not having software licenses. **Compromise**: gain in performance/loss of portability.

Figure 19.4. *Scope and limitations of virtual intermediate representations and TIC*

At this stage of our investigation, information can be extracted from CAD, with respect to the digital chain of design to create a representation of a virtual prototype. Virtual intermediate representations in 3D are capable of manipulation in an interactive manner and transmitted to a remote interlocutor for the evaluation of the proposed designs. But this information thus transmitted does not always allow efficient decision-making. In this sense, we are mobilizing the joint analysis approach used in marketing, by coupling it with previously obtained virtual intermediate representations via the Internet.

19.4. Developing a decision-making aid with joint analysis software

Joint analysis (JA) was introduced in the early 1970s by Green and Rao. Its applications have spread to marketing problems and industrial production. In France, the method is often quoted under the name of "trade-off". The name of this method more precisely indicates a certain type of application of analysis [DUS 98]. The method of joint analysis is defined as "a specific predictive technique (quantitative variables being able to be treated as random qualitative variables)" [SAP 03]. One of its specific characteristics is that it is "a model of compensatory choice: the consumer arbitrates between the qualities of a product which are not always compatible". While a choice is made, a compromise is also made between the advantages of one characteristic with relation to another. How are the preferences of the consumer to be explained between different objects based on attributes or characteristics which describe the object? The aim is to measure the joint effect of many independent (explanatory) variables, in the order of values taken by a dependent variable (preference). Among these criteria, we sometimes make a distinction between two types of evaluation. Certain attributes correspond to "naturally" digital evaluations, such as price, speed, costs, percentages (of market, profitability), known as "quantitative criteria". Other evaluations are not carried out naturally on a numeric scale, for example brand image, social risk, quality, are criteria as well, but for which there is no unit of measurement, and they are known as "qualitative criteria" [POM 93]. In this sense, the semantic differential and Kansei engineering, make it possible to distinguish analogical variables. The semantic differential[8] is used to characterize aspirations, desires [BAS 96]. This tool offers a capacity of measurement and of follow-up of the evolution of hedonistic aspects. Kansei is a methodological approach from Japan, aiming to explore associations between feelings (subjective criteria) and semantics, which are found in the bibliography under "Kansei science" or "Kansei engineering" [LEE 00]. We propose to carry out the integration of VIR and JA in a software.

19.4.1. *Software tools for joint analysis*

We have as our objective the proposition of a first level of software. We base our study on the evaluation of product designs in an organization of extensive concurrent engineering that is by considering the entire set of distant deciding factors. The tool should allow:

– compatibility on the Intranet and the Internet;

– creation of traceability in the process, specifications of the product;

– saving acquired knowledge, through the creation of archives of client needs and related activities having been used to identify these needs;

– diffusion of the understanding of client needs to all the actors participating in the designing process.

---

8 Introduced in 1957 [OSG 57], its use in the characterization of products is dated about 20 years.

The software tool for joint analysis makes it possible to treat data in a structured manner from different interlocutors involved in the designing process. We are not developing the genesis of the tool here [CHE 03c], but specifying its implementation through a five step method:

– choice of designs and specifications;

– collection of data;

– calculation of uses;

– simulation and synthesis;

– analysis and optimization.

19.5. Implementation of the software in SME of packaging creation

This SME, named Elyce innovation SA, designs every specific packaging with the patented technology of the Elyce open/close system as required by European and American customers. It includes the process: from the idea to the prototype. We have implemented this methodology in a SME for the creation of innovative packaging and described an application for bowls for food consumption.

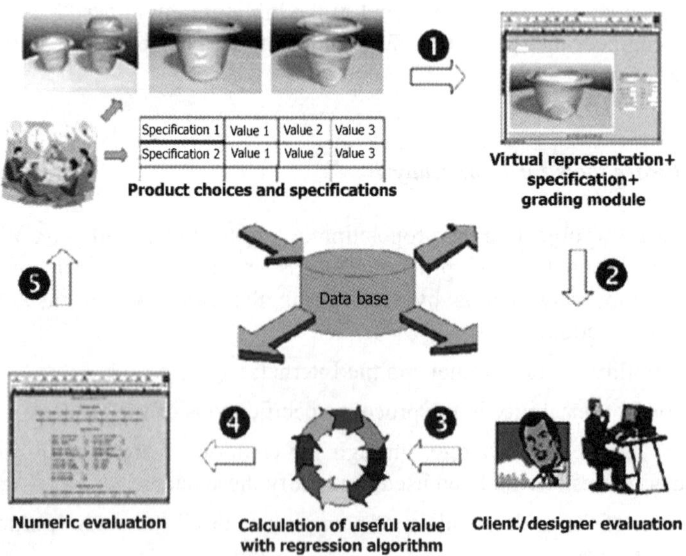

Table 19.1. *Implementation of joint analysis in the Elyce experiment*

19.5.1. *Choice of designs and specifications*

The design team has carried out a selection of designs for interesting food bowls, combining four structures for pots, with five structures of lids.

– The pots include motifs: waves/leaves/drops/thread.

– The lids are of the type: flat/mid-height /high/with tabs/without tabs.

The design team thus chose 18 sets of almost finalized design-products, without labels and virtualized them under the same viewing angle in an identical environment. The design team then defined the specifications which seemed important (described hereunder) with its associated values (values 1 to 3).

Brand	Brand I	Brand J	Brand K
Opening/closing	Easy	Difficult	Very difficult
Manufacturing	Slow (1 c/s)	Fast (3 c/s)	Very fast (5 c/s)
Labeling	Dry offset	Printed paper	Pad printing
Product vision	Insufficient	Average	Good
Consumption	Nomadic	Urban	Natural
Price	€100/1,000	€150/1,000	€200/1,000

Table 19.2. *List of specifications and possible values*

The project leader intervenes in the product design and the specification into the software.

Specifications								
Product	1	Brand I	Very difficult	Slow (1c/s)	Dry offset	Good	Nomadic	€100/1,000
Product	2	Brand I	Very difficult	Fast (3 c/s)	Printed paper	Average	Urban	€150/1,000
Product	3	Brand I	Very difficult	Very fast (5 c/s)	Pad printing	Insufficient	Natural	€200/1,000
Product	4	Brand I	Difficult	Slow (1c/s)	Dry offset	Average	Urban	€200/1,000
Product	5	Brand I	Difficult	Fast (3 c/s)	Printed paper	Insufficient	Natural	€100/1,000
Product	6	Brand I	Difficult	Very fast (5 c/s)	Pad printing	Good	Nomadic	€150/1,000
Product	7	Brand J	Easy	Slow (1c/s)	Printed paper	Good	Natural	€150/1,000
Product	8	Brand J	Easy	Fast (3 c/s)	Printed paper	Good	Natural	€100/1,000
Product	9	Brand J	Easy	Very fast (5 c/s)	Dry offset	Insufficient	Urban	€100/1,000
Product	10	Brand J	Very difficult	Fast (3 c/s)	Pad printing	Insufficient	Urban	€150/1,000
Product	11	Brand J	Very difficult	Fast (3 c/s)	Dry offset	Good	Natural	€200/1,000
Product	12	Brand J	Very difficult	Very fast (5 c/s)	Printed paper	Average	Nomadic	€100/1,000
Product	13	Brand K	Difficult	Slow (1c/s)	Printed paper	Insufficient	Nomadic	€200/1,000
Product	14	Brand K	Difficult	Fast (3 c/s)	Pad printing	Good	Urban	€100/1,000
Product	15	Brand K	Difficult	Very fast (5 c/s)	Dry offset	Average	Natural	€150/1,000
Product	16	Brand K	Easy	Slow (1c/s)	Pad printing	Average	Natural	€100/1,000
Product	17	Brand K	Easy	Fast (3 c/s)	Dry offset	Insufficient	Nomadic	€150/1,000
Product	18	Brand K	Easy	Very fast (5 c/s)	Printed paper	Good	Urban	€200/1,000

Table 19.3. *Definition of values of each specification, for each product*

The product designs and specifications being defined, it is possible to present them to the designers, so that the entire set may be assessed through an estimate.

19.5.2. *Collection of data*

One resorts to synthetic images along with a written list of characteristics. The image and specifications make it possible to get a common basis for communication between technicians, individuals from marketing, financiers, designers, etc. Five broad types of methods of collecting data can be identified [LIQ 00]. We are particularly interested in "complete profile techniques" and "grading test" due to their natural aspect and ease of implementation. The "grading test", consists of interval tests in a non-structured scale: Stevens' test to estimate size or "magnitude estimation". In this first experiment, the notation/marking test is done at non-structured intervals where the subject attributes a grade/mark from 1 to 10. The scale can further be reduced to 1 to 7 if required.

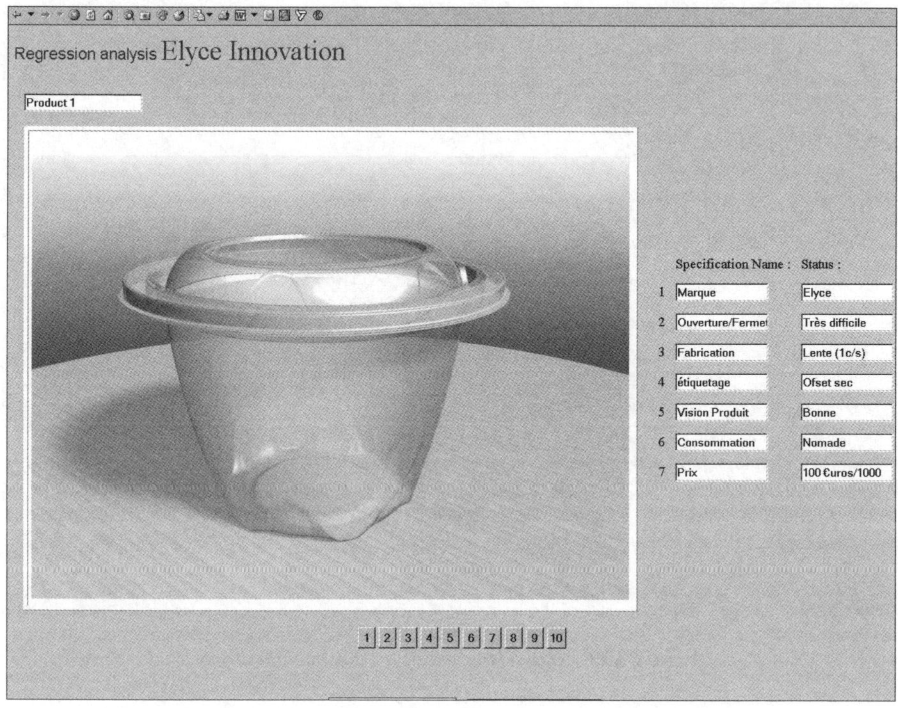

Figure 19.5. *Collection of data on joint analysis software*

In Figure 19.5, the choices are expressed in the form of numeric values, representing preferences for a product. Once these values are collected, it is necessary to analyze them further, to know which specifications held the most attention.

19.5.3. *Calculation of uses*

The calculation of uses can also be described as research on the merit of attributes. It makes it possible, on the basis of a group of X variables (specifications), to know which ones have the most influence on a Y variable (the expressed choice), determining the specifications which most influence its general appreciation.

D:\THESE\ILLUSTR\virtual_test\frcustform.html – Microsoft Internet Explorer

Beta Version v3.3

Customer answer

Product 1	Product 2	Product 3	Product 4	Product 5	Product 6	Product 7	Product 8	Product 9
8	7	6	5	10	8	7	3	5

Product 10	Product 11	Product 12	Product 13	Product 14	Product 15	Product 16	Product 17	Product 18
8	7	5	1	3	9	7	8	5

Regression result

Marque : Elyce Innovation	0		Ouv/Ferm : Très difficile	0
Marque : Fleury Michon	-1.22		Ouv/Ferm : Difficile	-1.44
Marque : Martinet	-2.44		Ouv/Ferm : Facile	-1.22
Fabrication : Lente (1 c/s)	0		Etiquetage : Offcet sec	0
Fabrication : Rapide (3 c/s)	-0.50		Etiquetage : Papier	-1.50
Fabrication : Très rapide (5 c/s)	0.33		Etiquetage : Tampographie	-1.65
Vision produit : Bonne	0		Conso : Nomade	0
Vision produit : Moyenne	-0.50		Conso : Urbain	0.5
Vision produit : Insuffisante	0.33		Conso : Nature	2.33
Prix: 100 €uros/1000	0			
Prix: 150 €uros/1000	1.83			
Prix: 200 €uros/1000	-2.00			

Rang Regression result

Prix	Marque	Consommation	Etiquetage	Ouverture/Fermeture	Fabrication	Vision produit
3.83	2.44	2.33	1.67	1.44	0.83	0.83

Figure 19.6. *Analysis of uses for each specification*

The calculation was carried out with the help of a regression algorithm. The results are presented in Figure 19.6. First of all, the effected/operating choices are summarized (for example here: 8 for product 1, 7 for product 2…5 for product 18). We then present the result of the regression and finally the classification of specifications in descending order of importance. Here, we observe in descending order: price (3.83), brand (2.44), mode of consumption (2.33) and lastly the product appearance (0.83).

19.6. Analysis of contributions of VIR with joint analysis in designing

This experiment makes it possible to present the quantified data to designers so as to make a choice between different designs of proposed products and then to refine the design as well. These elements are important at a time T for the designer, but are more important in terms of the evolution of the proposed product designs. An evaluation can thus be made and it is then possible to super-impose the graphics of one date on the other to keep track of the evolution of design.

In the experiment on packaging which have a strong market incidence, it is logical that notions relative to commercialization were mainly taken into account. In this example, labeling and practicality were least taken into account, manufacturing time was not considered and the product appearance was an even less important factor. In this sense, products having a high cost (with high lid), were rejected. In order to better understand these results, cognitive management constraints need to be discussed.

19.6.1. *Cognitive limitations*

It is always a difficult task to evaluate a virtual product without being able to manipulate it. However, this is often the case during the act of buying too (plastic packaging protection, sale through mail order, e-commerce). The single visual channel and the lack of participation of the other senses, particularly touch [CHE 94], limits global or general comprehension. As regards virtual representations, the images in 2D are aimed at simplifying the exchange and facilitating easy management of a data-base. On the other hand, the third dimension and virtual manipulation allow a general or global comprehension of the product in 3D. In response to this lack, the designing of a 3i model: information, infrastructure, interoperability, makes it possible to characterize VIR coupled with a numeric chain of design through TIC.

Limitations in colorimeters concern the colors or tints during the final finish of the product, compared to those seen on the screen. Considering the heterogenity of the peripherals used, the same color may not always have the same aspect. As for example, the color palettes for Macintosh and Windows are different. The colors are indexed differently and are hence restored differently according to the material of the client. On the other hand, the derived colorimetric can be important between two peripherals. We do not perceive colors in the same manner depending on the peripherals used [CHI 95]. In the case of the screen, the color is a luminous colored emission (synthetic additive) while in the case of a paper document the color is a result of the reflection of a luminous flow on the sheet of paper (synthetic subtractive). Hence, variations between each peripheral are important: RGB (red green blue) for display and CMYK (cyan magenta yellow black) for the print. The scanner and screen use the RGB coding, while the printers use the CMYK coding. At present there are no perfect CMYK/RGB conversion tables, which can pose problems during the transmission/passage of colors from the scanner to the screen and then to the printer. In this sense, the colors should be validated through real samples/scales, to be able to at least master the colorimetric chain. This requires, on a minimum weekly standardization of the peripheral factors.

19.6.2. *Limitations in terms of management of decision-making aids*

To make a choice, the decision maker uses many different areas of evaluation. If the choice is about a product, the determining factors will be, for example, the price, the quality, aesthetics, durability, etc. These areas of evaluation are the attributes for actions. When a minimum amount of information related to the preferences of the decision maker is added to these attributes, they become criteria. In other words, a criterion expresses more or less precisely, the preferences of the decision maker related to a given attribute. In this development, nothing is said about the independence or the links between different criteria. The model accepts the two distinct criteria, for example the fact that durability may be a component of quality. In the same way, the basic paradigm presupposes that all criteria work on the same level. The fact that some may be more important than others has not been considered. We have not yet integrated the fact that there are some that are more important than others: balancing attributes to products, balancing criteria integrating the decision makers.

This system, like all other systems that depend on measurement is subject to criticism. In fact, there are many different parameters which allow the definition of the activity of a group of individuals, without mentioning the parameters of capacity and fluctuation of individual performances! Therefore there is a sort of negative impact of research which in principle makes narration impossible. Any system of measurement of social activity is more complex to implement than a system of climatic measurement or of resistance of materials. We are aware of the objective limitations of our system which we presently compare with a system of survey. In the geological sense of the term – that which gives us indications about reality but never a description of reality. In an organization of extended concurrent engineering, many different tools are to be developed which may be easy to use by laymen and capable of being manipulated through numeric networks. We are looking to avoid arbitrary and unilateral choices made without consultation, with this type of tool. The limitations of different modes of investigations associated with social sciences are reinforced in our research through the introduction of virtual reality, which adds its own limitations to this approach (called affordance). With this point of view, we propose to observe the first perspectives of the evolution.

19.7. Perspectives

Our perspectives deal with an improvement of the software on the one hand and an introduction of VIR in interactive 3D rather than with fixed images, on the other. The improvement of software must be followed for the project leader and for the users.

For the project leader in charge of administration, the parameters can be customized, but demand a delicate manipulation of the code. Task automation through interfaces must be studied. The treatment of results needs to be improved. The connections with links like MS Access or another database need to be carried out. The development currently done in JavaScript, allows an interesting task as far as the client is concerned, but also leaves out the source codes from his side. Finally, work done through a PHP link, managed directly from the server will make it possible for a greater securitization of data, and hence a better capability of treatment.

For users, actors in designing, the evaluation module must be specified in a manner other than by numbers (from 1 to 10). It will be advantageous to complement it with a color code (for example from red to green). Limited understanding of fixed images has already been mentioned and is the object of development in mobile communication. We propose an integration of interactive 3D images, rather than fixed images. Moreover, it is possible to integrate an evaluation capability which can be linked to a database in this application so as to eventually carry out the desired statistical process. The development of this application with interactive 3D intermediate virtual representations is an important step [CHE 2003b]. Interactive 3D VIR are obtained by linking the software for surface modeling 3DSMax and integrating it with interactive Virtools. Three interfaces are suggested: a general interface for presentation, a specific interface for manipulation in 3D and an interface with marking system to carry out an evaluation.

Figure 19.7. *Evaluation of virtual products*

The evaluation module allows a qualitative feedback from the client so as to inform the designer about the evolution of products.

19.8. Conclusion

We have characterized the importance of innovation and new organizations in the field of designing which result from it. We have in a sense, brought out the importance of distributed concurrent engineering, and underlined the lack of tools to help in the decision making process between people who are geographically and culturally remote, in phases right from the beginning of the designing process. To answer to this, we have characterized intermediate virtual representations which may be exchanged over the Internet. We have also proposed the coupling of joint analysis and virtual representations through software help tools for decision. This software help tool has been experimented with in SME of packaging creation and makes it possible to arrive at a quantitative and qualitative decision.

In terms of quantity, the introduction of this systematic approach and software tool has helped the enterprise to reduce the time of design creation by 50% (6 months instead of 12 months) and to reduce the number of prototypes (3^9 instead of 5 on average for packaging). At the level of the organization of meetings, displacement of every 4 for 5 persons[10] thanks to different VIR, through rapid prototyping capacity directly to the concerned countries (by sending 3D files of rapid prototypes to local manufacturers).

In qualitative terms, tests have been carried out with a client and internal interlocutors. This has made it possible to validate the coupling between virtual prototyping and software tool for evaluation. Structured approaches and quantifications obtained in this phase have been very well received by the client. Being able to quantify product designs is the most important in the development of products, but besides, this allows the client to better justify time spent or budgets allocated for the designing activity.

In any case, specific recommendations connected with human impact must be respected regarding the use of VIR in the decision making process. It has been observed that the use of tools have their own set of deviances between the decision based on virtual prototypes and decisions based on physical prototypes. Nevertheless, even if these connections may be important, the current absence of decision tools in context of distance working is directly detrimental to the emergence of innovations through collaborative creation by an entire set of group members in designing and particularly the client. In its absence, promising designs can be preempted by a unilateral decision. Creation and creativity combine here to bring about a dialogue which allows and enriches a quasi interaction with the clients.

---

9 In case of an automobile manufacturer, the number of physical models was divided by 3 to 5 for designing the style of the vehicle according to Fuchs [2001].

10 For a client developing products for India, China and Brazil.

A systematic approach elaborated around specific tools is necessary. The onus is on the organization, like a construction phenomenon [WEI 79], through a management of organizations who are looking for reactivity, flexibility [WEI 95]. The advantage of concurrent work is seen when the organization can respond uniformly to the needs of the clients with more precision reactivity than any other.

In terms of evolution, it is clear that we are at the dawn of these developments. They are still linked with technological tools where manufacture is concerned, at the technical level, for tools for virtualization over the Internet and also sociological as far as human limitations are concerned. In fact, the client is not a designer, and does not always know what he wants but on the other hand, he knows what he does not want or what he prefers out of several alternatives. Finally it is the client who decides, and from then on, looking to integrate the decisions with ad hoc tools will become more and more an issue. We are aware of the novelty of this step and the improvements that need to be carried out. In this sense, two experiments jointly carried out, one with real products and the other with virtual products seems inevitable.

Chapter 20

Shapes, Knowledge and Innovation

20.1. Introduction

Innovation involves a process that is a result of exchanges of viewpoints and of taking decisions by way of conceptualizing and positioning a product in the market. But the result is only an assemblage of forms, lines, proportions, colors, mechanisms and technology that have little significance in and of themselves, if not of their exchange value that underlies their power of compensating for a social lack. Most often, innovation that is successful in the market shows optimization and harmony between performance and esthetics in all that is perceived and experienced by all those who are involved with the product. It is in this perspective, that a theory of form becomes inevitable to explain the contribution of chosen angular elements, certain notable proportions, their connections and complementary nature to esthetics, harmony, and efficiency.

In this chapter, we present a theory of form, its principal fundamentals, its implications in natural and artificial categories, and its method of application. We shall then describe the quantification of forms and proportions of two recent innovations in potable products. Finally, we shall bring the managerial contributions to bear upon the knowledge of theory and practice of form, based upon the chosen angles and remarkable proportions in the creative and analytical process linked to innovation.

Chapter written by Jean-Pierre MATHIEU, Michel LE RAY and Ilya KIRIA.

20.1.1. *Existence and theory of universal forms: chosen angles and sacred proportions*

The visual characteristics of a product, and also the tactile sensations produced by its coating, its weight, its ergonomic convenience, the animal or vegetal models that inspire its morphology and the artistic vibrations that it elicits, all make it an object of pleasure and desire. Thus, objects satisfy our senses, and sensory analysis, that is, "the use of human organs as measuring instruments" [MAC 97], appears to be appropriate for their conception and testing. However, the study of the consumers' and conceivers' relationship with their product brings us the inescapable notion of "esthetics", and, in this context, comparative criteria between art and design are established. Indeed, for us, the identification and subsequent choice of a product results from the perception of original forms [MAT 94]. For example, according to Fechner, beauty "designates all that is capable of evoking immediate pleasure" [FEC 1876]. This phenomenon of "natural" pre-selection is, in our opinion, based on perception of universal forms whose angular structures are formed by chosen angles [LER 80]. These have the capacity to radiate "the esthetic experience" created by the designer and decoded by the consumer, where the influence of certain angles and their associated forms (or design) as performance and esthetic criteria prove to be absolutely general in time and space [MAR 83].

Recent discoveries in physical sciences and many years of observations carried out by a team of scientists have enabled updating of chosen angular elements that contribute especially to "the esthetics, harmony and efficiency". The principles of these angles are those that have structured sacred proportions since Antiquity. Observations made first in physics and later in nature, art, architecture, and in numerous products [LER 80] make it possible to assert that "the inclusion of these angles is an essential criterion of performance, and that these angles play a vital role in adherence to the image" [MAT 99, 01, 02, 04]. Marketing [GAV 93] has shown the importance of the text/image relation and the results of his research explain the notions of congruence/redundancy between the image and the text of an advertising image. It is therefore particularly important from the sensory point of view, and especially commencing from the consumers' vision (the eye being the essential organ that relates the organism to its external world), to explain angular structures that underlie the forms of images and products. In this perspective, we shall describe briefly different areas of research in which these angles and other geometric forms are sources of dynamism, equilibrium, and harmony. It is supposed that these chosen forms facilitate the memorization by the serenity that these forms produce in the individual. The synthesis of contributions from quantum physics, neuro-physiology, cognitive psychology to the recognition of forms and hence to the processes of perception, identification, and memorization will aim at isolating major invariables from the complexity of explanatory phenomena and individual reactions to external stimuli.

The relation established between the microscopic level and macroscopic scale, due to the presence of a series of angles called "chosen" angles, has lead to the observation that in human creation there is a marked, deliberate, tendency to use these angles, which are sources of astonishment, concurrence, if not enthusiasm. Other geometric forms associated with these chosen angles favor sentiments of equilibrium, stability, and calm, even at the perception level of individuals.

20.2.1. *Notion of chosen angles developed by physical sciences and between microscopic and macroscopic scales*

This notion of chosen angles has been recognized at the microscopic scale since 1925-1930, and is based first at the level of the simplest atom (hydrogen) in which the rotation of the electron around the nucleus occurs, like that of a spinning top, around an axis which is itself situated on a cone whose generators create one of the chosen angles described later with an external magnetic field or with the axis of an external rotation. In 1972-1973, generalizing this fact at a much greater scale, Le Ray and his collaborators [LER 72, DER 73] showed the existence of vortexes in liquid super-fluid helium which obey, at a macroscopic level, the "macroscopic spatial quantification of the kinetic orbital moment", especially by the existence of spiral vortexes (local axis of rotation in a fluid) that make angles with their rotational axis given by the formula below, and only these. A further stage was attained [DER 75] in 1975-1976, when the major role of two families of angles, defined by the formula below, in the stability conditions of the vortex systems:

$$Cos\theta\{l,m\} = m \bigg/ \sqrt{l(l+1)} \qquad \{l,m\} \in I \quad m \leq l \text{ (l and m being integers)}$$

Let us remember that the cosine of an angle θ, written as cos θ, is defined as the ratio of the side of the right angle adjacent to the angle in question to the hypotenuse, which is the second side of the angle in question, in a triangle rectangle of which the angle considered is one of the two non-right angles.

These two families correspond to a simple relation between l and m and to the successive values of these two numbers or of one of them:

– for the first family (l = m = 1, 2, 3, 4, 5, 6, 7, 8, 9, etc.), these are angles equal to 45°; 35.3°; 30°; 26.6°; 24.1°; 22.2°; 20.7°; 19.4°; 18.4°, etc.;

– for the second (m = 2 ; l ≥ 2), the angles equal 35°, 3 (this angle being common to both families); 54.7°; 63.4°; 68.6°; 72°; 74.5°; 76.4°; 77.8°; 79°; 80°.

The existence of many combinations of these chosen angles must be observed: 45 = 26.6 + 18.4; 63.4 = 45 + 18.4 = 18.4 + 26.6 + 18.4; and, finally: 54.7 = 35.3 + 19.4; and many complementarities of two of these angles: 45 with itself; 54.7 with 35.3; 63.4 with 26.6.

20.2.2. *Golden angles and forms constructed by man*

Starting from aerodynamics, Le Ray and his collaborators asked themselves about the presence of these same chosen angles in certain natural structures subject to convolution by air or water currents, especially by natural wind. The straight or inflected lines of tree limbs and fine sand dunes like those in the Sahara showed chosen angles. Thousands of verifications showed, as early as 1976–1977, proof of the adaptation of these natural forms to angular conditions of equilibrium of the vortexes (in the case of sails, aircraft wings, of cars, the hulls and keels of ships, of locomotives of high-speed trains; see especially [LER 85]). In 1980 Le Ray [LER 80] published an article synthesizing analyses of thousands of traits of hundreds of documents (works of art, news photos, advertisements, etc.) that showed the existence of physically or psychologically important matching between lines according to the chosen angles now called golden angles.

Figures 20.1, 20.2, 20.3 and 20.4 mainly show the universality of the phenomenon of chosen angles in flows and forms created by man or by nature to adapt themselves to relative chosen positions (translated into chosen angles) of local axis of rotation, commonly called vortexes, in fluids (air or water, mostly), these local axis must be stable compared to the solid form to be circumvented, and moreover, in most cases, attach themselves to a ridge or a powerful curve of the form considered.

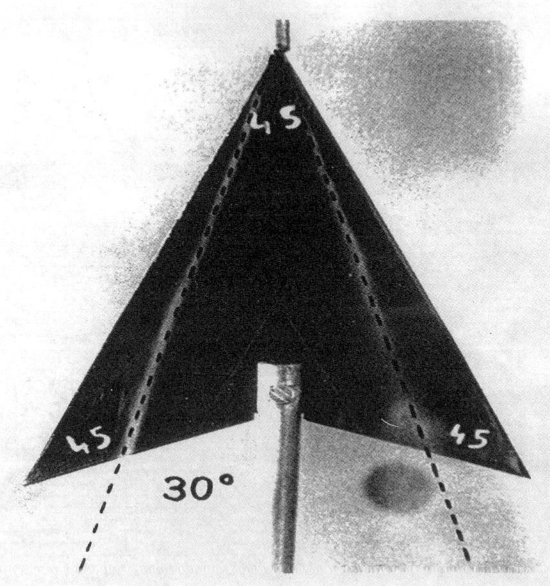

Figure 20.1. *Vortexes above a Delta Wing whose leading edges are at a chosen angle*

Figure 20.1 shows how, above a Delta Wing placed in a relative wind to which it is highly inclined (the inclination would be seen in profile and it is evidently meant to ensure an upward force called aerodynamic lift), vortexes called lift flaps are developed. Around these, the flow swirl while increasing its speed and decreasing the pressure exerted above the wing. If smoke is emitted in the immediate neighborhood of the front point of the wing, from where the vortexes are created, it splits and follows one or another of the vortexes (here rectilinear, as are the leading edges (front end) of this wing) because the length of the axis of each vortex, the rotation of fluid is nil, while it is intense around them. The smoke therefore shows up a couple of vortexes, thus revealing also that wings whose leading edges form a chosen angle $\theta 11 = 45°$ creates vortexes placed at $\theta 33 = 30°$ with respect to each other (Figure 20.1). A wing opened to $\theta 22 = 35.3°$, creates, for its part, vortexes angularly distanced by $\theta 55 = 24.1°$, just as a wing with an angle of opening called "apex angle" $\theta 33 = 30°$ produces an inter-vortex angle of $\theta 77 = 20.7°$ and that, finally, we witness "filiation" $\theta 44 = 26.6° \rightarrow \theta 99 = 18.4°$. Two other filiations are equally essential: $\theta 42 = 63.4° \rightarrow \theta 11 = 45°$, and $\theta 32 = 54.7° \rightarrow \theta 22 = 35.3°$. Thus were established [LER 85] the "laws of filiation" that will play, as will be shown later, an important role in the objects and the text-image matching.

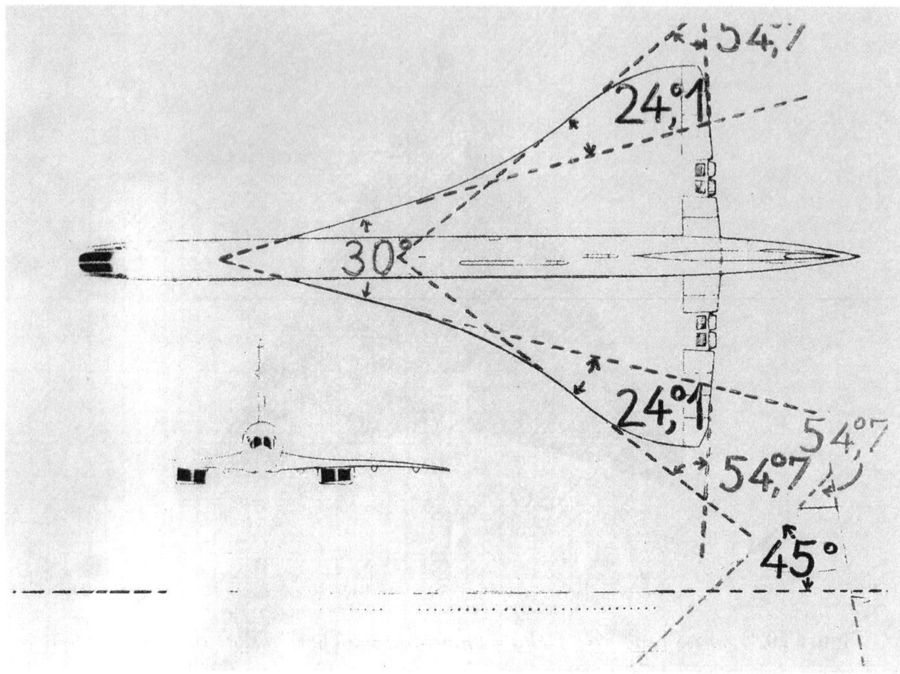

Figure 20.2. *Concorde and its angular coupling*

Figure 20.2 shows the leading edges of the Concorde with their 30° angle between the anterior portions on each wing, those of 24.1° between each anterior rectilinear portion and the following inflected portion on each wing, and finally that of 54.7° between the inflected portion of each leading edge and the rear edge called "lateral upper band". These angles have largely contributed to the stability and to the feeble flow noise around the aircraft (distinct noise made by the reactors) during the 34 years of flight of the most beautiful bird created by man to date.

Figures 20.3 and 20.4 illustrate the presence of chosen angles in natural structures subject to constant wind action, or which create in front of them, as efficiently as possible, relative wind (case of a bird's wing in flight):

– Very fine sand dunes in the Sahara (Figure 20.3), where the sand grains are so fine that the crests of dunes place their successive inflexions described by Roger Frison-Roche as "pure and quasi-abstract threads that link their ondulations between themselves" in a way that "the moving hills are so completely in proportion that one would not be able to give them dimensions and one would think that, in creating them, nature has divinely respected the Golden Number" [FRI 54].

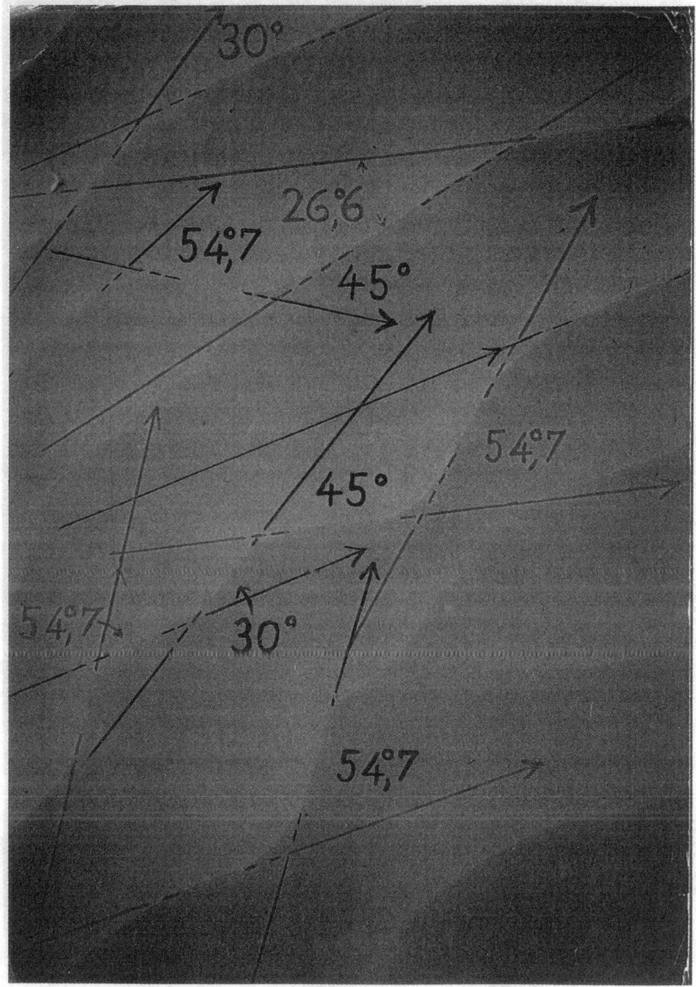

Figure 20.3. *Fine sand dunes in the Sahara, near Douiret (Southern Tunisia)*

– Successive elements of the leading edges of a bird's wing (Figure 20.4): the Golden Angle $\theta 42 = 63.4°$ gives an impressive majesty to the flight of a silver seagull, evoking the following affirmation by Gaston Bachelard: "the movement of flight gives, immediately, in remarkable abstraction, a dynamic, perfect, complete and total image" [BAC 43].

Figure 20.4. *Silver seagull in flight*

20.2.3. *Golden angles and other geometric forms*

These golden angles, along with other geometric forms (golden rectangles, root rectangles) seem to constitute a framework that is especially conducive to evocation of feelings of calm and stability. Numerous works of art from Antiquity show the use of these geometric forms to give rise to these sensations.

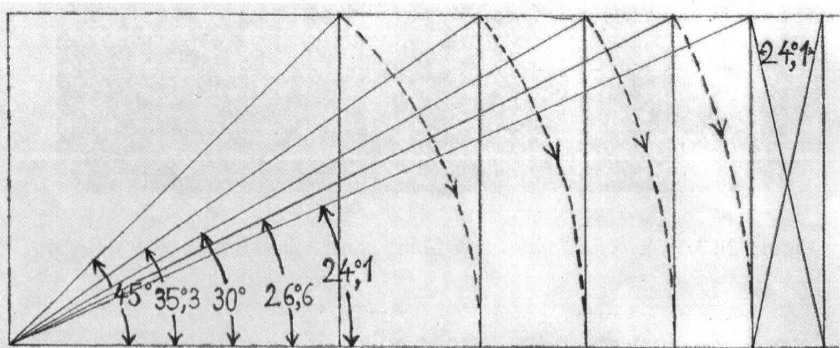

Figure 20.5. *Creation of dynamic rectangles from square and the relation between two consecutive dynamic rectangles and the one preceding this difference*

Since the early post-Pythagoras era, these rectangles and these numbers have been considered standards of beauty and harmony. Indeed, the chosen angles of the first family are also the angles between the longer side and diagonals of root rectangles, so named because the ratios of their sides are equal to the square roots of successive integers ($\sqrt{1} = 1$; $\sqrt{2} = 1.414$; $\sqrt{3} = 1,732$; $\sqrt{4} = 2$; $\sqrt{5} = 2.236$; etc.) and described by Plato as "dynamic rectangles". These rectangles are each obtained from the preceding one by dropping one its diagonals onto its longer side by extending it. The order of a dynamic rectangle is its number in the series which is just defined, the dynamic rectangle of the first order being the square. For example, the square of the first order, having characteristic angle of 45°, the dynamic rectangle of order 3 corresponds to an angle of 30°, and one of order 6 corresponds to 22.2°, in all cases between the longer side (or the short edge for the square) and diagonal (Figure 20.5). Moreover, it is very easy to establish mathematically (as, for example, is shown visually in Figure 20.1 in the case of the angle $\theta 55 = 24.1°$) that the difference (case of Figure 20.5) where the sum of two consecutive dynamic rectangles has the same angle between its diagonals as the angle between the diagonal and the longer side of the smaller one of the two dynamic rectangles. As to the chosen angles of the second family, they are, for the 35.3°, common with the first family, for the 54.7° and 63.4° complements of the 35.3° and 26.6° of the first family, the 63.4° being, also, the angle between the diagonals of the famous Golden Rectangle (Figure 20.6) whose ratio between sides is $\phi = (1+\sqrt{5})/2 = 1.618$ which is the famous Golden Number.

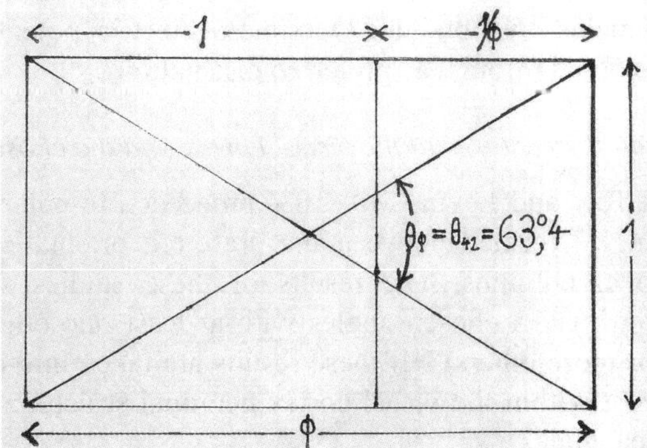

Figure 20.6. *Square, Golden Rectangle and Golden Number*

Let us remind ourselves that the Golden Rectangle is a rectangle such that if we remove from it a square constructed on three of its sides the remaining rectangle, which is also a Golden Rectangle, resembles the initial rectangle, thus fulfilling a "thirst for invariance of the Central Nervous System" [PAI 74].

20.2.4. *Contributions of neurophysiology*

As early as 1980, Le Ray spoke of the link between the existence of chosen angles as criteria of impact, memorization of images and the existence of cerebral mechanisms and structures that would explain the "thirst for invariance of the

Central Nervous System". The presence of "genetically pre-cabled detectors of form" was already linked to the functional specialization of the neurons of the occipital cortex (areas 17 and 18) studied since the 1960s by D. Hubel and T. Wiesel. These two neuro-physiologists of the Harvard Medical School in Boston, authors of about 20 publications since 1962, were awarded the Nobel Prize for Medicine and Physiology in 1981.

20.2.4.1. *Informative zones*

One of their most important discoveries concerns the existence of preferred orientations by neurons of the visual cortex or, more exactly, an orientation preferred by each of these neurons. Experiments were conducted in which a micro-electrode was introduced into the visual cortex of cats or monkeys obliquely to the external surface of the brain. The eyes, immobilized in advance, of these animals were subjected to stimuli consisting of black lines on white backgrounds or white lines on black backgrounds, or of black and white zones intersecting. In these three cases, the "informative zones" described by psycho-physiologists were involved. D Hubel and T Wiesel [HUB 77, 79] published some 15 articles giving figures which showed the directions preferred by neurons of one area of the visual cortex, this preference was signaled by obtaining maximum electric potential in this region when the stimulus passed through a "preferred orientation".

20.2.4.2. *The neuronal structures of the visual cortex and the chosen angles*

Martinache, Le Ray and Levin [MAR 83] proceeded to enlarge these diagrams and studied the angles formed between the preferred orientations of adjacent or neighboring neuronal columns. The results of these studies show that: "every preferred orientation makes chosen angles with at least one other orientation (but usually with two or more others)". If these results are taken into consideration, it is likely that there exist within the visual cortex neuronal structures that facilitate the perception of chosen angles.

20.2.4.3. *Angularly noticeable dipolar nature*

A dipole, generally, is a combination of a "source" and a "well" (source and attractor of a hydrodynamic and aerodynamic flow, positive and negative charge in electrostatic) placed very near one another and which define the axis of the dipole thus formed by the help of the segment which joins them. This structure in "source and well" is identical in its effects, that is, by the field of speed, to that of the very concentrated rotational loop of a fluid on itself, called a ring vortex, that makes the liquid pass into the interior of the ring and bringing it out along the lines of closed currents. Yet, the common form of the lines of fluid current created by a hydro or aerodynamic dipole and the electric or magnetic field lines created by an electrostatic or electric dipole (made of a small loop of electric current) shows properties remarkably like those of the angle characterizing the position of one point with respect to the dipole and its axis, and the angle characterizing the position of the speed of this point with respect to the line joining the dipole to the considered point, this line being the second side of the preceding angle. These observations make sense if we connect them to the work of neuro-physicists on the measurement external to the cranial box, in width, direction of magnetic fields coming from diverse cortex of the human brain, the measurement which enabled the reconstitution of a map of magnetic fields within the cortex, and of the electric currents responsible for their formation. The loops of electric current, thus revealed create a distribution of magnetic fields whose field lines are very similar to those created by an ideal dipole made of a loop of circular current. Both the dipolar point of view and the general vortex point of view explain the omnipresence of chosen angles.

20.2.5. *Contribution of cognitive psychology*

Although not explicitly stated, the chosen angles and the remarkable angular combinations appear indirectly in theories of cognitive psychology, especially in the recognition of forms and implementation of the attentional process where memory plays an important role. In the beginning, studies on recognition of forms were confused with more general perception processes, such as Associationism and Gestalt [GUI 37] but soon, two theories supported by two different conceptions of identification emerged:

– The first concept highlights pairing between two prototypes: being from the Gestalt thesis [WER 23], it opines that identification is based upon global perception. Secondly, Rosch [ROS 73, 76], introduced the idea that comparison between the example and the prototype could result in a judgment of closeness. Certain examples would be closer, more typical of, the prototype than others.

– The second, coming from "associationism" is based upon the description of images in lines, with the identification models [NEI 64] or componential [SMI 74].

Diverse contributions from quantum physics, neuro-physiology and cognitive psychology have rarely been used in fields so closely connected to marketing and design wherein forms, images constitute the apex of success of products sold to different consumers.

20.3. The spatial quantification of an object

In this section we shall present some analyses on two objects: the new Perrier bottle and a bottle of Coca-Cola. Previously, we referred the reader to some publications by the authors, or by one of them, and especially to the first of the works cited, which is the most thorough, with 16 graduated figures in domains from

Hydrodynamics to Pictorial Art, via natural forms [LER 80], as also to the most recent work, which is also quite comprehensive [MAT 04].

Every object, before any stylistic study [CHR 96], whether new or innovated, presented to a prospective consumer, must have an attractive shape, and as we have demonstrated [MAT 01, 02], in the case of wine bottles, this characteristic is linked to the presence of chosen angles and remarkable proportions, and to resonant combinations thereof, in different parts of the bottle and between themselves. In this perspective, and to complete the works cited above on bottles, we have chosen the new Perrier bottle that is identical in shape to the type "FLUO", and its equivalent in the market, the new bottle of Coca-Cola, that are current references in the soft drinks market. As we shall show in the following analysis, these two bottles possess fundamental directions that are dynamic and/or structuring.

In the Perrier bottle, the left-right and right-left crossings of the sides that do not correspond to angles of divergence greater than or less than 35.3° and 54.7° determine the angles 45° = (35.3°+54.7°)/2 and 45° = (54.7°+35.3°)/2 (Figure 20.7) which contributes to the unity of the form,

Figure 20.7. *50cl "FLUO" type Perrier bottle*

Moreover, the angle of divergence, this time towards the top of 24.1° (Figure 20.8) transforms the single filiation 54.7° → 35.3° into a double 54.7° →35.3° →24.1°.

Figure 20.8. *50cl "FLUO" type Perrier bottle*

On its part, the Coca-Cola bottle, *a priori* more classical and more austere, perhaps less spontaneous dynamically, shows, in contrast, a resonant structuring that is more elaborate, wherefrom, literally, emerges an impression of authority. Figure 20.9 (with the angles 45°, 54.7° and 35.3°) is particularly subtle, complete and stable.

Figure 20.9. *50cl Coca-Cola bottle*

Figure 20.10 shows the initial divergence-convergence at 45° (compared to "Bourgogne Tradition" bottles) that was seen from the outset. This divergence-convergence must obviously be taken up again in Figure 20.11, already referred to for explaining the effects of perpendicularity and resonance with elements of the main body of the bottle.

Figure 20.10. *50cl Coca-Cola bottle*

Figure 20.11 with the angles of 63.4° and 26.6° and the effects of perpendicularity associated with it shows the phenomena linked to two "Golden wing-backs" (with diagonals at 63.4° between them), both of vertical axis, one sideways, the other flat. These different angular sets of diagonals indicate a pertinence of the placing of the "waist" (the narrowest part) of the bottle, in the framework of the general architecture of the bottle, defined in Figure 20.11. The appearance of 63.4° and 26.6° completes almost totally (with the exception of the 30°) the use of significant chosen angles in the geometry of the bottle.

Figure 20.11. *50cl Coca-Cola bottle*

The shape of the main body of the bottle, from the top bulge to the bottom of the divergent portion, up to the widest part of the lower portion (section characterized by an identifiable seam in the photograph, but clearly visible on the object) appears as a Rectangle whose diagonals are at an angle of $\theta_{32} = 54.7°$ between them. This corresponds to a longer side to shorter side ratio, which in this case is, height upon width (diameter of the cylinder enveloping the bottle) equal to $x = (\sqrt{3}+1)/\sqrt{2} = \sqrt{2}/(\sqrt{3}-1) = 1{,}932$. This ratio can be split up as $x = \sqrt{2}+(\sqrt{3}-1)/\sqrt{2}$, or $x = \sqrt{2}+1/x$ (A), a ratio that is very similar to: $\phi = 1+1/\phi$ (B), defining the Golden Number and the accompanying Golden Rectangles, whose difference is a square, and whose inter-diagonal angles are equal to $\theta_{42} = 63.4°$. Here, the role played in (B) by the unit is played in (A) by $\sqrt{2} = 1.414$. The relation (A) shows clearly that the large rectangle that is lying vertically, and whose inter-diagonal angle is 54.7° is the succession:

– of a rectangle of relative height $\sqrt{2}$ which is clearly a dynamic rectangle of the second order, characterized by its angle between longer side and diagonal θ22 = 35.3° and its angle, complementary to the first one, between smaller side and diagonal equal to θ32 = 54.7°;

– and of a rectangle of relative height $1/x$ = $(\sqrt{3}-1)/\sqrt{2}=\sqrt{2}/(\sqrt{3}+1)$ = 0.518, similar, of course, to the large rectangle of proportion x, but "flat" (that is, with large horizontal sides), however, the first one is "on its side" (that is, with longer sides vertical). This new rectangle, similar to the entire rectangle, has, also, diagonals at 54.7° between themselves, each of these diagonals being perpendicular to a diagonal of the similar, vertical large rectangle (as seen in Figure 20.11), making an angle θ22 = 35.3° (not labeled in the figure, but easily identifiable in the rectangle triangles containing the angle θ32 = 54.7° = 90° − 35.3°) with the other diagonal of the large vertical rectangle.

Finally, in its turn, the dynamic rectangle of proportion $\sqrt{2}$ (with its angles θ22 = 35.3° and θ32 = 54.7° between diagonals, and respectively, its longer and shorter sides) can be reduced to a square (with, obviously, its 45° angles between sides and diagonals) and a rectangle constituted by the difference between the initial rectangle and the square, hence by the difference between the consecutive dynamic rectangles of proportion $\sqrt{2}$ and $\sqrt{1}$ − 1. Indeed, we know, from the general theorems about the sum and the difference of two consecutive dynamic rectangles that the angle between the diagonals of the difference rectangle is, here too, equal to θ11 = 45°.

The angular analysis of the bottle (Figure 20.11) makes clear all these successive elements from the various elements that have just been described in detail. The sides and diagonals of each element of the general combination are, for the most part, remarkably associated with morphological or graphic elements of the bottle and its label.

The lower square extends from the maximal lower section cited at the beginning of this analysis up to a first increase in the bottle's contour just below the label. Above it is the rectangle whose diagonals are at a 45° angle. Its lower edge is obviously the upper side of the previous square, that is, the bulge or excrescence that has just been mentioned.

The upper edge of this rectangle, which is, evidently, the lower side of the next rectangle of the decomposition, is mixed with the lower level of the Coca-Cola logo on the label, spectacularly highlighting this label and the famous logo.

This highlighting is even more striking because the diagonals at 45° between themselves are each, on the one hand perpendicular to the side of the angle, which itself is 45°, and which frames the upper divergence of the bottle, and on the other, at 45° with the other side of the angle of this divergence. Thus, we are faced with a fundamental whole of stable and resonant elements, with an aerial and dynamic character that we may compare, as we have done earlier in the case of certain wine bottles, to the esthetic and aerodynamic qualities of a Delta Wing, especially a "swallow tail" Delta Wing made up of chosen angles, especially angles of 45° [LER 85; MAT 01, 02].

Finally, this lower level of the Coca-Cola logo is not only the upper side of the rectangle with 45° between its diagonals and of that lying vertically with 35.3° and 54.7° between its diagonals, and, respectively, larger sides vertical and smaller sides horizontal, but it is also the lower side of the rectangle bound on its upper side by the uppermost bulge of the bottle, a rectangle whose diagonals are at the same angle of 54.7° like the angle that exists between diagonals of the big rectangle lying vertically and which envelops the main body of the bottle, from the widest section of the rounded lower portion to the top bulge, at the lower limit of the conical portion (or divergent – convergent portion) of the bottle. This results, as we have seen in the general study of similar rectangles with large and small sides perpendicular to each other, that each of the diagonals of these two rectangles makes, with one diagonal or another, either a right angle, or a chosen angle of 35.3° = 90° – 54.7°. One can easily imagine, even in this case, the contributions made by such phenomena to the image of the universal brand of the product.

20.4. Overall finding

We have presented a current status of the art of highlighting chosen angles and remarkable proportions which inscribe themselves naturally in the field of "Gestalt" while making an explanatory, supplemental contribution, an operationality and an application of this last, to the quantification of new or innovative forms. To this end, the present chapter suggests, in particular, three paths of reflection and application for systematically taking into account the analyzed phenomena for more effective innovation:

– first, update of work likely to help understand better the influence of chosen angular components on perception, categorization, and memorization of forms, and hence, objects that will enable such processes, for a better sensitivity towards research conducted in marketing design, innovation and creation, especially within the framework of positioning strategies;

– contribute to demonstrate the importance of angular analyses, translating spatial quantification of images and forms in the creation and evaluation of design and design management, and especially in the study of stylistics of objects and their execution in connection, particularly, with the creation, analysis, maintenance, and launches of these objects or products and communications around the innovative aspects of the latter.

– finally, suggest the importance of training, experience, or at least good artistic and stylistic knowledge to designers, directors, and all involved in innovation.

Bibliography

[ACH 98] ACHARD P., BERNAT J.-P., *L'intelligence économique: mode d'emploi*, Paris, ADBS Editions, 1998.

[AFA 97] AFAV, Qualité en conception – La rencontre Besoin-Produit-Ressources, Paris, AFNOR Editions, 1997.

[AFI 02] AFIC, PRICE WATERHOUSE COOPER, "Rapport sur l'activité du capital investissement en France 2002".

[AFI 03] AFIC, Evaluation, reporting et statistiques, les nouveaux standards de la profession, AFIC, 2003, available on www.afic.asso.fr.

[AFN 96] AFNOR, Norm NF-EN 1325-1 "Vocabulaire du management par la valeur, de l'analyse de la valeur et de l'analyse fonctionnelle – Partie 1: Analyse de la valeur et analyse fonctionnelle", Paris, AFNOR Editions, 1996.

[AFN 98] AFNOR, collection of standards "De l'analyse de la valeur au management par la valeur", Paris, AFNOR Editions, 1998.

[AFN 00] AFNOR, Norm NF-EN 12973 "Management par la valeur – Principes – Mise en œuvre – Outils", Paris, AFNOR Editions, July, 2000.

[AGH 04] AGHION P., COHEN E., *Education et croissance*, La documentation française, 2004.

[ALB 74] ALBERS J., *Interaction of Color*, New Haven, London, 1936 (French translation, *L'interaction des couleurs*, Paris, 1974).

[ALI 98] ALIX J.-P., KRIEGER E., "La confiance, clé de voûte des réseaux d'innovation", in *La Revue Française de Géo-économie*, no. 8, winter 98/99, Paris, Economica, pp. 125-138, quotation p. 129.

[ALT 02] ALTER N. (Ed.), *Les logiques de l'innovation. Approche pluridisciplinaire*, La Découverte Recherches, 2002.

[ALT 88] ALTSHULLER G., *Creativity as an Exact Science*, NY Gordon, Breach, 1988.

[AMI 96] AMIDON D.M., "The challenge of fifth generation R&D", *Research Technology Management*, July-August, 1996.

[AMI 97] AMIDON D.M., The Ken Awakening: Management Strategies for Knowledge Innovation, Butterworth-Heinemann Publishing, 1997.

[AND 94] ANDERSON E.W., FORNELL C., LEHMAN D.R., "Customer satisfaction, market share, and profitability: findings from Sweden", *Journal of Marketing*, vol. 58, pp. 53-66, July 1994.

[ART 94] ARTUS P., Increasing Returns and Path Dependence in the Economy, U. Mich. Press, 1994.

[ART 01] ARTUS P., *La nouvelle économie*, Repères, La Découverte, 2001.

[ATK 00] ATKINSON T., BOYER R., HERZOG R., JACQUET P., *et al.*, *Questions européennes*, La documentation française, Paris, 2000.

[AUR 92] AURIFEILLE J.-M. and VALETTE-FLORENCE P., "A chain-constrained clustering approach in means-end analysis. An empirical illustration", in K. GRUNERT and D. FUGLEDE (Eds.), "Marketing for Europe – Marketing for the Future", *Annual Conference of the European Marketing Acadamy*, vol. 21, Aarhus, Denmark, pp. 49-64, 1992.

[AUS 91] AUSTIN J.L., *Quand dire, c'est faire*, Seuil, 1991.

[BAC 43] BACHELARD G., *L'Air, les Songes, Essai sur l'Imagination du Mouvement*, Chapitre II: "La poétique des Ailes", p. 79, José Corti, Paris, 1943.

[BAR 00] BARBIERI MASINI E., *Penser le futur. L'essentiel de la prospective, de ses méthodes*, Dunod, Paris, 2000.

[BAR 93] BARBIERI MASINI E., *Why Futures Studies?*, Grey Seal Books, London, 1993.

[BAR 86] BARNEY J.B., "Strategic factor markets: expectations, luck and business strategy", *Management Science*, no. 42, 1986.

[BAS 96] BASSEREAU J.F., LEGEY T., DUCHAMP R., "Du différentiel sémantique au profil sensorial", *Journal of Design Sciences and Technology*, STC Europia production, vol. 5 no. 2, 1996.

[BAT 99] BATSCH L., *Finance et stratégie*, Economica, 1999

[BAU 91] BAUMARD P., *Stratégie, surveillance des environnements concurrentiels*, Masson, Paris, 1991.

[BAU 96] BAUMARD P., Organisations déconcertées. La gestion stratégique de la connaissance, Masson, Paris, 1996.

[BEA 01] BEAURAIN N., JEZ E., *Les noms de domaine et l'Internet*, Litec, Paris 2001.

[BER 57] BERGER G., "Sciences Humaines et prévision", *La Revue des deux mondes*, 1 February 1957, pp. 3-12.

[BER 59] BERGER G., "L'éducation dans un monde en accélération", *L'Encyclopédie française*, vol. XX: Le monde en devenir, 1959, p. 20/54-15/56/02.

[BER 60] BERGER G., "En guise d'avant-propos: méthodes et résultats", *Prospective*, no. 6, November 1960, pp. 1-14.

[BER 61] BERGER G., "Le chef d'entreprise, philosophe en action", (Colloquium 8 March 1955 at the department of general studies at the Centre de Recherche, d'Etudes des Chefs d'Entreprise), *Prospetive*, no. 7, 1961, pp. 47-66.

[BER 64] BERGER G., *Phénoménologie du temps, prospective*, PUF, 1964, quote p. 233.

[BER 99] BERGER P., *R&D*, L'Harmattan, Paris, 1999.

[BER 01] BERGER P., *Traité de systemique digitale*, documento de trabajo, Paris, 2001.

[BER 00] BERGMAN E. (Ed.), *Information Appliances and Beyond*, Morgan Kaufmann, 2000.

[BER 00] BERNARD-BOUSSIÈRES, J., *Aide à l'élaboration du cahier des charges fonctionnel*, AFNOR Editions, Paris, 2000.

[BER 68] BERTALANFFY L.V., *General System Theory: Foundations, Development, Applications*, George Braziller, New York, 1968.

[BER 73] BERTALANFFY L.V., *Théorie générale des systèmes*. Paris, Dunod, 1973.

[BES 96] BESSON B., POSSIN J.-C., *Du Renseignement à l'intelligence économique*, Dunod, 1996.

[BES 98] BESSON B., PONSIN J.-C., *L'audit de l'intelligence économique. Mettre en place, optimiser un dispositif coordonné d'intelligence collective*, Dunod, 1998.

[BIE 94] BIENAYME, L'économie des innovations technologiques, Que sais-je, 1994.

[BIEN 94] See [BIE 94].

[BLA 02] BLANCO, *et al.*, "Une expérience de conception collaborative à distance", *Revue Mécanique, Industrie*, Elsevier, 2002, pp.153-161.

[BOC 98] BOCQUET J.-C., "Modèles fonctionnels", in M. TOLLENAERE (Ed.), *Conception de produits mécaniques*, Hermes Science Publications, Paris, 1998, pp. 397-411.

[BOI 93] BOISSELIER P., L'investissement immatériel; gestion et comptabilisation, Edition de Boeck, 1993.

[BOL] BOLY V., Ingénierie de l'innovation: organisation et méthodologies des entreprises innovantes, Hermes Sciences/Lavoisier, Paris, 2004.

[BON 00] BONFOUR A., "La valeur dynamique du capital immatériel", *Revue Française de Gestion*, no. 130, September-October, 2000, pp. 111-124.

[BOR 00] BORJA DE MOZOTA B., *Design Management*, Editions d'Organisation, Paris, 2000.

[BOS 97] BOSSARD, "Origine, définition de l'ingénierie concorrante", in P. BOSSARD, C. CHANCHEVRIER, P. LECLAIR (Eds.), *L'ingénierie Concourante – de la technique au social*, Editions Economica, Paris 1997.

[BOU 97] BOUCHARD C., "Modélisation du processus de style automobile – Méthode de veille stylistique adaptée au design du composant d'aspect", PhD Thesis, ENSAM no. 1997-14, Paris, 1997.

[BOU 99] BOUCHARD C., CHRISTOFOL H., ROUSSEL B., AUVRAY L., AOUSSAT A., "Identification., Integration of Product Design Trends", in *Proceeding of International Conference on Engineering Design ICED '99*, Munich, August 24-26, 1999.

[BOU 02] BOULIER D. and PAVARD B., *Acceptabilité, ergonomie et usage*, CNRS RTP, May 2002.

[BOU 00-a] BOULTON R.E.S., LIBERT B.D. and SAMEK S.M., "A business model for the new economy", *The Journal of Business Strategy*, vol. 21, Issue 4, 2000, pp. 29-35.

[BOU 00-b] BOURNOIS F., ROMANI P.-J., *L'intelligence économique, stratégique dans les entreprises françaises*, IHEDN, Economica, 2000.

[BRI 91] BRIDGES W., *Managing Transitions – Making the Most of Change*, Addison-Wesley, Massachusetts, 1991, pp. 2-5.

[BRU 01] BRUN G., CONSTANTINEAU F., *Le Management par la valeur – Un nouveau style de management*, AFNOR Pratique, Paris, 2001.

[BRY 00] BRYSON J. and ANDERSON S., "Applying Large-Group Interaction Methods in the Planning and Implementation of Major Change Efforts", *Public Administration Review*, vol. 60, Issue 2, 2000, pp. 143-163.

[BUC 99] BUCHANAN D. and BADHAM R., *Power, Politics and Organizational Change: Winning the Turf Game*, Sage Publications, London, 1999.

[BUD 01] BUDHWANI K., "Becoming part of the e-generation", *CMA Management*, vol. 75, Issue 4, 2001, pp. 24-27

[BUZ 93] BUZAN T., *The Mind Map Book*, BBC Books, London, 1993.

[BWK 02] ENTREVISTA a TIM B.-L., "The Web's Weaver Looks Forward", BusinessWeek Online 27-03-2002.

[CAP 01] CAPPELLI P., NEUMARK D., "External Job Churning, Internal Job Flexibility", *document de travail du NBER*, no. 81-11, 2001.

[CAR 03] CARAYON B., (Report to the French Prime Minister), "Intelligence économique, compétitivité, cohésion sociale", www.bcarayon-ie.com.

[CAR 95] CARNELL C., *Managing Change in organizations*, Prentice Hall International Ltd, UK, 1995.

[CAR 01] CARNEY M., "The Development Of A Model To Manage Change: Reflection on a Critical Incident in a Focus Group Setting. An Innovative Approach", *Journal of Nursing Management*, vol. 8, Issue 5, 2001, pp. 265-273.

[CAR 99] CAROLI E., "Flexibilité interne versus flexibilité externe du travail: quels éléments peut-on tirer de la firme en termes de competences", *Report INRA-LEA*, 1999.

[CAR 91] CARTER J., "Implementing Supplier Bar Codes", *Production and Inventory Management Journal*, 1991, Fourth Quarter, pp. 42-47.

[CAS 88] CASPAR P., AFRIAT C., *L'investissement intellectuel. Essai sur l'économie de l'immatériel*, Economica-CPE, 1988.

[CAS 98] CASTELLS M., *La société en réseaux*, Fayard, 1998.

[CAS 99] CASTELLS M., *The Information Age: Economy, Society, Culture*, Blackwell, 1999.

[CAV 97] LUTZ P., "TRIZ, un concept nouveau de résolution des problèmes d'innovation", *2nd International Conference on Industrial Engineering*, Albi, 1997.

[CAV 99] CAVALLUCCI D., "Contribution à la conception de nouveaux systèmes mécaniques par intégration méthodologique", PhD Thesis in Mechanics, Louis Pasteur University, Strasbourg, December 1999.

[CHA 95] CHABBAL R., Caractéristiques des politiques d'innovation, 1995.

[CHA 02] CHAFFEY D., *E-Business and E-Commerce Management*, Prentice Hall, 2002.

[CHA 99] CHAMBRON G., www.art-of-design.com, 1999.

[CHA 03] CHANG W.-C., VAN Y.-T., "Researching design trends for the redesign of product form", in *Design Studies*, vol. 24, no. 2, pp. 173-180, 2003.

[CHA 98] CHAVANNE A., BURST J.-J., *Droit de la propriété industrielle*, Dalloz, 5th edition, Paris, 1998.

[CHE 99] CHECKLAND P., SCHOLES J., *Soft Systems Methodology in Action*, John Wiley, Chichester, 1999.

[CHE 97] CHEDMAIL P., BOCQUET J.-C., DORFELD D., *Integrated Design, Manufacturing in Mechanical Engineering*, Kluwer Academic Publishers, Dordrecht, 1997.

[CHE 01] CHEDMAIL P., MAILLE B., RAMSTEIN E., "Etat de l'art sur l'accessibilité en réalité virtuelle, application à l'étude de l'ergonomie", *Primeca Colloquium*, La Plagne, 5 April 2001.

[CHE 94] CHÉNÉ E., "Recherche d'un outil de communication industriel sur le toucher", *Design Recherche*, no. 6, 1994.

[CHE 03a] CHÉNÉ E., SAMIER H., RICHIR S., "Processus virtuel de décision dans la chaîne numérique de conception en PME en ingénierie concourante par Internet", in *International Journal of Information Sciences for Decision Making*, no. 8, May 2003.

[CHE 03b] CHÉNÉ E., RICHARD P., CHRISTOFOL H., RICHIR S., "Virtual Reality Application for Product Design Evaluation: Virtual Phone Design" *Virtual Concept 2003*, Biarritz, 2003.

[CHE 03c] CHÉNÉ E., "Développement d'une méthode, d'outils d'aide à la décision en ingénierie concourante étendue, intégrant le prototypage virtuel en conception de produits", PhD Thesis, ISTIA Innovation, 2003.

[CHR 95] CHRISTOFOL H., "Modélisation systémique du processus de conception de la coloration du produit", PhD Thesis, ENSAM, Paris, 1995.

[CRH 96] CHRISTOFOL H., AOUSSAT A., DUCHAMP R., "L'analyse de Contenu Iconique, un outil de représentation pour le concepteur de la coloration d'un produit", *Design Recherche*, no. 8, Quadrature, 25-34, 1996.

[CHR 00] CHRISTOFOL H., BOUCHARD C., ROUSSEL B., AOUSSAT A., "Analogue Reasoning a fundation of Stylistic, ergonomic, technologival creativity", in *Proceeding of 3rd International Conference on Integrated Design, Manufacturing in Mechanical Engineering (IDMME'2000)*, Presses Internationales Polytechnique, Montreal, 2000.

[CHR 01] CHRISTOFOL H., BOUCHARD C., ROUSSEL B., RICHIR S. and TARAVEL B., "Des valeurs humaines à la valeur du produit: méthodes de l'innovation collaborative dans les phases préliminaires du processus de conception", in *Proceeding of International Conference AFAV*, 6-7 November, Paris, pp. 77-85, 2001.

[CHR 04] CHRISTOFOL H., SAMIER H., "La veille tendances: comment surveiller les secteurs d'influences des métiers, des produits", *VSST Colloquium*, Toulouse 2004.

[CLA 97] CLARKE A., GARSIDE J., "The Development Of A Best Practice Model For Change Management", *European Management Journal*, 1997, vol. 15, Issue 5, pp. 537-545.

[CNI 00] CNISF, FMOI, CNAM, *TRANSinfo 3 – L'entreprise et l'effet réseau*, Paris, ADBS Editions, 2000

[CNI 97] CNISF, FMOI, CNAM, "Modèles de communication et stratégies d'entreprises – problèmes d'organisation ou problème de management?" *Actes du colloque international TRANSinfo 96*, Paris, ADBS Editions, 1997.

[COC 01] COCPIT, "Contribution des tendances prescriptrices des styles intérieurs automobiles à la conception couleur-matière d'un démonstrateur automobile", Report, SERAM, Paris 2001.

[COC 99] COCPIT, "Formalisation des tendances en vue du design d'une innovation technologique – Projet DD", Report, SERAM, Paris, 1999.

[COH 04] COHEN C., *Veille et intelligence stratégiques*, Hermes Sciences/Lavoisier, Paris, 2004.

[COM 01] COMMISSION EUROPENNE DG RECHERCHE, The Norway Seminar, Workshop 2001.

[COM 92] COMBS R.E., MOORHEAD J.D., *The Competitive Intelligence Handbook*, The Scarecrow Press, 1992.

[CON 03] CONCEIÇÃO P., HEITOR M.V. and VELOSO F., "Infrastructures, Incentives, and Institutions: Fostering Distributed Knowledge Bases for the Learning Society" *Technological Forecasting and Social Change*, 70(7): 583–617, 2003.

[COR 90] CORIAT B., *L'atelier et le robot*, Christian Bourgeois Editeur, 1990.

[COR 01] CORSI P. and DULIEU M., *The Innovation Drive – The Marketing of Advanced Technologies*, Knowledge 2 Know-How Ed., 2001.

[COR 03] CORSI P. *et al.*, *Piloter le changement avec les cybertechnologies, Innover pour rester compétitif*, Hermes Sciences/Lavoisier, Paris, 2003.

[COR 05] CORSI P., LEVY-DREYFUS J.M., *Introducing the Next Big Thing: The Transactional Internet;* "IST at the Service of a Changing Europe by 2020: Learning from World Views Environments", *FISTERA IPTS International Conference*, Seville, Spain, 16-17 June 2005.

[COU 02] COUTROT T., Innovation, stabilité des emplois, Review made by the Commissariat général du Plan, presented on 17 January 2002.

[CRO 96] CROSS N., CHRISTIAANS H., DORST K., *Analysis Design Activity*, John Wiley, England, 1996.

[CRU 02] CRUBLEAU P., "L'identification des futures générations de produits industriels. Proposition d'une démarche utilisant les lois d'évolutions de TRIZ", PhD Thesis, Industrial Engineering, Angers University, December 2002.

[DAI 03] DAILY NEWS, Informations for the Global GRID Community, October 6, 2003: Vol. 2 no. 40, http://www.gridtoday.com/gridtoday.html.

[DAR 07] DARSES, "L'ingénierie concourante: un modèle en meilleure adéquation avec les processus de conception", in BOSSARD P., CHANCHEVRIER C., LECLAIR P. (Eds.), *L'ingénierie Concourante – de la technique au social*, Economica, Paris, 1997.

[DAV 99] DAVIDSON W., "Beyond Re-Engineering: The Three Phases of Business Transformation", *IBM Systems Journal*, Vol. 38, Issue 2/3, pp. 485-499.

[DE 95] DE P.K. and HUEFNER B., "Technological Competitiveness of Germany: A Post-Second World War Review", *Technology Management*, Vol. 2, no. 6, 1995 pp. 265-274.

[DEF 02] DEFAUX M., "Travail collaboratif: entre rêve et réalité", *Harvest*, no. 69, pp. 31-38, April 2002.

[DEF 85] DEFORGE Y., *Technologie, génétiques des objets industriels*, Maloine, Paris 1985.

[DEL 00] DELTOUR F., *L'innovation dans l'organisation: dépasser les ambiguïtés du concept*, les cahiers de la recherche au CLAREE – CNRS 8020 – 2000.

[DEN 03] DENIS M., "Cognition humaine, Exemples d'Activités de Recherche du groupe, LIMSI", http://www.limsi.fr/RS2002FF/CHM2002FF/CH2002FF/, last accessed on 16 July 2003.

[DER 73] DEROYON M.J., DEROYON J.P., LE RAY M.," Experimental evidence of microscopic spatial quantization of Angular Momentum in rotating Helium", *Physics, Letters-A 45*, pp. 237-238, 1973.

[DER 75] DEROYON J.P., DEROYON M.J., LE RAY M., "Influence on normal fluid on the quantized vortices of the helium 2, vortices in the wake of bubbles in a classical fluid", *Proceedings of the 14th international conference on low temperature Physics LT 14* (Helsinki) pp. 279-282, North Holland Publishing Company, Amsterdam, 1975.

[DER 75] DEROYON J.P., DEROYON M.J., LE RAY M., Tourbillons quantifiés dans l'hélium, *Résumé de la communication no. 6-C-4: 2nd French colloquium on Mechanics*, Toulouse, 1975.

[DES 96] DESHAYES P., "The duality environment/teleology in modelling process of conception", in *CESA'96 Multiconference Computational Engineering in Systems Applications*, Lille, France, 1996.

[DEV 02] DEVARAJ S. and KOHLI R., *The IT Payoff*, Prentice Hall, New Jersey, 2002.

[DIE 00] DIESE M., NOWICKOV C., KING P. and WRIGHT A., *Executives Guide to e-Business: From Tatics to Strategy*, John Wiley and Sons, New York, 2000.

[DOD 00] DODGE M., KITCHIN R., *Mapping Cyberspace*, Routledge, 2000.

[DOD 01] DODGE M., KITCHIN R., *The Atlas of Cyberspace*, Addison-Wesley, 2001.

[DOU 95] DOU H., Veille technologique, compétitivité, Dunod, 1995.

[DOY 02] DOYLE M., "Selecting managers for transformational change", *Human Resource Management Journal*, Vol. 12, Issue 1, pp. 3-16, 2002.

[DRU 66] DRUCKER P., *The Effective Executive*, Harper & Row, 1966.

[DUK 98] DUKAN L., *Psychologie de l'innovation*, article, conférence, Forum Mondial de l'Innovation, Paris, 1998.

[DUK 00] DUKAN L., *Les freins psychologiques à l'innovation*, article, conference, Institut des cadres dirigeants, ICAD, 2000.

[DUK 03] DUKAN L., *Piloter le changement avec les cybertechnologies*, M. SAADOUN (Ed.), Hermès Science/Lavoisier, Paris, 2003.

[DUS 98] DUSSAIX A-M., SAPORTA G., CARLE P., DARMON R-Y., GRIMMER J-F., MORINEAU A., *L'analyse conjointe, la statistique, le produit idéal, Méthodes, applications*, 1998.

[EDV 97] EDVINSSON L., MALONE M., Intellectual Capital Realizing Your Company's True Value by Finding its Hidden Brainpower, Harper Collins Publishers, 1997.

[EDV 99] EDVINSSON L., MALONE M., Le Capital Immatériel de l'Entreprise. Identification, Mesure, Management, Maxima, 1999.

[EME 41] EMERSON R.W., *Self-reliance* [*Essays*, 1841-1844]

[EME 00] EMERSON R.W., *La confiance en soi* [*Essays*, 1841-1844], Petite Bibliothèque, Payot & Rivages, Paris, 2000.

[EPI 99] EPICE Project D21 "Guidelines for Programme Management in the Virtual Enterprise". Reference guide for establishing, operating Project Management among trading partners collaborating together within a Virtual Enterprise. 21 September 1999, EPICE ESPRIT Project n°27089, 1999.

[ERI 90] ERICKSON T.J., MAGEE J.F., ROUSEL P.A. and SAAD K., "Managing Technology as a Business Strategy", *Sloan Management Review*, Spring, pp 73-77, 1990.

[ERN 00] ERNST and YOUNG, European life sciences 2000.

[ERN 02] ERNST and YOUNG, La performance du capital investissement en France, *Etude 2002*, available on www.afic.asso.fr

[ETT 00] ETTIGHOFFER D., VAN BENEDEN P., Mét@-Organisations, Les modèles d'entreprise créateurs de valeur, Village Mondial, Paris, 2000.

[ETT 90] ETTKIN, L. P., HELMS, M. M. and HEYNES, P. J., "People: A Critical Element of New Technology Implementation", *Information Management*, September-October, pp.27-29, 1990

[ETT 92] ETTIGHOFFER D., *L'entreprise virtuelle ou les nouveaux modes de travail*, Odile Jacob, Paris, 1992.

[EUR 99] EUROPEAN COMMISSION (Fifth Research Framework Programme), *Innovation Management – Building competitive skills in SMEs*, Luxembourg Office for Publications of the European Communities 1999.

[EVE 03] EVENSON R.E. and GOLLIN D., *Crop Variety Improvement and Its Effect on Productivity: The Impact of International Agricultural Research*, Wallingford, UK: CABI Publishing, 2003.

[FAH 01] FAHEY L., SRIVASTAVA R., SHARON J.S., SMITH D.E., "Linking E-Business And Operating Processes: The Role Of Knowledge Management", *IBM Systems Journal*, Vol. 40, Issue 4, pp. 889-908, 2001.

[FAL 70] FALLON C., "Value Analysis". Washington: Miles Value Foundation.

[FAR 01] FARHOOMAND A. and LOVELOCK P., *Global e-commerce: Text and Cases*, Prentice Hall, Singapore, 2001.

[FEC 1876] FECHNER G.T., *Vorschule der Asthetik*, Hildesheim, New York, 1876.

[FIS 86] FISCH R., WEAKLAND J.H., SEGAL L., *Tactiques du changement*. Seuil, Paris, 1986.

[FLO 90] FLOCH J-M., *Sémiotique, marketing, communication – sous les signes les stratégies*, Formes sémiotiques, Vendôme, PUF, Paris, 1990.

[FON 97], FONG P.S.W., "Establishing value engineering as an academic discipline worldwide", *1997 SAVE Proceedings*, USA, pp. 82-90, 1997.

[FON 04], FONG P.S.W., "A critical appraisal of recent advances and future directions in value management", *European Journal of Engineering Education*, vol. 29, no. 3, 2004, pp. 377-376.

[FOR 99] FORAY D., MAIRESSE J. (Ed.), *Innovations et Performances, Approches interdisciplinaires*, EHESS Editions, 1999.

[FOS 01] FOSTER I., KESSELMAN C., TUEKE S., The Anatomy of the grid, Enabling Scalable Virtual Organization, 2001, www.globus.org/research/papers/anatomy.pdf.

[FOS 02] FOSTER I., KESSELMAN C., NICK J., TUEKE S., "The Physiology of the Grid: An Open Grid Services Architecture for Distributed Systems Integration", *the Fourth Global Grid Forum*, Toronto, Canada, 2002, www.globus.org/research/papers/ogsa.pdf.

[FOU 66] FOURASTIE J., *Essais de morale prospective. Vers un nouveau comportement*, Paris, Editions Gonthier, 1966.

[FRE 04] French-Japan Workshop on Grid Computing March 8-9, 2004, in http://www-direction.inria.fr/international/ASIE_OCEANIE/japon.html.

[FRI 54] FRISON-ROCHE R., *Le rendez-vous d'Essendilène*, p. 201, Arthaud, Paris, 1954.

[FUC 01] FUCHS P., MOREAU G., PAPIN J.-P., *Le Traité de la réalité virtuelle*, Les Presses de l'Ecole des Mines de Paris, 2001.

[FUC 03] FUCHS P. *et al.*, *Le Traité de la réalité virtuelle, version 2*, Les presses de l'Ecole des Mines de Paris, 2003.

[GAR 85] GARDNER H., *Les intelligences multiples*, RETZ, 1985.

[GAR 99] GARRON M., "Une expérience de conception coopérative, simultanée via l'outil informatique", Thesis CNAM, Nancy, June 1999.

[GAU 78] GAUDIN T., *L'écoute des silences*, Paris, UGE, collection 10/18, 1978.

[GAU 98] GAUDIN T., 2100, De l'Innovation, www.2100.org/text_innovation.htm.

[GAU 98] GAUDIN T. and AUBERT J.-E., *De l'innovation*, La Tour-d'Aigues, Editions de l'Aube, 1998.

[GAU 03] GAUDIN T., L'YVONNET F., *Discours de la méthode créatrice*, Le Relié, Ose savoir, Gordes, 2003.

[GAU 05] GAUDIN T., *La Prospective*, Que Sais-je?, PUF, 2005.

[GAV 83] GAVARD-PERRET M.L., "La présence Humaine dans l'image, facteur d'efficacité de la communication publicitaire", *Recherche, applications en Marketing*, 8, 2, pp. 1-22, 1993.

[GAZ 98] GAZEAU M., Etat de veille, www.sesame-ouvre-toi.com.

[GEA 01] GEAR R., "Peer-to-peer (P2P) computing", *Computer Weekly*, 18/06/2001. http://www.computerweekly.co.uk .

[GEN 02] GENTZSCH W., What is grid computing?, in www2.cio.com/ask/expert/expert.cfm?EXPERT=180.

[GIA 93] GIARD V., MIDLER C., *Pilotages de projet, entreprises*, Economica, Paris, 1993.

[GIG 98] GIGET Marc, *La dynamique stratégique de l'entreprise*, Paris, Dunod, 1998.

[GIL 03] GILAD B., Early Warning: Using Competitive Intelligence to Anticipate Market Shifts, Control Risk, Create Powerful Strategies, 2003.

[GIL 92] GILDER G., *Into the Fibersphere*, Forbes, 7 December 1992.

[GIN 87] GINGER S., *La gestalt, une thérapie du contact*, Hommes & Groupes Editeurs, 1987.

[GIR 05] GIROD DE L'AIN M., *2010, Futur Virtuel*, M2 éditions, 2005.

[GOL 92] GOLD B., "Senior Managements' critical role in strengthening technological competitiveness", *Int Journal of Technology Management, Special Issue on Strengthening Corporate and National Competitiveness Through Technology*, 1992,Vol. 7 nos 1,2,3, pp. 5-15.

[GRA 01] GRANDHAYE J.P., GUIDAT C.L., TANI M., "Le Management par la Valeur., l'impulsion par les hommes – Vers une méthodologie robuste", *Revue Française de Gestion Industrielle*, vol. 20, no.2, pp. 75-86, 2001.

[GRA 01] GRACC (Groupe de recherche sur l'Activité de Conception Collaborative), Une expérience de conception collaborative à distance, www.3s.hmg.inpg.fr/ci/GRACC/, 2001.

[GRE 96] GREENAN N., MAIRESSE J., Computers, Productivity in France: some evidence, NBER, Working Paper, no. 5836, 1996.

[GUI 37] GUILLAME P., *La psychologie de la forme*, Flammarion, 1937.

[GUT 82] GUTMAN J., "A means-end chain model based on consumer categorization processes", *Journal of marketing*, 46(1), 1982.

[HAC 00] HACKBARTH G. and KETTINGER W., "Building and e-business strategy", *Information systems Management*, Summer 2000, pp.78-93.

[HAM 98] HAMBRICK D., NADLER D. and TUSHMAN M., *Navigating Change: How CEOs, Top Teams and Boards Steer Transformation*, Boston, Harvard Business School Press, 1998.

[HAM 80] HAMMING R-W., *Coding, Information Theory*, Prentice-Hall, 1980.

[HAR 92] HARBULOT C., *La machine de guerre économique: Etats Unis, Japon, Europe*, Economica, 1992.

[HAR 97] HARTLEY J., BENINGTON J. and BINNS P., "Researching the role of internal change agents in the management of change", *British Journal of Management*, vol. 8, Issue 1, pp. 61-73, 1997.

[HAS 97] HASSID L., MOINET N., JACQUES-GUSTAVE P., *Les PME face au défi de l'intelligence économique*, Dunod, 1997.

[HAT 00] HATCHUEL A., LE MASSON P., WEIL B., "Innovation/projet: des liens complexes", *La cible*, no. 88, pp.1-5, 2000.

[HER 01] ERMINE J.-L., "Les processus de gestion des connaissances", *Extraction des connaissances, apprentissage*, n° 1-2, Hermès, 2001.

[HIG 95] HIGGINS J.M., *Innovate or Evaporate*, The New Management Publishing Company, 1995.

[HIP 98] HIPEL V., *Sources of innovation*, Oxford University Press, 1998.

[HIR 98] HIRIGOYEN G., CABY J., Histoire de la valeur en finance d'entreprise, *Actes des XIVèmes journées nationales des IAE*, Nantes, 1998, vol. 1.

[HOA 01] HOARAU C., TELLIER R., *Création de valeur et management de l'entreprise*, Vuibert, Collection Entreprendre, 2001

[HUB 63] HUBEL D., WIESEL T., Shape, arragement of columns in cat's striate cortex, *Journal of Physiology*, 165, pp. 559-568, 1963.

[HUB 77] HUBEL D., WIESEL T., Fonctionnal architecture of macaque monkey visual cortex, *Proceedings of the Royal Society*, London B,198, pp. 1-59, Figure 15, 23, 1977.

[HUB 79] HUBEL D., WIESEL T., Les mécanismes cérébraux de la vision – Le cerveau, special issue, *Pour la science* (French edition of *Scientific American*), Figure 10, p. 86, 1979.

[HUN 90] HUNT C., ZARTARIAN V., *Le renseignement stratégique au service de votre entreprise*, Editions First, 1990.

[IDE 99a] IDEATION International Inc., *Tools of classical TRIZ*, 1999.

[IDE 99b] IDEATION International Inc., *TRIZ in progress*, 1999.

[INS 00] Institute of Directors, *E-Business – Helping Directors to Understand and Embrace the Digital Age*, Director Publications, London, 2000.

[ITU 03] INTERNATIONAL TELECOMMUNICATION UNION, "World telecommunication development: Access indicators for the information society", *Report 2003* Geneva.

[JAG 93] JAGOU P., "Concurrent Engineering – La maîtrise des coûts, des délais, de la qualité", *Systèmes d'Informations*, Hermès, Paris, 1993.

[JAK 91] JAKOBIAK F., *Pratique de la veille technologique*, Edition d'Organisation, Paris, 1991.

[JAK 98] JAKOBIAK F., *L'intelligence économique en pratique*, Editions d'Organisation, Paris, 1998.

[JAO 94] JAOUI H., *La créativité mode d'emploi*, 2nd edition, ESF, 1994.

[JEA 98] JEANTET A., "Les objets intermédiaires dans la conception, éléments pour une sociologie des processus de conception" 1998, in *Sociologie du travail* vol. 3, pp. 291-316, 1998.

[JOL 89] JODELET D., *Les représentations sociales*, Sociologie d'aujourd'hui, Presses Universitaires de France, Paris 1989.

[JON 99] JONES P., PAWAR K.S., RIEDEL J.-C., PALLOT M., "Electronic Commerce for Programme management Information sharing in the Concurrent Enterprise (EPICE): Opening up the black box". *Proceedings of the 5th International Conference on Concurrent Enterprising, ICEE'99 The concurrent Enterprise in Operation*, The Hague, The Netherlands, March 1999.

[JUM 92] JUMA C., CLARK N., *Long-run economic growth: An evolutionary approach to economic growth*, London, Pinter Publishers, 1992.

[KAT 98] EISENHARDT K.M., BROWN S.L., "Time spacing: competing in markets that won't stand still", in *Harvard Business review*, March-Avril, pp. 59-69, 1998.

[KIL 81] KIDDER T., *The Soul of a New Machine*, Atlantic-Little, Brown, 1981.

[KLI 04] KLINIOTOU M., "Identifying, Measuring and Monitoring Value during Project Development ", *European Journal of Engineering Education*, vol. 29, no. 3, 2004.

[KOE 99] KOENIG G. (Ed.), *De nouvelles théories pour gérer l'entreprise*, Economica, 1999.

[KOR 33] KORZYBSKI A., *Science and Sanity: an Introduction to Non-Aristotelian Systems and General Semantics*, Institute of General Semantics, 1996 (reprint), 1st ed. in 1933.

[KUM 95] KUMAR A. and MOTWANI J., "A methodology for assessing time-based competitive advantage of manufacturing firms", *International Journal of Operations & Production Management*, Vol 15 no. 2 pp 36-53, 1995.

[LAC 96] LACHMAN J., *Financer l'innovation des PME*, Economica, 1996

[LAM 04] LAMY, *Droit de l'Informatique et des Réseaux*, Editions Lamy 2004.

[LAR 90] LARROCHE M., *Mes cellules se souviennent*, Tredaniel, 1990.

[LAS 03] LASO BALLESTEROS I., "3rd Wave of Internet to allow collaborative working environments", http://www.e-global.es, 2003.

[LAV 01] LAVAL C., LOCHOT M., "Management par la Valeur et développement durable – un apport pour les collectivités locales", *Revue Française de Gestion Industrielle*, vol. 20, no. 2, pp. 51-60, 2001.

[LEB 01] LEBORGNE C., "proposition d'une démarche anthropocentrée de conception de produits nouveaux basée sur l'usage et destinée a une meilleure intégration, par l'ergonome, des besoins et des attentes des usagers.", PhD Thesis, ENSAM, Paris, 2001.

[LEC 92] LE COQ M., "Approche intégrée en conception de produits", PhD Thesis, ENSAM, 1992.

[LEE 00] LEE S., HARADA A., STAPPERS P.-J., "Pleasure with Products. Design based on Kansei", Industrial Design Engineering, Delft University of Technology, Netherlands & Art & Design Institute, University of Tsukuba, Japan, 2000.

[LEE 01] LEE S. (2001) Change management at the desktop *InfoWorld*, vol. 23, Issue 30.

[LEE 98] BERNERS-LEE T., The fractal nature of the web, 1998, last accessed 28/11/2001 in http://www.w3.org/DesignIssues/Fractal.html.

[LEM 90] LE MOIGNE, J.-L., *La modélisation des systèmes complexes*, Afcet Système, Dunod, Paris, 1990.

[LEM 02] LE MASSON P., "De la R&D à la RID, de nouvelles formes organisationnelles pour répondre aux enjeux de la conception innovante", support de cours du Master Management de l'Innovation., Ingénierie de la Conception, ENPC, Paris, 2002.

[LEN 02] LENHARDT V., *Les responsables porteurs de sens*, Insep Consulting Editions, 2002.

[LER 72] LE RAY M., DEROYON J.P., DEROYON M.J., FRANÇOIS M., VIDAL F., "New results on the vortex lattice", *Proceedings of the 13th International Conférence on Low Temperature Physics LT 13 Boulder*, Colorado, USA, Plenum Press, New-York, Tome 1, p.96-100, 1972.

[LER 80] LE RAY M., *Dialogue du physicien, de l'esthète*, Communication, langages, Paris, 45, p.49-69., 1980.

[LER 85] LE RAY M., DEROYON J.P., DEROYON M.J., MINAIR C., "Critères angulaires de stabilité d'un tourbillon hélicoïdal ou d'un couple de tourbillons rectilignes, rôle des angles privilégiés dans l'optimisation des ailes, voiles, coques des avions, navires", *Bulletin de l'Association technique maritime, aéronautique*, Paris, pp. 511-529, 1985.

[LER 94] LE RAY M., MARTINACHE M., *Formats, composition des couvertures de magazine: effets esthétiques*, Lez Valenciennes, pp. 16, 113-128, 1994.

[LES 94] LESCA H., *Veille stratégique: l'intelligence de l'entreprise*, Astier, 1994.

[LEV 01] LEVASSEUR, R.E., "People skills: Change Management tools – Lewin's change model", *Interfaces*, vol. 31, Issue 4, pp. 71-73, 2001.

[LEV 01] LEVET J.-L., *L'intelligence Economique, mode de pensée, mode d'action*, Economica, 2001.

[LEV 96] LEVITT T., *The Marketing Imagination*, The Free Press, 1996.

[LEV 99] LEVAN S., *Le projet Workflow*, Eyrolles, Paris, 1999.

[LHO 01] LHOMME Y., "Le financement de l'innovation technologique dans l'industrie", *Le 4 pages des statistiques industrielles*, SESSI, no. 156, novembre 2001.

[LIE 01] LIEBMAN L., "Change management gets critical", *Communications News*, vol. 38, Issue 10, pp. 76, 2001.

[LIM 03] LIM D., "Modélisation du processus de conception centrée utilisateur, basée sur l'intégration des méthodes et outils de l'ergonomie cognitive: Application à la conception d'IHM pour la télévision interactive", PhD Thesis, ENSAM, Paris, 2003.

[LIQ 00] LIQUET and BENAVENT, *L'analyse conjointe, ses applications en marketing*, IAE Lille, 2000.

[LOR 95] LORINO P., *Comptes et Récits de la Performance. Essai sur le Pilotage de l'Entreprise*, Editions d'Organisation, Paris, 1995.

[LOW 03] LOWE R.K., "Animation and learning: Selective processing of Information in Dynamic Graphics", *Learning and Instruction*, 13, 157-176, 2003.

[LUC 01] LUCAS D., TIFFREAU A., *Guerre économique, information*, Ellipses, Paris, 2001.

[LUC 98] LUCAS A., *Droit d'auteur et numérique*, Litec, Paris, 1998.

[MAC 96] MAC LEOD P., extract from "Quand l'industrie joue le sensoriel", *L'usine nouvelle*, 2622, p.48-57, 1996.

[MAI 92] MAITRE P., MIQUEL J.D., *De l'idée au produit*, Eyrolles, 1992.

[MAI 99] MAITRE B., *Les Business models de la Nouvelle Economie*, Dunod, 1999.

[MAL 00] MALLET-POUJOL N., *La création multimédia et le droit*, Litec, Paris, 2000.

[MAR 83] MARTINACHE M., LE RAY M., LEVIN M., Les angles privilégiés dans la lecture, la rédaction des images, *Journées d'études de la Société des électriciens, électroniciens, radioélectriciens*, Rennes, pp. 27-49, 1983.

[MAR 88] MARION A., "La place de l'investissement immatériel dans l'évaluation des entreprises", *Revue Française de Gestion*, no. 67, January-February, pp. 6-12, 1988.

[MAR 94] MARTRE H., (Ed.), *Intelligence économique, stratégie des entreprises*, Commissariat Général du Plan, La Documentation française, Paris, 1994 .

[MAR 95] MARTINET B., MARTY M., *L'intelligence économique, les yeux, les oreilles de l'entreprise*, Editions d'Organisation, Paris, 1995.

[MAR 96] MARTORY B., PIERRAT C., *La gestion de l'immatériel*, Nathan, Paris, 1996.

[MAS 02] MASTERS W.A., "Research Prizes: A Mechanism to Reward Agricultural Innovation in Low-Income Regions". *AgBioForum*, Vol. 6, no.1 and 2, Article 14, 2002.

[MAS 06] MASSOTTE P., CORSI P., *La complexité dans les processus de décision et de management*, Hermes Science/Lavoisier, Paris, 2006.

[MAT 01] MATHIEU J.-P., KIRIA I., LE RAY M., "The Influence of certain angles, shapes, their relevance for marketing, design", *5th congrès international de l'Association Russe de Marketing: Marketing in Russia; World Practice & Russian Experience*, 2001.

[MAT 02] MATHIEU J.P., LE RAY M., "The influence of privileged angles, remarkable proportions in shapes, their relevance for marketing., product design", *The 11th International Forum on Design Management*, Education at Northeastern University, Boston, Massachusset, USA, 2002.

[MAT 02] MATHIEU J.P., LE RAY M., LAMBERT J.L., "Marketing des vins, quantification spatiale des bouteilles", *Premier Colloque Marketing Spécialisé, Thématiques pour le Marketing, Bayonne*, pp. 39-52, 2002.

[MAT 04] MATHIEU JP., LE RAY M., "Universalité dans les Formes, Angles Privilégiés, Proportions Remarquables: Application aux images, aux objets", *Premières Journées de Recherche sur le Marketing., le Design*, Nantes, edited by AUDENCIA Nantes-Ecole de Management, Economica, Paris, 2004.

[MAT 94] MATHIEU J.P., "Apport des angles privilégiés au marketing", Thesis, ESA-Grenoble, 1994.

[MAT 99] MATHIEU J.P, THIEBLEMONT R., GOUSTY Y., "The influence of certain angles, shapes, their relevance for marketing, industrial design", *PICMET'99 Technology, Innovation Management*. Portland, Oregon, USA, CD-ROM, 1999.

[MCC 97] MCCRIMMON M., *The Change Masters: Managing and Adapting to Organizational Change*, London: Institute of Management/Pitman Publishing, 1997.

[MCG 01] MCGARVEY R., New corporate ethics for the new economy, *World Trade*, vol. 14, Issue 3, p. 43, 2001.

[McK 01] MCKNIGHT L.W., VAALER P.M., KATZ R.L. (Eds.), *Creative Destruction; Business Survival Strategies in the Global Internet Economy*, MIT Press, 2001.

[MEY 48] MEYER F., *L'accélération évolutive*, Presses Universitaires de France, Paris, 1948.

[MEY 54] MEYER F., *Problématique de l'évolution*. Presses Universitaires de France, Paris, 1954.

[MEY 74] MEYER F., *La surchauffe de la croissance. Essai sur la dynamique de l'évolution*, Paris, Fayard, 1974.

[MIC 96] MICHEL J., "L'analyse de la valeur adaptée à la reconfiguration de petites et moyennes organisations", *La Valeur des produits, procédés et services*, no. 69, pp. 17-20, July 1996.

[MIC 99a] MICHEL J., "Sustainable Value Management Versus Managing Sustainable Value", *Proceedings of the International Conférence 1999 of the Hong-Kong Institute of Value Management*, Hong-Kong, May 1999.

[MIC 99b] MICHEL J., "Analyse de la Valeur et Management de l'Information – Vers la Value Information", *La Valeur des produits, procédés et services*, no. 79, pp. 19-22, Janvier 1999.

[MIC 00a] MICHEL J. "Nouvelles transformations de la formation des ingénieurs induites par les TIC", *Les Cahiers du Numérique*, Hermès, vol. 1, no. 2, pp. 67-86, 2000.

[MIC 00b] MICHEL J. "L'information et documentation – Un domaine d'activité professionnelle en mutation", *Les Cahiers du Numérique*, Hermès, vol. 1, no. 3, pp. 47-64, 2000.

[MIC 01a] MICHEL J., "Le Management par la Valeur, démarche de coaching pour l'accompagnement du changement dans les organisations", *Revue Française de Gestion Industrielle*, vol. 20, no. 2, p. 75-86, 2001, pp. 61-73

[MIC 01b] MICHEL J., "Le knowledge management, entre effet mode et (ré)invention de la roue…", *Documentaliste - Sciences de l'Information*, vol. 38, nos. 3-4, pp, 176-186, 2001.

[MIC 01c] MICHEL J. "Management par la Valeur, création de Valeur, chaîne de Valeur, … Parle-t-on de la même Valeur? - Proposition d'un cadre conceptuel pour la valeur généralisée et contribution au développement de la Valorique", *La Valeur des produits, procédés et services*, N°90, p. 2-7, octobre 2001.

[MIC 03a] MICHEL J., "Les documentalistes: l'urgence d'une reconnaissance sociale", *Les Journalistes ont-ils encore un pouvoir* - Hermès, pp. 185-193, 2003.

[MIC 03b] MICHEL J., "L'Infopolis, une nécessaire utopie pour mieux articuler veille informative, gestion documentaire, partage et dynamisation des connaissances. Des organisations en quête de cohérence et d'efficacité", *Proceedings du Colloque IERA'2003*, Nancy, April 2003.

[MIC 04] MICHEL J., "Le management par la valeur pour définir la stratégie de communication d'une grande organization fédérale", *Proceedings du Congrès Francophone du Management de Projet 2004*, Paris, December 2004.

[MIC 97] MICAELLI J. P., PERRIN J., "Les trois âges de la capitalisation des connaissances", in FOUET J.M. (Ed.), *Connaissances et savoir-faire en entreprise. Intégration et capitalisation*, Hermes, Paris, pp. 23-40, 1997.

[MID 93] MIDLER C., *L'auto qui n'existait pas*, InterEditions, Paris, 1993.

[MID 98] MIDLER C., *L'auto qui n'existait pas*, Dunod, Paris, 1998.

[MID 93] MIDLER C., *L'auto qui n'existait pas – Management des projets, transformation de l'entreprise*, Economica, Paris, 1993.

[MIL 61] MILES L.D., *Techniques of Value Analysis and Engineering*, New York, 1961.

[MIL 89] MILES L.D., *Techniques of Value Engineering and Value Analysis*, 3rd ed., USA, Miles Value Foundation, 1989

[MIL 65] MILLER H., BIERI J., "Cognitive Complexity as a Function of the Stimulus Objects Being Rated", *Psychological Reports*, 1965.

[MIL 99] MILLER W.L., MORRIS L., *4th Generation R&D: Managing Knowledge, Technology, Innovation*, john Wiley and Sons, 1999.

[MIL 03] MILLET P., *L'Etude des marchés qui n'existent pas encore*, Editions d'Organisation, Paris, 2003.

[MOK 02] MOKYR, *Gifts of Athena: Historical origins of the knowledge economy*, Princetonn N.J., Princeton University Press, 2002.

[MOL 70] MOLES A., ROLAND C., *Créativité, méthodes d'innovation dans l'entreprise*, Fayard-Mame, 1970.

[MOL 91] MOLES A., "Pour un néo-fonctionnalisme in Informel", vol. 4 no. 1, *Montreal and conferences attended at the technological university of Compiègne*, 18 October 1991.

[MOR 00] MORIN E., *Les 7 savoirs nécessaires à l'éducation du futur*, Seuil, Paris, 2000.

[MOR 81] MORIN E., *La méthode*, vol. 1-4, Seuil, Paris, 1981.

[MOR 99] MORIN E., LE MOIGNE J.-L., *L'intelligence de la complexité*, Paris, L'Harmattan, 1999.

[MUC 91] MUCCHIELLI A., *Les méthodes qualitatives*, PUF, Paris, 1991.

[NEI 64] NEISSER U., "Visual Search", *Scientific American*, 210, pp. 94-102, 1964.

[NEI 67] NEISSER U., *Cognitive Psychology*, New York: Appleton-Century Crofts 1967.

[NFW 91] 91NF X 50-151, "Analyse de la valeur, Analyse fonctionnelle: expression fonctionnelle du besoin, cahier des charges fonctionnel", Paris, 1991.

[NIE 93] NIELSEN J., *Usability engineering*, Boston, Academic Press, 1993, 358p.

[NIG 04] NIGHTINGALE P., "Echnological Capabilities, Invisible Infrastructure and the Unsocial Construction of Predictability: The Overlooked fixed costs of Useful Research", *Paper presented at the DRUID Summer Conference 2004 on Industrial Dynamics, Innovation and Development*, Elsinore, Denmark, June 14-16, 2004.

[NON 97] NONAKA I., TAKEUCHI H., La connaissance créatrice. La dynamique de l'entreprise apprenante, DeBoeck University, 1997.

[NOO 90] NOORI, H., *Managing the Dynamics of New Technology: Issues in Manufacturing Management*, Prentice Hall New Jersey, 1990.

[NOR 01] NORTON D.P., KAPLAN R.S., *The Strategy Focus Organization*, Harvard Business School Press, 2001.

[NOR 96] NORTON D.P., KAPLAN R.S., *The Balanced Scorecard*, Harvard Business School Press, 1996.

[OHM 01] OHMAE K., *Géographie secrète de la nouvelle économie*, Editions Village, Paris, 2001.

[PAI 74] PAILLARD J., *Le traitement des informations spatiales, Association de Psychologie scientifique de langue française (de l'espace culturel à l'espace écologique)*, Paris, PUF, pp. 8-54, 1974.

[PAL 00] PALLOT M., MAIGRET J.-P., BOSWELL J., DE JONG E., "EPICE: Realising the Virtual Project Office", *Proceedings of the e-Business for Industry Conference, Flagship Conference of UK Council for Electronic Business*, London, 2000.

[PAL 93] PALLOT M., "Enabling Interaction using Meta-data within a Concurrent Engineering environment", *Proceedings of the 6ʰ Annual Conference, Exposition CALS International Expo*, Atlanta, USA, 1993.

[PAL 96] MARC M., "Suitability of available IT & Tools for Concurrent Engineering", *Proceedings of the 3rd International Conference on Concurrent Engineering, Electronic Design Automation*, Robinson College, Cambridge, UK, 1996.

[PAL 98] PALLOT M., SANDOVAL V., *Concurrent Enterprising: Toward the Concurrent Enterprise in the era of the Internet, Electronic Commerce*, Kluwer Academic Publishers, USA, 1998.

[PAS 01] PASINI W., FRANCESCATO D., *Le courage de changer*, Odile Jacob, Paris, 2001.

[PAT 00] PATON R. and MCCALMAN J., *Change Management: A Guide to Effective Implementation*, 2ⁿᵈ Edition, London, Paul Chapman Publishing, 2000.

[PAT 98] PATEYRON E., *La veille stratégique*, Economica, Paris, 1998.

[PER 01a] PERRIN J., *Concevoir l'innovation industrielle; Méthodologie de conception de l'innovation*, Paris, CNRS Editions, 2001.

[PER 01b] PERRIN J., "Analyse de la valeur et valeur économique des biens et services", *Revue Française de Gestion Industrielle*, vol. 20, no. 2, pp. 9-20, 2001.

[PER 81] PERRIAULT J., *Mémoires de l'ombre et du son. Une archéologie de l'audiovisuel.* Paris, Flammarion, 1981.

[PEY 04] PEYRAT-GUILLARD D., SAMIER N., "TIC, Implication et Climat social: vers une autonomie sous contraintes des salariés", *Revue de Gestion des Ressources Humaines*, no. 51, pp. 39-58, January-February-March 2004.

[PEY 95] PEYREFITTE A., *La Société de Confiance. Essai sur les origines et la nature du développement*, Paris, Odile Jacob, 1995.

[PIA 79] PIAGET J., CHOMSKY N., *Théories du langage, théories de l'apprentissage*, Seuil, Paris, 1979.

[PIN 93] PINSONNEAULT A., KRAEMER K.L., "The Impact of Information Technology on Middle Management", *Management Information System Quarterly*, September, pp. 271-292, 1993.

[POM 93] POMEROL J.-C., BARBARA-ROMERO S., *Choix multicritère en entreprise*, Hermes, Paris, 1993.

[POR 04] PORTNOFF A-Y., *Sentiers de l'innovation*, Futuribles international, Paris, 2004.

[PRA 90] PRAHALAD C.K., HAMEL G., "The core competence of the Corporation, *Harvard Business Review*, no. 68, 1990.

[PRE 95] PRESCOTT J.E., GAURUAB B., "Competitive intelligence practices", *Competitive Intelligence Review*, vol. 6, no. 2, pp. 4-14, 1995.

[PUL 02] PULLI P., ANTONIAC P., PALLOT M., STANESCU A., "Modelling Scenarios of Innovative Mobile Services", *Proceedings of the 8th International Conference on Concurrent Enterprising, ICEE'2002 Ubiquitous Engineering in the collaborative Economy*, Roma, Italy, June 2002.

[RAO 94] RAO K.V.S. and DESHMUKH S.G., 'Strategic Framework for Implementing Flexible Manufacturing Systems in India', *International Journal of Production Management*, Vol 14 no. 4 pp. 50-63, 1994

[REG 04] REGAS C., "Un aperçu des technologies dans la Silicon Valley", French Consulat in San Francisco, Mission pour la Science, la Technologie,, www.france-science.org, 2004.

[REN 02] RENAUDEAU F., Brochure de présentation du Micad 2002, www.birp.com/micad, 2002.

[REY 93] REYNE Maurice, *Les prévisions technologiques. Matériaux, procédés et produits de demain*, Paris, Éditions d'Organisation, Hommes et Techniques, 1993.

[RIC 03] RICHIR S., SAMIER H., SANDOVAL V., *Les cybertechnologies dans les entreprises industrielles*, Hermes, Paris, 2003.

[RIF 00] RIFKIN J., *L'âge de l'accès: la révolution de la nouvelle économie*, La découverte, Paris, 2000.

[RIL 03] RILEY R.A., PEASON T.A., TROMPETER G., "The value relevance of non-financial performance variables and accounting information: the case of the airline industry", *Journal of Accounting & Public Policy*, vol. 22, pp. 231-254, 2003.

[ROB 02] ROBERTS E.B. (Ed.), *Innovation; Driving Product, Process, and Market Change.* MIT Sloan Management Review, Jossey Bass, 2002.

[ROG 95] ROGERS E.M., *Diffusion of Innovations.* 4th edition, The Free Press, 1995.

[ROK 73] ROKEACH M. J., 1973, *The Nature of Human Values*, New York, The Free Press

[ROS 73] ROSCH E., *Natural categories, Cognitive Psychology*, 4, pp. 328-350, 1973.

[ROS 76] ROSCH E., MERVIS C.B, GRAY W.D, JOHNSON D.M., BOYES-BRAEM B., "Basic objects in natural categories", *Cognitive psychology*, no. 8, pp. 382-439, 1976.

[ROS 99] ROSENFELD R. and WILSON D., *Managing Organizations: Texts, Readings and Cases*, 2nd edition, London, McGraw-Hill, 1999.

[ROU 96] ROUACH D., *La veille technologique, l'intelligence économique*, PUF, 1996.

[ROU 96] ROUSSEL B., "Proposition d'une méthode centrée sur la formulation de principes de solution dans le processus interdisciplinaire de conception de produits". Thesis ENSAM, no. 14874, 1996.

[ROW 02] ROWE F., "Communication, coopération à distance", in ROWE F. (ed.), *Faire de la recherche en systèmes d'information*, Paris, Vuibert, pp. 173-199, 2002.

[SAA 00] SAADOUN M., *Technologies de l'information et management*, Hermès, 2000.

[SAA 02] SAADOUN M., *Les systèmes d'information: Art, pratiques, ouvrage collectif*, Editions d'Organisation, 2002.

[SAA 03] SAADOUN M. (Ed.), *Piloter le changement avec les cybertechnologies*, Hermès, 2003.

[SAA 04] SAADOUN, M., "Virtual Spaces, Shaping Business Strategy in a Networked World", *Proceedings of ICEB2004*, Vol. I and II, December 5-9, Beijing, China, 2004.

[SAA 05] SAADOUN M., Behind a concept… a reality: Impact of ICT, *CISTM*, July 24-26, New Delhi, 2005.

[SAA 92] SAAD K.N., BOHNIN N.H., VAN OENE F., *R&D de troisième génération*, Editions d'Organisation, Paris, 1992.

[SAA 96] SAADOUN M., *Le projet groupware*, Eyrolles, 1996.

[SAA 98] SAADOUN M., *Avec le temps*, Editions d'Organisation, 1998.

[SAA 00] SAADOUN M., *Technologies de l'information, du management*, Hermès Science publications, Paris, 2000.

[SAL 98] SALMON R., *L'intelligence compétitive*, Economica, 1998.

[SAL 99] SALAMATOV Y., *The right solution at the right time: a guide for innovative problem solving*, Edited by Valeri Souchkov, 1999.

[SAM 04] SAMIER H., "La veille sur les Weblog", *VSST Colloquium*, Toulouse 2004.

[SAM 02] SAMIER H., SANDOVAL V., *La veille stratégique sur internet*, Hermes Sciences, Paris, 2002.

[SAM 99] SAMIER H., SANDOVAL V., *La recherche intelligente sur l'internet, l'intranet*, Hermès Science Publications, Paris 1999, 2nd edition, 1999.

[SAN 02] SANDERS T., Love is the Killer Application, How to Win Business, Influence Friends, Crown Business, 2002.

[SAN 95] SANDOVAL V., *Les autoroutes de l'information*, Hermes, Paris, 1995.

[SAP 03] SAPORTA G., "Principes, méthodes de l'analyse conjointe", *5th EDF R&D conference "Pour capter les préférences du client, l'analyse conjointe"*, May 2003.

[SAR 92] SARAPH J. and SEBASTIAN R., "Human Resource Strategies for Effective Introduction of Advanced Manufacturing Technologies", *Production and Inventory Management Journal*, First Quarter, pp 64-70, 1992.

[SAU 00] SAUNDERS C., *Virtual teams: piecing together the puzzle,* in *Framing the domain of IT Management*, Zmud R. (ed.), Cincinnati: Pinneflex, 2000.

[SCH 02] SCHUH G. MUELLER M. and TOCKENBUERGER L.W., Successful change management from strategy to transformation: Results of an EC project *International Journal of Manufacturing Technology and Management*, vol. 4, Issue 1, 2 pp. 90-95, 2002.

[SCH 43] SCHUMPETER J.A., *Capitalism, Socialism and Democracy*, (1943) 5th edition. George Allen & Unwin Ltd, London, 1976.

[SCH 61] SCHUMPETER J.A., *Théorie de l'évolution économique*, 1st edition in French (1935), Dalloz, 1961.

[SCHR 02] SCHROEMBERGER, *The internet of things*, Forbes, 2002.

[SEN 01] SENI D., "Desiderata for an idealized scenario: towards an international masters program in value management. Workshop on towards – A VM and VE Educational Framework at Masters Level", *Proceedings of the MVF and VEAMAC International Conference in Fort Lauderdale*, USA, The Miles Value Foundation and VEAMAC, 2001.

[SES 02] SESSI, DIGITIP, INSEE, INPI, *Tableau de bord de l'innovation 2002*, 7th edition, April 2002, available on www.industrie.gouv.fr.

[SES 98] SESSI, "Les compétences pour innover", *Le 4 Pages*, no. 85, January 1998.

[SHA 00] SHARMA P., "E-transformation basics: key to the new economy", *Strategy & Leadership*, vol. 28, Issue 4, pp. 27-31, 2000.

[SHA 01] SHARMA S.K. and GUPTA J.N.D., *E-commerce Opportunities and Challenges, in E-commerce Diffusion: Strategies and Challenges*, edited by MOHINI Singh and THOMPSON Teo, Heidelberg Press, pp. 21-42, 2001.

[SHA 03] SHARMA S.K. and GUPTA J.N.D., *Creating Business value through E-commerce*, edited by NAMCHUL S., Idea Group Publishing, pp. 166-186, 2003.

[SHA 04] SHARMA S., "A Change Management Frame Work for E-Business Solutions", in *E-Business Innovation and Change Management*, edited by SINGH M. and WADDELL D., IDEA Group Publishing, Hershey, USA pp. 54-69, 2004.

[SHA 89] SHARPE J., Building and Communicating the Executive Vision, *IBM Australia, Working paper* 89-001, Melbourne, 1989.

[SIM 69] SIMON H.A., *The science of the artificial*, Cambridge, Mass., the MIT Press, 1969.

[SIN 00] SINGH, M., "Electronic Commerce in Australia: Opportunities and Factors Critical for Success", *proceedings of the 1st World Congress on the Management of Electronic Commerce (CD-ROM)*, 19-21 January, Hamilton, Ontario, Canada, 2000.

[SIN 04] SINGH M., "Innovation and Change Management", in *E-Business Innovation and Change Management*, edited by SINGH M. and WADDELL D., IDEA Group Publishing, Hershey, USA, pp. 1-19, 2004.

[SIN 97] SINGH M., "Effective Implementation of New Technologies in the Australian Manufacturing Industries", PhD Thesis, Monash University, Australia, 1997.

[SIN 98] SINGH M., "An Implementation Model for New Technologies", *Proceedings of the World Innovation & Strategy Conference*, 2-5 August, Sydney, Australia, pp. 649-654, 1998.

[SMI 74] SMITH E.E., SHOBEN E.J. and RIPS L.J., "Structure, process in semantic memory: a featural model for semantic memory", *Psychological Review*, Vo. 81, no. 3, 1974, pp. 214-241.

[SMI 89] SMITH G.V., PARR R.L., *Evaluation of intellectuel property and intangible assets*, John Wiley and Sons, 1989.

[STA 98] STALLMAN R., *The GNU Manifesto*, www.gnu.org/gnu/thegnuproject.html.

[STE 00] STEWART T.A., "Three rules for managing in the real-time economy", *Fortune*, vol. 141, Issue 9, pp. 332-334, 2000.

[STU 95] STUART R., "Experiencing organizational change: triggers, processes and outcomes of change journeys", *Personnel Review*, vol. 24, Issue 2, pp. 5-53, 1995.

[SUH 88] SUH N., "Keynote papers, Basic concepts in Design for Productibility", *Annals of the CIRP*, Vol. 37121, 1988.

[SWA 01] SWANSON S., "More than technology: User attitudes make a difference", *Information week*, Issue 826, p. 48, 2001.

[TAK 03] TAKÁCS B., KISS B., "Virtual Human Interface: a Photo-realistic Digital Human", *IEEE Computer Graphics and Applications, Special Issue on Perceptual Multimodal Interfaces*, September-October 2003.

[TAK 05] TAKÁCS B., "Special Education and Rehabilitation: Teaching and Healing with Interactive Graphics", *IEEE Computer Graphics and Applications*, September-October 2005.

[TAP 01] TAPSCOTT D., TICOLL D., LOWY A., *Capital réseaux*, Editions Village Mondial, Paris, 2001.

[TEZ 94] TEZENAS DU MONTCEL H., "L'avenir appartient à l'immatériel dans l'entreprise", *Revue Française de Gestion*, September-October, pp. 97-101, 1994.

[THO 00] THORNHILL A., LEWIS P., MILLMORE M. and SAUNDERS M., *Managing Change: A Human Resource Strategy Approach*, London, Financial Times/Prentice-Hall, 2000.

[TIC 95] TICHKIEWITCH S., CHAPA E., "Un modèle produit multi-vues pour la conception intégrée", *PRIMECA Colloquium*, Nancy, 7 December 1995.

[TIC 97] TICHKIEWITCH S., "Relecture de l'estampage à la lumière de la mécanique", in *Connaissances, savoir-faire en entreprise*, Hermes, Paris, 1997.

[TIC 98] TICHKIEWITCH S., "Introduction. Les enjeux des nouvelles techniques de conception", in TOLLENAERE M. *et al.*, *Conception de produits mécaniques – méthodes, modèles, outils*, Hermès, Paris, 1998, pp. 19-26.

[TYS 90] TYSON K.W.M., *Competitor intelligence manual, guide: gathering, analysing, using business intelligence*, Englewood Cliffs, N.J., Prentice Hall, 1990.

[TYS 98] TYSON K.W.M., *The Complete Guide to Competitive Intelligence*, Ring-bound, 1998.

[UES 00] UESHIBA K., *L'esprit de l'Aikido*, Budo Editions, 2000.

[ULR 89] ULRICH D., "Assessing Human Resource Effectiveness: Stakeholder, Utility and Relationship Approaches", Human Resource Planning, vol. 12, no. 4, pp. 301-316, 1989.

[VAD 96] VADCARD P., "Aide à la programmation de l'utilisation des outils des outils en conception de produits", PhD Thesis, ENSAM, November 1996.

[VAD 97] VADCARD P., "La créativité", in LE COQ M. *et al.*, *La conception de produits – approche méthodologique*, ENSAM Paris, pp. 72-90, 1997.

[VAL 91] VALETTE-FLORENCE P. and RAPACCHI B. "Improvements in Means-End Chain Analysis Using Graph Theory and Correspondence Analysis", *Journal of Advertising Research*, 31, 1, 30-45, 1991.

[VAL 95] VALETTE-FLORENCE P., "Introduction à l'analyse des chaînages cognitifs", *Recherche, Application en Marketing*, Vol. 9, no. 1, 1995, pp. 93-118.

[VAU 00] DE VAUCLEROY, G. *European Business Summit*, Brussels, June 2000.

[VEL 00] VELTZ P., *Le nouveau monde industriel*, Gallimard, 2000.

[VEN 00] VENKATARAMAN N., "Five steps to a dot-com strategy: how to find your footing on the web", *Sloan Management Review*, spring 2000, pp. 15-28.

[VER 03] VERNIMMEN P., *Finance d'entreprise*, Dalloz, 5th edition, also available on www.vernimmen.net.

[VIL 91] VILLAIN J., *L'entreprise aux aguets: information, surveillance de l'environnement, propriété, protection industrielles, espionnage, contre-espionnage au service de la compétitivité*, Paris, Masson, 1991.

[VIV 89] VIVANT M., "La création générée par ordinateur", *Cahiers Lamy*, February 1989.

[WAR 01] WARUSFEL B., *La propriété intellectuelle et l'Internet*, Flammarion 2001.

[WAT 75] WATZAWITCK P., WEAKLAND J.H., FISCH R., *Changements: paradoxes, psychothérapie*, Seuil, Paris, 1975.

[WAT 86] WATZAWITCK P., *Le langage du changement.* Seuil, Paris, 1986.

[WEI 99] WEISER M., http://www.ubiq.com/hypertext/weiser/weiser.html, 1999

[WEL 98] WELLS S., *Choosing the Future. The Power of Strategic Thinking.* Boston, Butterworth-Heinemann, 1998.

[WER 23] WERTHEIMER M., KÖHLER S., "Untersuchungen zur Lehre von der Gestalt II", *Psychologische Forschung*, 1923, Vol. 4, pp. 301-305.

[WHO 01] WORLD HEALTH ORGANIZATION "Macroeconomics and Health: Investing in Health for Economic Development", *Report of the Commission on Macroeconomics and Health*, Geneva, 2001.

[WIN 88] WINTER R.I., BERTEND H.E., SULSARCZUL M.M.G., "The role of concurrent engineering in weapon system acquisition", *IDA Report R-388*, Institute of Defence system Analysis, Alexandria, VA, 1988.

[WOL 99] WOLTON D., *Internet, et après*. Flammarion, Paris, 1999.

[WOO 01] WOODHEAD R., DOWNS C., *Value Management; Improving Capabilities*. London, Thomas Telford Publishing, 2001.

[WOO 02] WOODHEAD R., McCUISH J.D., *Achieving Results: How to create value*. London, Thomas Telford Ltd, 2002.

[WOO 04] WOODHEAD R., BALL F., LI X., "Silk flowers are just as artificial as plastic ones: Value Engineering in the university context", *European Journal of Engineering Education*, vol. 29, no. 3, 2004.

[WOR 02] WORLD BANK, *Constructing Knowledge Societies: New Challenges for Tertiary Education*, Washington D.C, 2002.

[WOR 04] WORLD BANK. *World Development Indicators 2004*, Washington, D.C., 2004.

[XIE 04] XIE W. and WHITE S., "Sequential Learning in a Chinese Spin-off: The Case of Lenoro Goup Limit", *R&D Management*, Vol. 34, no. 4, pp 407-422, September 2004.

[YAN 01] YANNOU B. Préconception de Produits Mémoire d'HDR Ecole centrale Paris, June 2001.

[YAN 04] YANNOU B., BIGAND M., "A curriculum of value creation and management in engineering", *European Journal of Engineering Education*, vol. 29, no.3, 2004, pp. 355-366

[YAN 98] YANNOU B., "Analyse fonctionnelle., analyse de la valeur", in TOLLENAERE M. (Ed.), *Conception de produits mécaniques*, Hermes, Paris, pp. 77-104, 1998.

[YOU 75] YOUNG S. and FEIGIN B., "Using the benefit chain for improved strategy formulation", *Journal of Marketing*, vol. 39, July 1975.

[ZAI 92] ZAIRI M., *Management of Advanced Manufacturing Technology*, Sigma Press, Wimslow, UK, 1992.

[ZIG 96] ZIRGLER P, HARTLEY A, "The effect of acceleration techniques on product development time", *IEEE transactions on engineering management*, pp. 143-152, 1996.

[ZLO 99] ZLOTIN B., ZUSMANN A., "Directed evolution of technological systems". University of Minnesota, Ideation International Inc, 1999.

[ZU 90] ZU SUN., *L'art de la guerre*, Economica, Paris 1990.

List of Authors

Sylvain ALLANO (ENS Cachan, Paris) was appointed as Professor in 1989 and is presently Head of the SATIE Laboratory with research activities in electrical engineering and bio-microsystems.

Carole BOUCHARD (ENSAM, Paris) is Assistant Professor at the Laboratory of New Products Design and Innovation at the Ecole Nationale Supérieure des Arts et Métiers in Paris. She works on Kansei engineering, applied creativity and innovation, and design watch.

Pascale BRENET (University of Paris V) is Assistant Professor and Director of a professional Masters on entrepreneurship. She works on academic high tech spin-offs and assists with the University of Franche-Comté for training students intending to start up their own firm.

Emmanuel CHÉNÉ (IAE Nantes, France) was a founder of two technology start-ups and is now Associate Professor at the Institute of Business Administration (IAE) at the University of Nantes, President of www.innovation.fr and a consultant in innovation management.

Hervé CHRISTOFOL (ISTIA, Angers) is Associate Professor and coordinates the Technology Transfers and European Projects Master at ISTIA, University of Angers. He is a founder of the research group on anticipation and integration of the trends in product design process.

Patrick CORSI (Consultant, Brussels) worked for IBM and later went through the full successful cycle of a start-up in AI in Paris. He renewed and directed an AI Department at SYSECA/THOMSON-CSF and worked at the European Commission in charge of monitoring the European AI portfolio of projects. He is now an international consultant and a frequent keynote speaker in innovation engineering and is Associate Professor at ISTIA Innovation, University of Angers and Laval.

Pascal CRUBLEAU (ENSAM, Angers) is Associate Professor at ISTIA, Angers University and is a TRIZ consultant.

Marc De FOUCHÉCOUR (ENSAM Paris) is Professor and Director of the knowledge management program at ENSAM Engineering School in Paris, and is partner of ICCE.

François DRUEL (ENSAM, Paris) started as an IT consultant, and then served as a junior analyst for Atelier Télématique. He then co-founded Business Village and during this period he also co-headed l'Afa. Nowadays he is a PhD student in Presence Innovation Laboratory (ENSAM Angers) and also works for a major French telecom operator.

Laurent DUKAN (Consultant, Paris) specializes in innovation and human changes associated with technology development. A consultant in innovation, he also teaches at the ISTIA, University of Angers.

Fabienne GOUX-BAUDIMENT (ProGective, Paris) is a foresight specialist and President of proGective. She has taught foresight studies at the ISTIA, University of Angers since 1995.

Christopher Burr JONES is currently CEO and Senior Futurist for neoFutures. He taught for three years as Visiting Professor for the MS Program in studies of the futures at the University of Houston-Clear Lake and prior to that as Professor of Political Science at Eastern Oregon University, USA. A Vice-President of the World Futures Studies Federation (2001-2005), he is currently a WFSF Executive Board member.

Ilya KIRIA (Moscow State University) is Assistant Professor at the Faculty of Journalism, Moscow State University and Co-Director of the Franco-Russian MS in Journalism.

Michel LE RAY (Valenciennes) is Emeritus Professor of the University of Valenciennes having formerly been Director of the Laboratory of Hydrodynamics, Aerodynamics and Energy.

Dokshin LIM (Samsung Telecommunication Network Division, Seoul) is Doctor of Industrial Engineering. He works for user interface design at Mobile Human Interface Design Team.

Darrell MANN (European TRIZ Association) is one of the world's leading TRIZ researchers, teachers and users. Over the course of the last 10 years he has taught TRIZ and other systematic innovation methods to senior management teams of blue-chip organizations from all around the world including Rolls-Royce, BAE Systems, Exxon, MoD, Mars, Sara Lee, Hoover, Intel, Heidelberg Digital, United Utilities and Volvo. He is currently President of the European TRIZ Association.

Jean-Pierre MATHIEU (Audencia, Nantes) is Professor of Marketing and Director of the MS in Marketing, Design and Creation at Audencia, Nantes.

Olaf MAXANT (EDF, France) is a researcher at Electricité de France R&D and works on the CREATEAM innovative project.

Jean MICHEL (VM Consultant, Paris) is a civil engineer, an international consultant in value management and information and knowledge management. Vice-President of the AFAV, he is presently Chairman of the Certification Board of ADBS, and the editor of the EJEE (European Journal of Engineering Education).

Hiroshi MITZUTA (Virtual Enterprise Center, Tokyo) is Head of the Virtual Enterprise Center in Tokyo, Japan.

Marc PALLOT (MPC, Paris) is an international consultant on collaborative innovation, concurrent engineering and business process improvement. Formerly Vice-President for Product Marketing & Engineering at Win Technology, he is co-founder and has been co-organizer of the annual ICE international conference since 1994.

Kulwant PAWAR (Nottingham University, UK) is Professor of Operations Management and Director of the Centre for Concurrent Enterprise at Nottingham University Business School. He is involved in several national, European and International research and consultancy projects and is the Editor-in-Chief of the International Journal of Logistics: Research & Application.

Gérald PIAT (EDF R&D, France) is a researcher at Electricité de France R&D and manages the CREATEAM innovative project.

André-Yves PORTNOFF (Futuribles, Paris) is a consultant, researcher on intangible factors management and Director of Studies on intangibles at the Futuribles Group in Paris.

Simon RICHIR (ISTIA, Angers) is Professor at ENSAM Engineering School in Angers and is Director of the Virtual Reality Masters in Laval (France).

Albert "Skip" RIZZO (ICT and USC, Los Angeles) is a research scientist at the Institute for Creative Technologies in Marina del Rey and is Research Assistant Professor at the School of Gerontology at the University of Southern California.

Benoît ROUSSEL (ENSGSI, Nancy) is Associate Professor at the ENSGSI Institut National Polytechnique in Nancy, France.

Mélissa SAADOUN (Consultant, Paris) is Managing Director of INEDIT. Teaching ICT and Project Management in different universities, she has also extensive experience consulting in management, organization and ICT.

Henri SAMIER (ISTIA, Angers) is Associate Professor and Director of the Innovation Department at ISTIA Innovation, where he is also and Head of the Technology Innovation Masters.

Nathalie SAMIER (Angers University) is Associate Professor and Head of the International Management of Human Resources Master at Angers University.

Victor SANDOVAL (ENPC, Paris) is a researcher in ICT and Virtual Spaces at the Ecole Nationale des Ponts et Chaussées, Paris. He currently works on principles, laws and nature of hyperspaces and its power and challenges for human activities.

Michel SINTES (Audiconsult, Geneva) is President and founder of Audioconsult (Vision-Passion-Action), Creativity Consultant and gives lectures at universities, enterprises and public institutions.

Barnabás TAKÀCS (Digital Elite, Los Angeles) is President of Digital Elite Inc., which he founded in 2000. In 1999 he was credited for creating the first "digital clone" while later he co-invented a method for digital plastic surgery.

Roy Mark WOODHEAD (Oxford Brookes University, Oxford) is Senior Lecturer in Technology Management at the School of Technology, Oxford Brookes University, UK. He is also Vice-President for the Society of American Value Engineering (SAVE) International.

Lin YANNING (REALIZER, Beijing) is a lawyer at the Bar of Beijing in the law firm REALIZER specializing in law of new information technology. Her main interests of practice and research focus on the legal aspects aroused by the application of ICT in the new economy. She currently carries out a research in e-banking at the University of Paris I.

Index

R, S, T

R&D 15, 89-94, 132, 226, 298, 299
StoryoFil 235, 236, 238
TRIZ 6, 22, 63, 181, 215, 217-219, 221

V

Virtual environments 44, 48, 49, 199, 204, 206, 207, 210, 211
Virtual reality 37, 43, 44, 48, 49, 199, 203-206, 211, 272